Fractal Functions,
Fractal Surfaces,
and Wavelets

Fractal Functions, Fractal Surfaces, and Wavelets

Peter R. Massopust

Sam Houston State University
Department of Mathematics
Huntsville, Texas

Academic Press

San Diego   New York   Boston   London   Sydney   Tokyo   Toronto

Copyright © 1994 by ACADEMIC PRESS, INC.

Academic Press, Inc.
A Division of Harcourt Brace & Company
525 B Street, Suite 1900, San Diego, California 92101-4495

United Kingdom Edition published by
Academic Press Limited
24-28 Oval Road, London NW1 7DX

Library of Congress Cataloging-in-Publication Data

Massopust, Peter Robert, date.
    Fractal functions, fractal surfaces, and wavelets / by Peter R.
Massopust.
        p.   cm.
    Includes index.
    ISBN 0-12-478840-8
    1. Fractals.   I. Title.
QA614.86.M32   1994
514' .74--dc20   94-26551
                                                     CIP

PRINTED IN THE UNITED STATES OF AMERICA
94 95 96 97 98 99 BC 9 8 7 6 5 4 3 2 1

Contents

Preface

This monograph gives an introduction to the theory of fractal functions and fractal surfaces with an application to wavelet theory. The study of fractal functions goes back to Weierstraß's nowhere differentiable function and beyond. However, it wasn't until the publication of B. Mandelbrot's book (cf. [123]) in which the concept of a fractal set was introduced and common characteristics of these sets were identified (such as nonintegral dimension and geometric self-similarity) that the theory of functions with fractal graphs developed into an area of its own. Seemingly different types of nowhere differentiable functions, such as those investigated by Besicovitch, Ursell, Knopp, and Kiesswetter, to only mention a few, were unified under the fractal point of view. This unification led to new mathematical methods and applications in areas that include: dimension theory, dynamical systems and chaotic dynamics, image analysis, and wavelet theory.

The objective of this monograph is to provide essential results from the theory of fractal functions and surfaces for those interested in this fascinating area, to present new and exciting applications, and to indicate which interesting directions the theory can be extended. The book is essentially self-contained and covers the basic theory and different types of fractal constructions as well as some specialized and advanced topics such as dimension calculations and function space theory.

The first part of the book contains background material and consists of four chapters. The first chapter introduces the relevant notation and terminology and gives a brief review of some of the basic concepts from classical analysis, abstract algebra and probability theory that are necessary for the remainder of the book. The reader who is not quite familiar with some of the material presented in this first chapter is referred to the bibliography where most of these concepts are defined and motivated. However, efforts were made to keep the mathematical requirements at a level where a graduate student with a solid background in the afore-mentioned areas will be able to work through most of the book.

The second chapter introduces same basic constructions of fractal sets. The first is based upon the approach by J. Hutchinson [98] and M. Barnsley and S. Demko [9] using what is now called an iterated function system. This method is then generalized and compared to M. Dekking's [45] construction of so-called recurrent sets associated with certain semigroup endomorphisms and C. Bandt's approach [5, 6] via topological Markov chains. Finally, a graph-directed fractal construction due to D. Mauldin and S. Williams [134] is presented. The emphasis is on iterated function systems and their generalizations; however. In this chapter the foundations for the rigorous treatment of univariate and multivariate fractal functions are laid.

Next, the concept of dimension of a set is introduced. This is done by first reviewing the different notions of dimension that are used to characterize and describe sets. The last two sections in this chapter are devoted to the presentation of dimension results for self-affine fractal sets.

A short chapter dealing with the fascinating theory of dynamical systems follows. The emphasis is on the geometric aspects of the theory and it is shown how they can be used to describe attractors of iterated function systems.

In the second part of this book, univariate and multivariate fractal functions are discussed. The fifth chapter introduces fractal functions as the fixed points of a Read-Bajraktarević operator. This approach differs from that undertaken by M. Barnsley [8] who introduced fractal functions for interpolation and approximation purposes. It is also shown how M. Dekking's approach to fractals can be used to define fractal functions and the iterative interpolation process of S. Dubuc and his co-workers is presented. The remainder of the chapter deals with different classes of fractal functions and discusses several of their properties.

Chapter 6 is devoted to dimension calculations. Formulae for the box dimension of the graphs of most of the fractal functions introduced in the previous chapter are presented here. The second part of the chapter deals with an interesting relationship between certain classes of smoothness spaces and the box dimension of the graphs of affine fractal functions.

In Chapter 7, the basic concepts and notions of wavelet theory are introduced, and it is demonstrated how a certain class of fractal functions generated by iterated function systems can be used to generate a multiresolution analysis of $L^2(\mathbb{R})$. This class of fractal functions then provides a new construction of continuous, compactly supported and orthogonal scaling functions and wavelets.

The next chapter introduces multivariate fractal functions. The graphs of these functions are called fractal surfaces. Properties of fractal surfaces are then discussed and formulae for the box dimension derived.

In order to construct multiresolution analyses based on the fractal surfaces defined in Chapter 8, the theory of Coxeter groups needs to be employed. This is done in Chapter 9, after some rudimentary concepts of this theory are introduced.

Because of the limited scope of this monograph, certain topics could not be covered. This includes a more in-depth presentation of the geometric theory of dynamical systems and the role fractals play in this theory. Furthermore, some of the work of T. Lindstrøm on nonstandard analysis, iterated function systems, fractals, and especially Brownian motion on fractals is beyond the limits of this book. The interesting work of J. Harrison dealing with geometric integration theory and fractals could also not be described. However, references pertaining to these as well as other topics are listed in the bibliography. The bibliography also contains research papers and books not explicitly used or mentioned in this monograph. They were included to give the reader a more well-rounded perspective of the subject.

This book grew out of the work of many mathematicians from several areas of mathematics, and the author has greatly benefited from numerous conversations and discussions with my colleagues. Special thanks go to Doug Hardin and Jeff Geronimo, who have influenced and shaped some of my thoughts and ideas. In particular, I am grateful to Doug Hardin for allowing me to use his *Mathematica* packages to make some of the figures in this monograph. I also wish to thank Patrick Van Fleet for introducing me to the theory of Dirichlet splines and special functions.

Working with Academic Press was a pleasure. I would like to especially express my gratitude to Christina Wipf, who gave me the idea of writing this monograph, and to Peter Renz, who guided me through the final stages.

Last but not least, I wish to thank my wife Maritza and my family for their continuous support and encouragement during the preparation of this monograph.

Peter R. Massopust

Part I

Foundations

Chapter 1

Mathematical Preliminaries

This chapter provides most of the mathematical preliminaries necessary to understand the results in the following chapters. It is a mere collection of definitions and theorems given without a proof (the only exceptions are the Banach Fixed-Point Theorem, and the Existence Theorems for free semigroups and free groups). The bibliography contains a list of references in which all these results are motivated and proved. In a sense, this first chapter compiles notation and terminology and serves as a reference guide for the remainder of the book.

The relevant material is discussed in three sections: analysis and topology, probability theory, and algebra. The first section covers such basic topics as linear spaces, normed and metric spaces, point-set topology, measures, and the different notions of convergence encountered in analysis. In the second section, probability measures, distribution functions, random variables, and their interconnections are considered. Then the Lebesgue spaces are defined, the Riesz Representation Theorem is stated, and a brief overview of Markov processes and Markov chains is given. The last section deals with diagrams, semigroups, groups, and semigroup and group endomorphisms and introduces free semigroups and free groups. A brief review of category theory and direct and inverse limits is also provided.

1.1 Analysis and Topology

Throughout this monograph, $\mathbb{N} := \{1, 2, 3, \ldots\}$ denotes the set of natural numbers, $\mathbb{Z}$ the ring of integers, and $\mathbb{R}$ the field of real numbers. Let $\mathbb{K}$ be a subfield of $\mathbb{C}$, the field of complex numbers, and suppose that the mapping

$\alpha : \mathbb{C} \to \mathbb{C}$, $z = x + iy \mapsto \bar{z} = x - iy$ maps $\mathbb{K}$ into itself (α is called an *involuntary automorphism of* $\mathbb{C}$).

Suppose X and Y are linear spaces over $\mathbb{K}$.

Definition 1.1   1. A mapping $\varphi : X \to \mathbb{K}$ is called semilinear or a semi-linear form iff

(a) $\forall x, x' \in X : \varphi(x + x') = \varphi(x) + \varphi(x')$.

(b) $\forall x \in X \; \forall k \in \mathbb{K} : \varphi(kx) = k\varphi(x)$.

If $\mathbb{K} \subseteq \mathbb{R}$ all semilinear forms are linear.

2. A mapping $\varphi : X \times X \to \mathbb{K}$ is called sesquilinear or a sesquilinear form iff

(a) $\forall x, x', y \in X : \varphi(x + x', y) = \varphi(x, y) + \varphi(x', y)$.

(b) $\forall k \in \mathbb{K} : \varphi(kx, y) = k\varphi(x, y)$.

(c) $\forall x, y, y' \in X : \varphi(x, y + y') = \varphi(x, y) + \varphi(x, y')$.

(d) $\forall k \in \mathbb{K} : \varphi(x, ky) = \bar{k}\varphi(x, y)$.

If $\mathbb{K} \subseteq \mathbb{R}$, all sesquilinear forms are bilinear.

3. A sesquilinear form φ is called Hermitian iff $\forall x, y \in X : \varphi(x, y) = \overline{\varphi(y, x)}$ (if $\mathbb{K} \subseteq \mathbb{R}$ a Hermitian form is called symmetric).

4. A sesquilinear form φ is called positive definite, respectively positive semidefinite, iff $\forall x \in X$, $x \neq 0$, $\varphi(x, x) > 0$, respectively $\forall x \in X$, $\varphi(x, x) \geq 0$.

Definition 1.2 An inner product on a linear space X over $\mathbb{K}$ is a positive definite Hermitian sesquilinear form $\varphi : X \times X \to \mathbb{K}$.
The pair (X, φ) is called an inner product space. If X is a linear space over $\mathbb{R}$, respectively $\mathbb{C}$, (X, φ) is called an Euclidean space, respectively unitary space.

Notation. Instead of writing $\varphi(x, y)$, $x, y \in X$, the shorter notation $\langle x, y \rangle$ is sometimes used.

An inner product φ on X can be used to define the norm of an element $x \in X$, and the distance between two elements $x, y \in X$. More precisely, the *norm* of $x \in X$ is defined as

$$\|x\|_\varphi := \sqrt{\varphi(x, x)}, \tag{1.1}$$

and the *distance* between $x, y \in X$ by

$$d_\varphi := \|x - y\|_\varphi = \sqrt{\varphi(x - y, x - y)}. \tag{1.2}$$

Proposition 1.1 (Cauchy-Schwartz Inequality) *Let (X, φ) be an inner product space over $\mathbb{K}$. Then, for all $x, y \in X$,*

$$|\varphi(x, y)| \le \sqrt{\varphi(x, x)} \sqrt{\varphi(y, y)} = \|x\|_\varphi \|y\|\|_\varphi, \tag{1.3}$$

with equality iff there exists a $k \in \mathbb{K}$ such that $x = ky$. ∎

Definition 1.3 Suppose that X is a linear space over $\mathbb{K}$. A non-negative functional $\| \cdot \| : X \to \mathbb{R}$ is called a norm on X iff the following conditions hold:

(a) $\forall x \in X : \|x\| \ge 0, \|0\| = 0$.

(b) $\forall x \in X \, \forall k \in \mathbb{K} : \|kx\| = |k| \, \|x\|$.

(c) $\forall x, y \in X : \|x + y\| \le \|x\| + \|y\|$.

(d) $\|x\| = 0 \Rightarrow x = 0$.

If only properties (a) — (c) are satisfied, $\| \cdot \|$ is called a semi-norm on X. The pair $(X, \| \cdot \|)$ is called a normed (linear) space.

Proposition 1.2 *Suppose (X, φ) is an inner product space over $\mathbb{K}$. Then $\| \cdot \|_\varphi : X \to \mathbb{K}$ as defined in (1.1) is a norm on X.* ∎

Definition 1.4 Suppose M is a set. A mapping $d : M \times M \to \mathbb{R}$ is called a metric on M iff the following conditions are satisfied:

(a) $\forall x, y \in M : d(x, y) \ge 0, \quad d(x, x) = 0$.

(b) $\forall x, y \in M : d(x, y) = d(y, x)$.

(c) $\forall x, y, z \in M : d(x, z) \le d(x, y) + d(y, z)$.

(d) $d(x, y) = 0 \Rightarrow x = y$.

If only properties (a) — (c) hold, then d is called a semi-metric on M. The pair (X, d), where d is a (semi-)metric on the set X, is called a (semi-) metric space.

Proposition 1.3 *Suppose* $\| \cdot \|$ *is a norm, respectively a semi-norm, on a linear space* X *over* $\mathbb{K}$*. Then*

$$d(x,y) := \|x - y\|, \ x, y \in X, \tag{1.4}$$

defines a metric, respectively a semi-metric, on X. ∎

A norm $\| \cdot \| : X \to \mathbb{R}$ on a linear space X over $\mathbb{K}$ induces in a canonical way a topology on X, the so-called *norm* or *strong topology*. At this point the definition of topology on a set M is recalled.

Definition 1.5 Let M be an arbitrary set and let $\mathcal{T}$ be a collection of subsets of M. Then $\mathcal{T}$ is called a topology on M provided

(a) For all i in some index set I, $T_i \in \mathcal{T} \Rightarrow \bigcup_{i \in I} T_i \in \mathcal{T}$.

(b) $T_1, \ldots, T_n \in \mathcal{T} \Rightarrow \bigcap_{i=1}^{n} T_i \in \mathcal{T}$.

(c) $M \in \mathcal{T}$, $\emptyset \in \mathcal{T}$.

The elements of $\mathcal{T}$ are called open sets and the pair $(X, \mathcal{T})$ a topological space.

The norm topology on X is then defined as follows: Let $A \subseteq X$, and let $B_r(a) := \{x \in X | \|x - a\| < r\}$ denote the ball of radius $r > 0$ centered at $a \in X$. The set A is called *open* iff for each $a \in A$ there exists a ball $B_r(a)$, $r > 0$, contained entirely in A. It is easy to show that $\mathcal{T}_{\| \cdot \|} := \{A \subseteq X | A$ is open$\}$ is a topology on X.

The topological space $(X, \mathcal{T}_{\| \cdot \|})$ is also *Hausdorff*.

Definition 1.6 A topology $\mathcal{T}$ on a set M is called Hausdorff iff two distinct points $x, y \in M$ can be separated by two disjoint sets U and V in $\mathcal{T}$, i.e., $\forall x, y \in M$, $x \neq y$, $\exists U, V \in \mathcal{T}$ such that $x \in U$, $y \in V$, and $U \cap V = \emptyset$.

Suppose that X is a linear space over $\mathbb{K}$, and $\| \cdot \|_i : X \to \mathbb{R}$, $i = 1, 2$, are arbitrary norms on X. $\| \cdot \|_1$ and $\| \cdot \|_2$ are called *equivalent*, written $\| \cdot \|_1 \approx \| \cdot \|_2$, iff there exist positive real numbers c_1 and c_2 such that for all $x \in X$,

$$\|x\|_1 \leq c_1 \|x\|_2, \quad \text{and} \quad \|x\|_2 \leq c_2 \|x\|_1. \tag{1.5}$$

Proposition 1.4   *1. All norms on* $\mathbb{R}^n$ *are equivalent.*

2. *All norms on* $\mathbb{R}^n$ *generate the same topology.* ∎

Definition 1.7 Suppose $(X, \mathcal{T})$ is a topological space. $\mathcal{B} \subseteq \mathcal{T}$ is called a basis of $\mathcal{T}$ iff every open set is a union of elements of $\mathcal{B}$:

$$T \in \mathcal{T} \Rightarrow T = \bigcup_{B \in \mathcal{B}'} B,$$

with $\mathcal{B}' \subseteq \mathcal{B}$.

The following result gives necessary conditions for a subset $\mathcal{B}$ of $\mathcal{T}$ to be a base.

Proposition 1.5 *Let* $(X, \mathcal{T})$ *be a topological space and let* $\mathcal{B} \subseteq \mathcal{T}$. *If* $\mathcal{B}$ *is a base of* $\mathcal{T}$, *then*

(a) $X = \bigcup_{B \in \mathcal{B}} B$;

(b) $\forall B_1, B_2 \in \mathcal{B} \; \forall x \in B_1 \cap B_2 \; \exists B_x \in \mathcal{B} : x \in B_x \subseteq B_1 \cap B_2.$ ∎

The concept of topology allows one to precisely define notions such as *distance, convergence, and continuity.*

Definition 1.8 Let $\mathbf{D}$ be a non-empty set. A relation $\preceq$ on $\mathbf{D}$ is called directed iff it has the following properties:

(a) Reflexivity: $\forall \alpha \in \mathbf{D} : \alpha \preceq \alpha$.

(b) Transitivity: $\forall \alpha, \beta, \gamma \in \mathbf{D} : \alpha \preceq \beta, \; \beta \preceq \gamma \Rightarrow \alpha \preceq \gamma$.

(c) $\forall \alpha, \beta \in \mathbf{D} \; \exists \gamma \in \mathbf{D} : \alpha \preceq \gamma, \; \beta \preceq \gamma$.

A directed set is a set with a directed ordering.

Remark. Some authors define a directed set as a non-empty partially ordered set satisfying condition (c) above.

Let $(X, \mathcal{T})$ be a topological space. A *net* in X consists of a directed set $\mathbf{D}$ and a mapping $\delta : \mathbf{D} \to X$. It is common to write the image of $\alpha \in \mathbf{D}$ under δ in X as x_δ instead of $\delta(\alpha)$. Nets are then denoted by $\{x_\alpha\}_{\alpha \in \mathbf{D}}$, or simply by $\{x_\alpha\}$ if it is understood which directed set is meant. Clearly, every sequence in X is a net in X: take $\mathbf{D} = \mathbb{N}$ and $\preceq := \leq$.

Recall that a set $N \subseteq X$ is called a *neighborhood of* $x \in X$ iff N is a superset of an open set containing x.

***Definition* 1.9** Let $(X, \mathcal{T})$ be a topological space and let $\{x_\alpha\}$ be a net in X. A point $x \in X$ is called a limit point of $\{x_\alpha\}$ iff for any neighborhood N of x there exists an $\alpha_0 \in \mathbf{D}$ such that all x_α with $\alpha_0 \preceq \alpha$ are points in N.

Notation. If x is a limit point of a net $\{x_\alpha\}$, then one writes for short: $x_\alpha \to x$ (in $\mathcal{T}$).

The classical characterization of convergence (in the strong topology) is obtained by choosing X to be a normed linear space, $\mathcal{T} = \mathcal{T}_{\|\cdot\|}$, and $\mathbf{D} = \mathbb{N}$:

$$x_n \to x \Longleftrightarrow \forall \varepsilon > 0 \; \exists n_0 \; \forall n \geq n_0 : \; \|x_n - x_0\| < \varepsilon.$$

Let $(X, \mathcal{T})$ and $(X', \mathcal{T}')$ be two topological spaces, and $F : X \to X'$ a mapping of sets. Then F is *continuous* iff $x_\alpha \to x$ in $(X, \mathcal{T})$ implies $F(x_\alpha) \to F(x)$ in $(X', \mathcal{T}')$, for every net $\{x_\alpha\}$ in X. A mapping $F : X \to X'$ is called a *homeomorphism* iff F is bijective and F and its inverse F^{-1} are continuous.

***Definition* 1.10** Let $(X, \mathcal{T})$ be a topological space and let $\overline{\mathbb{R}} := \mathbb{R} \cup \{\pm\infty\}$ be the completed real line. A function $f : X \to \overline{\mathbb{R}}$ is called upper semi-continuous, respectively lower semi-continuous, at $x_0 \in X$ iff for all $\alpha \in \overline{\mathbb{R}}$ with $\alpha > f(x_0)$, respectively $\alpha < f(x_0)$, there exists a neighborhood N of x_0 in X such that for all $x \in N$ one has $\alpha > f(x)$, respectively $\alpha < f(x)$.
A function f is called upper semi-continuous, respectively lower semi-continuous, on X iff it is upper semi-continuous, respectively lower semi-continuous, at each $x_0 \in X$.

It is clear that if f is upper semi-continuous then $-f$ is lower semi-continuous. Also, if a function f is both upper and lower semi-continuous at $x_0 \in X$ then f is continuous at x_0. The next proposition characterizes lower semi-continuous functions.

Proposition 1.6 *A function f from a topological space $(X, \mathcal{T})$ into the completed real line $\overline{\mathbb{R}}$ is lower semi-continous iff for all $\alpha \in \overline{\mathbb{R}}$ the set*

$$f^{-1}(\alpha, \infty]$$

is open in X, or equivalently, the set

$$f^{-1}[-\infty, \alpha]$$

is closed in X.   ∎

***Definition* 1.11** Let $(X, \mathcal{T})$ be a topological space and let $\{x_\alpha\}$ be a net in X. A point $x \in X$ is called an accumulation point of $\{x_\alpha\}$ iff for every neighborhood N of x and any α_0 there exists an $\alpha \succeq \alpha_0$ such that $x_\alpha \in N$.

Nets provide a simple characterization of compactness in a topological space $(X, \mathcal{T})$.

Theorem 1.1 *Let $(X, \mathcal{T})$ be a topological space. X is compact iff every net in X has an accumulation point.* ∎

The next theorem is of great importance in other areas of mathematics as well.

Theorem 1.2 (Tychonov) *The (cartesian) product of any collection of compact topological spaces is compact.* ∎

Nets are also a useful tool for characterizing the closure $\overline{A}$ of a set $A \subseteq X$ in a topological space $(X, \mathcal{T})$.

Theorem 1.3 $\overline{A}$ *consists precisely of the limit points of all the nets in A. A set $A \subseteq X$ is closed iff it contains every limit point of every net in A.* ∎

The characterization of completeness in terms of so-called *Cauchy nets* requires the concept of a *uniform space*.

Let X be a set and let U be a subset of $X \times X$. Define U^{-1} as the set of all pairs $(x, x') \in X \times X$ such that $(x', x) \in U$, and let $D := \{(x, x) \in X \times X \mid x \in X\}$. If $U, U' \in X \times X$, define $UU' := \{(x, x') \in X \times X \mid \exists x'' \in X: (x, x'') \in U \wedge (x'', x') \in U'\}$.

Definition 1.12 Let X be a set and let $\mathcal{U}$ be a collection of subsets of $X \times X$. The pair $(X, \mathcal{U})$ is called a uniform space (with uniformity $\mathcal{U}$) iff the following conditions hold:

  (a) $\bigcap_{U \in \mathcal{U}} = D$.

  (b) $\forall U \in \mathcal{U}: U^{-1} \in \mathcal{U}$.

  (c) $\forall U \in \mathcal{U} \, \exists U' \in \mathcal{U}: U'U' \subseteq U$.

  (d) $\forall U, U' \in \mathcal{U}: U \cap U' \in \mathcal{U}$.

  (e) $\forall U \in \mathcal{U} \, \forall U' \supseteq U: U' \in \mathcal{U}$.

The elements of $\mathcal{U}$ are called uniformities or entourages.

Clearly, every metric space and thus every normed linear space is a uniform space (the uniformities are given by the neighborhoods of D in $X \times X$).

Let $(X, \mathcal{U})$ be a uniform space. The definition of net in a uniform space is just as above, with *topological* replaced by *uniform*.

A *Cauchy net* $\{x_\alpha\}_{\alpha \in \mathbf{D}}$ in $(X, \mathcal{U})$ is a net having the following property: $\forall U \in \mathcal{U} \, \exists \alpha \in \mathbf{D}$ such that $(x_{\alpha'}, x_{\alpha''}) \in U$ whenever $\alpha \preceq \alpha'$ and $\alpha \preceq \alpha''$.

Theorem 1.4 *A uniform space $(X, \mathcal{U})$ is complete iff every Cauchy net converges to a point of X.* ∎

The next definition introduces two important classes of complete linear spaces.

Definition 1.13 A complete inner product space $(X, \langle , \rangle)$ is called a Hilbert space, and a complete normed linear space $(X, \| \cdot \|)$ is called a Banach space.

Next mappings between linear spaces are considered. For this purpose let X and Y be linear spaces over $\mathbb{K}$. A mapping $F : D \subseteq X \to Y$, $D \neq \emptyset$, is called a *function* on X with domain D. If $X = Y$, then F is usually called an *operator* on X with domain D, and if $Y = \mathbb{K}$, F is called a *functional*.

Now assume that X and Y are normed linear spaces. A function $F : D \subseteq X \to Y$ is called *Lipschitz continuous* iff there exists an $L > 0$, called the *Lipschitz constant*, such that $\|F(x) - F(x')\| \leq L\|x - x'\|$, $\forall x, x' \in D$. If $L < 1$, F is called *contractive* or a *contraction*. Let $T : D \subseteq X \to X$ be an operator. A point $x \in X$ is called a *fixed point* of T iff $Tx = x$. In case $TD \subseteq D$, one defines the *powers* or *iterates* of T by

$$T^0 := id_D,$$

$$T^n := TT^{n-1}, \ \forall n \in \mathbb{N}.$$

(Here id_D denotes the identity function on D).

The next theorem will be used extensively in the following chapters. Its proof uses a technique that will be encountered numereous times.

Theorem 1.5 (Banach Fixed-Point Theorem) *Suppose X is a normed linear space over $\mathbb{K}$, and T a Lipschitz continuous operator with complete domain $D \subseteq X$. Assume that $TD \subseteq D$. Furthermore, let L_n denote the Lipschitz constant of T^n, $n \in \mathbb{N}_0$, and assume that*

$$\sum_{n=0}^{\infty} L_n < \infty.$$

Then T has exactly one fixed point $x^* \in D$. Moreover, if $x_0 \in D$ is chosen arbitrarily and if $x_{n+1} := Tx_n$, $n \in \mathbb{N}_0$, then the sequence $\{x_n\}$ converges to x^*, independently of x_0.

Proof. EXISTENCE: Let $\ell, m, n \in \mathbb{N}_0$ be such that $m < n \le \ell$. Let $x_0 \in D$ be arbitrary. Then

$$\|x_m - x_n\| \le \|\sum_{k=n}^{m-1} x_{k+1} - x_k\| \le \sum_{k=n}^{m-1} \|T^{k+1}x_0 - T^kx_0\|$$

$$\le \sum_{k=n-\ell}^{m-\ell-1} \|T^k x_{\ell-1} - T^k x_\ell\| \le \left(\sum_{k=n-\ell}^{\infty} L_k\right) \|x_{\ell+1} - x_\ell\|.$$

Now, $\forall \varepsilon > 0 \ \exists N \in \mathbb{N}$ such that $(\sum_{k=n-\ell}^{\infty} L_k)\|x_1 - x_0\| \le \varepsilon$. Therefore, setting $\ell = 0$ yields

$$\|x_m - x_n\| \le \left(\sum_{k=n}^{\infty} L_k\right) \|x_1 - x_0\| < \varepsilon.$$

The Cauchy sequence $\{x_n\}$ has, by the completeness of D, a limit point $x^* \in D$. Letting $n \to \infty$ and using the fact that T is Lipschitz continuous gives $Tx = x$.

UNIQUENESS: Suppose x and x' are fixed points of T. Then, $\forall n \in \mathbb{N}_0$,

$$\|x - x'\| = \|T^n x - T^n x'\| \le L_n\|x - x'\|.$$

Hence, $L \ge 1$, for all $n \in \mathbb{N}_0$. This contradiction now yields uniqueness.

The fact that $x_n \to x^*$, independent of x_0, follows readily from the foregoing arguments. ∎

Now some of the properties of functions between normed spaces are presented. Let X and Y be normed spaces over $\mathbb{K}$, and let $F : D \subseteq X \to Y$ be a continuous linear function on $D \neq \emptyset$ (note that D is necessarily a linear subspace of X). It is a well-known fact that each such function has a smallest Lipschitz constant $L \ge 0$, i.e., there exists a least non-negative number L with the property

$$\|Fx\| \le L\|x\|, \ \forall x \in D, \ x \neq 0.$$

This smallest Lipschitz constant is equal to the supremum of the set

$$\{M \in \mathbb{R} \mid M = \|Fx\|/\|x\| \wedge x \in D \wedge x \neq 0\},$$

in the case of dim $D > 0$; otherwise it is 0 (here dim denotes the dimension of the linear subspace D). It is easy to show that under the usual definition of addition and scalar multiplication of functions, and equipped with the norm of the smallest Lipschitz constant

$$\|F\| := \begin{cases} 0 & \text{if } \dim D = 0 \\ \sup_{x \in D}\{\|Fx\|/\|x\| \mid x \neq 0\} & \text{if } \dim D > 0 \end{cases},$$

the set of all continuous linear functions $F : D \subseteq X \to Y$ forms a normed linear space over $\mathbb{K}$. This normed linear space will be denoted by $\mathcal{L}(D, Y)$. In the case where Y is a Banach space, $\mathcal{L}(D, Y)$ is also a Banach space.

Uniform convergence in $\mathcal{L}(D, Y)$ implies norm convergence, which then implies pointwise convergence.

Definition 1.14 Suppose X and Y are normed linear spaces over $\mathbb{K}$. A continuous function $F : D \subseteq X \to Y$ is called equicontinuous iff for every bounded sequence $\{x_n\} \subseteq X$ the sequence $\{Fx_n\}$ contains a convergent subsequence.

The following fixed-point theorem is due to Schauder.

Theorem 1.6 (Schauder Fixed-Point Theorem) *Let X be a Banach space and let $T : D \to D$ be an equicontinuous operator with bounded, closed, and convex domain $D \subseteq X$. Then T has a fixed point in D.* ∎

Recall that a subset C of a linear space X over $\mathbb{K}$ is called *convex* iff $x, x' \in C$ implies $\lambda x + (1 - \lambda)x' \in C$, for all $\lambda \in [0, 1]$.

The next result is of great importance in real analysis as well as functional analysis. Only its normed space version is stated.

Theorem 1.7 (Hahn-Banach Extension Theorem) *Let X be a normed linear space over $\mathbb{K}$ and let $U \subseteq X$ be a linear subspace. Then every continuous linear functional $\varphi : U \to \mathbb{R}$ can be linearly and continuously extended to all of X in such a way that it preserves the norm, i.e., there exists a continuous linear functional $\psi : X \to \mathbb{R}$ with the properties*

$$\forall u \in U : \ \psi u = \varphi u, \tag{1.6}$$

and

$$\|\varphi\| = \|\psi\|. \tag{1.7}$$

∎

Remark. The extension of a continuous linear functional — as guaranteed by the Hahn-Banach Extension Theorem — may not be unique.

Now suppose X is a normed space over $\mathbb{K}$. The set of all continuous linear functionals $\varphi : X \to \mathbb{K}$ is called the *dual (space)* X^* of X. Clearly, if addition and scalar multiplication on X^* is defined in the usual way and X^* is equipped with the norm of the smallest Lipschitz constant, then $X^* = \mathcal{L}(X, \mathbb{K})$, and hence X^* is a Banach space.

The elements of X^* are usually denoted by x^*.

Proposition 1.7 $\dim X = \dim X^*$. ∎

Let $\varepsilon > 0$ be given, and let $x_1^*, \ldots, x_n^*$ be a finite set of elements of X^*. One can define a topology on X as follows:

For each $x_0 \in X$ let

$$N(x_0; \varepsilon, x_1^*, \ldots, x_n^*) := \{x \in X \mid \|x_i^*(x) - x_i^*(x_0)\| < \varepsilon, \ i = 1, \ldots, n\}$$

be a basis of neighborhoods.

Clearly, this defines a locally convex topology on X, the *weak topology*.

Let $\{x_n\}_{n \in \mathbb{N}}$ be a sequence in X. This sequence is called *weakly convergent* to an $x \in X$ iff

$$\forall x^* \in X^* : \ \lim_{n \to \infty} x^*(x_n) = x^*(x). \tag{1.8}$$

The element x is called the *weak limit* of the sequence $\{x_n\}$. Instead of (1.8), the shorter notation $x_n \overset{w}{\to} x$ is used.

A subset U of a normed linear space X over $\mathbb{K}$ is called *weakly compact* if every sequence $\{x_n\} \subseteq U$ has a subsequence which converges weakly to an element of U.

Definition 1.15 A subset U of a normed linear space X is called *separable* iff it satisfies the following equivalent conditions:

(a) There exists a sequence $\{x_n\} \subseteq X$ such that each $u \in U$ is the limit of a subsequence of $\{x_n\}$.

(b) There exists a sequence $\{x_n\} \subseteq X$ that is dense in X.

Definition 1.16 Let X be a normed linear space X over $\mathbb{K}$. A sequence $\{e_n\} \subseteq X$ is called a *Schauder basis* of X iff every element $x \in X$ has a unique representation of the form

$$x = \sum_{n \in \mathbb{N}} \alpha_n e_n, \tag{1.9}$$

for $\{\alpha_n\} \subseteq \mathbb{K}$.

Remark. The sum in (1.5) is understood as a limit in the norm topology.

The following result will be used later.

Proposition 1.8 *Finite dimensional normed linear spaces and infinite di-mensional normed linear spaces with Schauder basis are separable.*   ∎

Let X be a normed linear space over $\mathbb{K}$ and let X^* be its dual (considered as a Banach space). One can define the dual of X^*, X^{**}, to be the collection of all continuous linear functionals $\Phi : X^* \to \mathbb{K}$,

$$(\Phi\varphi)(x) := \varphi x, \qquad \forall x \in X. \tag{1.10}$$

It is straightforward to show that $\|\Phi\| = \|x\|$.

One can give the dual of X a topology that is even weaker than the weak topology on X. This topology is called the *weak* topology* on X:
The basis of neighborhoods of a point $x_0^* \in X^*$ is defined by

$$N(x_0^*; \varepsilon, x_1, \dots, x_n) := \{x^* \in X^* | \, \|x^*(x_i) - x_0^*(x_i)\| < \varepsilon, \, i = 1, \dots, n\},$$

where $\varepsilon > 0$ and $x_1, \dots, x_n$ is a finite set of elements of X. This obviously defines a locally convex topology on X^*.

A sequence $\{x_n^*\} \subseteq X^*$ is called *weak* convergent* to $x^* \in X^*$ iff

$$\lim_{n\to\infty} x_n^*(x) = x^*(x), \, \forall x \in X. \tag{1.11}$$

One calls x^* the weak* limit of the sequence $\{x_n^*\}$ and writes: $x_n^* \xrightarrow{w^*} x^*$.

A subset $U^* \subseteq X^*$ is called *weak* compact* iff every sequence $\{x_n^*\} \subseteq U^*$ has a weak* convergent subsequence whose limit is in U^*.

In general, $X \neq X^{**}$. Banach spaces for which $X = X^{**}$ are called *reflexive*.

A topological space $(X, \mathcal{T})$ is called *metrizable* if its topology can be defined by a metric on X, and *locally compact* if every point of X possesses a compact neighborhood.

Let X be a compact and metrizable topological space. The algebra of all continuous functions $f : X \to \mathbb{K}$ together with $\|f\| := \sup_{x \in X} |f(x)|$ is a Banach space. It will be denoted by $C_{\mathbb{K}}(X)$. A *measure* μ on X is an element of the dual of $C_{\mathbb{K}}(X)$, i.e., a continuous linear functional $f \mapsto \mu(f)$ on $C_{\mathbb{K}}(X)$ satisfying

$$|\mu(f)| \leq a\|f\|, \tag{1.12}$$

for some $a \in \mathbb{R}$ and all $f \in C_{\mathbb{K}}(X)$.

Now suppose X is a separable, metrizable, and locally compact topological space. For every compact subset $K \subseteq X$, denote by $\mathcal{K}_{\mathbb{K}}(X|K)$ the linear subspace of $C_{\mathbb{K}}(X)$ generated by those functions whose support is contained in K. Let $\mathcal{K}_{\mathbb{K}}(X)$ be the linear space of all functions $f : X \to \mathbb{K}$ with compact support, i.e.,

$$\mathcal{K}_{\mathbb{K}}(X) := \bigcup_{K \subseteq X} \mathcal{K}_{\mathbb{K}}(X|K).$$

A measure on X is a linear functional μ on $\mathcal{K}_{\mathbb{K}}(X)$ with the property that for each compact subset $K \subseteq X$, there exists a non-negative constant a_K, depending on K, such that

$$\forall f \in \mathcal{K}_{\mathbb{K}}(X) : |\mu(f)| \leq a_K \|f\|. \tag{1.13}$$

Lebesgue measure λ on $\mathbb{R}$ is, for instance, obtained as follows: Let $f \in \mathcal{K}_{\mathbb{R}}(X)$ and consider the linear functional $f \overset{\lambda}{\mapsto} \int_{\mathbb{R}} f(t)\, dt$ on $\mathcal{K}_{\mathbb{K}}(X)$ (note that since f is compactly supported, the integral is well-defined). Letting $K = [a, b]$, $a, b \in \mathbb{R}, a < b$, yields

$$\left| \int_{\mathbb{R}} f(t)\, dt \right| \leq (b - a)\|f\|,$$

for all $f \in \mathcal{K}_{\mathbb{K}}(X|\mathbb{R})$.

Clearly, if μ and ν are measures on X, and if $a \in \mathbb{K}$, then $\mu + \nu$ and $a\mu$ are also measures on X. Hence, the collection of all measures on X is a linear subspace of $\mathbb{K}^{\mathcal{K}_{\mathbb{K}}(X)}$. This subspace is denoted by $\mathcal{M}_{\mathbb{K}}(X)$.

Since $\mathcal{M}_{\mathbb{K}}(X)$ is a linear subspace of $\mathbb{K}^{\mathcal{K}_{\mathbb{K}}(X)}$, it can be endowed with a weak topology. Therefore, a sequence $\{\mu_n\}$ of measures in $\mathcal{M}_{\mathbb{K}}(X)$ is called weak* convergent to a measure $\mu \in \mathcal{M}_{\mathbb{K}}(X)$ if $\mu_n(f) \to \mu(f)$ in $\mathbb{K}$ for all $f \in \mathcal{K}_{\mathbb{K}}(X)$.

Let $\mathcal{J}_*(X)$ be the set of all lower semi-continuous functions on X with values in the completed real line $\overline{\mathbb{R}}$ that are minorized by functions from $\mathcal{K}_{\mathbb{R}}(X)$, i.e., if $f \in \mathcal{J}_*(X)$ then there exists a $g \in \mathcal{K}_{\mathbb{R}}(X)$ so that $g(x) \leq f(x)$ for all

$x \in X$. The set $\mathcal{J}_*(X)$ is clearly non-empty, for any lower semi-continuous non-negative function is an element of $\mathcal{J}_*(X)$. Similarly, one defines the set $\mathcal{J}^*(X)$ of all upper semi-continuous functions on X with values in $\overline{\mathbb{R}}$ that are majorized by functions from $\mathcal{K}_{\mathbb{R}}(X)$. Note that $\mathcal{J}^*(X) = -\mathcal{J}_*(X)$. For each function $f \in \mathcal{J}_*(X)$ define

$$\mu^*(f) := \sup_{g \leq f, \, g \in \mathcal{K}_{\mathbb{R}}(X)} \mu(g). \qquad (1.14)$$

The number $\mu^*(f)$ is an element of $(-\infty, \infty]$. It is easy to see that if $f \in \mathcal{K}_{\mathbb{R}}(X)$ then $\mu^*(f) = \mu(f)$, and that μ^* is a subadditive linear functional on $\mathcal{J}_*(X)$. (The scalars are assumed to be positive reals.) Analogously, one defines

$$\mu_*(f) := \inf_{g \geq f, \, g \in \mathcal{K}_{\mathbb{R}}(X)} \mu(g). \qquad (1.15)$$

Obviously,

$$\mu_*(f) \leq \mu^*(f),$$

for all functions $f : X \to \overline{\mathbb{R}}$.

Definition 1.17 A function $f : X \to \overline{\mathbb{R}}$ is said to be μ-integrable on X iff $\mu_*(f)$ and $\mu^*(f)$ are finite and equal; their common value is called the integral of f with respect to μ and is denoted by $\int_X f \, d\mu$.

Remark. The integral of f with respect to μ is sometimes also denoted by $\mu(f)$, $\langle \mu, f \rangle$, or $\int_X f(x) \, d\mu(x)$.

This section is closed out with a few remarks about Hilbert spaces. If $(X, \langle \, , \, \rangle)$ is a Hilbert space over $\mathbb{K}$ and if $x \in X$, then

$$\varphi := \langle \, \cdot \, , x \rangle \qquad (1.16)$$

defines a linear functional in X^* with $\|\varphi\| = \|x\|$.

Conversely, every $\varphi \in X^*$ is of the form (1.16). The element x is uniquely determined by the conditions $\langle u, x \rangle = 0$, $u \in \ker \varphi$, and $\varphi(x) = \|\varphi\|^2$. (Here $\ker \varphi = \{x \in X | \varphi(x) = 0\}$.)

Let T be a continuous and linear, i.e., bounded, operator on $(X, \langle \, , \, \rangle)$. Denote the set of all such operators by $B(X)$.

Proposition 1.9 *For each operator $T \in B(X)$ there exists a unique bounded operator T^*, called the adjoint of T, such that $\langle x', T^*x \rangle = \langle Tx, x' \rangle$, for all $x, x' \in X$.* ∎

If $X := \mathbb{R}^n$, then a bounded linear operator T on $\mathbb{R}^n$ can be represented by its associated $n \times n$-matrix $A = A(T)$. An $n \times n$ diagonal matrix will be denoted by $\mathrm{diag}(a_i)$, $a_i \in \mathbb{R}$, $i = 1, \ldots, n$.

Definition 1.18 1. A matrix A is called positive iff all its elements are positive. If A is positive one writes $A > 0$.

2. A vector $v \in \mathbb{R}^n$ is called positive iff all its components are positive, and non-negative if all its components are non-negative. In the former case one writes $v > 0$, and in the latter $v \geq 0$.

The next two theorems will be used in the following chapters (cf. [21]).

Theorem 1.8 (Perron-Frobenius Theorem) *Suppose that A is a positive matrix. Then there exists a unique eigenvalue $\lambda = \lambda(A)$ of A which has greatest absolute value. This eigenvalue is positive and simple, and its associated eigenvector may be taken to be positive.* ∎

Theorem 1.9 *Let A be a positive matrix and let $\lambda(A)$ be defined as above. Denote by $S(\lambda)$ the set of all non-negative $\lambda \in \mathbb{R}$ for which there exist non-negative vectors $x \in \mathbb{R}^n$ such that $Ax \geq \lambda x$, and by $T(\lambda)$ the set of positive $\lambda \in \mathbb{R}$ for which there exist vectors $y \in \mathbb{R}^n$ such that $Ay \leq \lambda y$. Then*

$$\lambda(A) = \left\{ \begin{array}{l} \max\{\lambda | \lambda \in S(\lambda)\} \\ \min\{\lambda | \lambda \in T(\lambda)\} \end{array} \right.$$

∎

Now let $T \in B(X)$. The *spectrum $\sigma(T)$* of T is defined as

$$\sigma(T) := \{\lambda \in \mathbb{K} | \lambda I - T \text{ is singular}\}. \tag{1.17}$$

Proposition 1.10 *Let $T \in B(X)$. Then $\sigma(T)$ is a non-empty compact subset of $\mathbb{K}$, and $\sigma(T) \subseteq \{\lambda \in \mathbb{K} | |\lambda| \leq \|T\|\}$.* ∎

Definition 1.19 For $T \in B(X)$, the spectral radius, $r(T)$, of T is defined by

$$r(T) := \sup\{|\lambda| \, | \, \lambda \in \sigma(T)\}. \tag{1.18}$$

The following theorem gives Gel'fand's formula for the spectral radius.

Theorem 1.10 (Spectral Radius Theorem) *If $T \in B(X)$ then $r(T) = \lim_{\nu \to \infty} \|T^\nu\|^{1/\nu}$.* ∎

1.2   Probability Theory

In Section 1.1 measures were introduced as elements of the dual of a function space. However, there is another approach to measures, namely via numerically valued set functions. This approach is usually undertaken when measures are first introduced in probability theory.

In this section probability measures are defined as numerically valued set functions and some of their basic properties are presented. Once this has been achieved, relations between probability measures, random variables, and distribution functions are considered. This then leads to the Lebesgue spaces $L^p(\Omega, \mathbb{R}, \mu)$ and the Riesz Representation Theorem. Markov processes are then defined via non-expansive positive operators on $L^1(\Omega, \mathbb{R}, \mu)$. Finally, the concept of a sequence space is used to introduce countable Markov chains.

Definition 1.20 Let Ω be a non-empty set. A collection $\mathcal{F}$ of subsets of Ω is called a σ-algebra or Borel field iff

1. $\emptyset \in \mathcal{F}, \Omega \in \mathcal{F}$.

2. $B \in \mathcal{F} \Rightarrow B^c \in \mathcal{F}$.

3. $B_j \in \mathcal{F}, j \in \mathbb{N} \Rightarrow \bigcup_{j \in \mathbb{N}} B_j \in \mathcal{F}$.

Definition 1.21 Let Ω be a non-empty set and $\mathcal{F}$ a Borel field of subsets of Ω. A function $\mu : \mathcal{F} \to \overline{\mathbb{R}}$ is called a measure on $\mathcal{F}$ provided

1. $\mu\emptyset = 0$.

2. $B_i, B_j \in \mathcal{F}, B_i \cap B_j = \emptyset, i, j \in \mathbb{N}, i \neq j \Rightarrow \mu(\bigcup_{j \in \mathbb{N}}) = \sum_{j \in \mathbb{N}} \mu B_j$.

A function $\pi : \mathcal{F} \to \overline{\mathbb{R}}$ is called a probability measure on Ω iff the following additional conditions are satisfied:

1. $\forall B \in \mathcal{F} : \pi B \geq 0$.

2. $\pi\Omega = 1$.

The triple $(\Omega, \mathcal{F}, \mu)$, respectively $(\Omega, \mathcal{F}, \pi)$, is called a measure space, respectively probability space. The sets in $\mathcal{F}$ are called μ-*measurable* , respectively π-*measurable*.

A measure space $(\Omega, \mathcal{F}, \mu)$ is called $\sigma - finite$ iff it is the countable union of μ-measurable sets of finite measure.

A set $A \subseteq \Omega$ is μ-*measurable* iff for each set $B \subseteq \Omega$,

$$\mu B = \mu(B \cup A) + \mu(B \setminus A). \tag{1.19}$$

The *support* of a measure μ is defined to be the closed set

$$\mathrm{supp}(\mu) := \Omega \setminus \bigcup \{O \,|\, O \text{ is open} \wedge \mu O = 0\}. \tag{1.20}$$

For any non-empty collection $\mathcal{C}$ of subsets of Ω there exists a smallest Borel field containing $\mathcal{C}$. This minimal Borel field is said to be generated by $\mathcal{C}$.

A measure μ is called *complete* if, whenever $A \subseteq B$, $B \in \mathcal{F}$ and $\mu B = 0$, A is μ-measurable and $\mu A = 0$. A measure space $(\Omega, \hat{\mathcal{F}}, \hat{\mu})$ is said to be the *completion of a measure* μ iff $\hat{\mu}$ is complete, each μ- measurable set B is $\hat{\mu}$-measurable with $\hat{\mu} B = \mu B$, and each element $\hat{B} \in \hat{\mathcal{F}}$ is given by $\hat{B} = B \cup Z$, where $B \in \mathcal{F}$ and Z is a subset of a set of μ-measure zero.

It is easy to see that every measure μ has a completion. Indeed, given the measure space $(\Omega, \mathcal{F}, \mu)$, define

$$\hat{\mathcal{F}} := \{E \subseteq \Omega \,|\, E = B \cup Z \wedge E \in \mathcal{F} \wedge Z \subseteq Z' \in \mathcal{F} \wedge \mu Z' = 0\}.$$

Then it follows almost immediately that $(\Omega, \hat{\mathcal{F}}, \hat{\mu})$ is a measure space where if $B \subseteq E$ and $E \setminus B$ is a subset of a set of μ-measure zero, $\hat{\mu} E = \mu B$.

An important example of a probability space is given by $(\mathbb{R}, \mathcal{B}^1, \mu)$ where $\mathcal{B}^1$ is the (linear) Borel field generated by $\mathcal{C} := \{(a, b] | -\infty < a < b < +\infty\}$. Note that Borel-Lebesgue measure λ is not a probability measure on $\mathbb{R}$, however, the measure space $(\mathbb{R}, \mathcal{B}^1, \lambda)$ is σ-finite.

The next definition introduces certain classes of measures that admit good approximations of various types.

Definition 1.22   1. The smallest σ-algebra of $\mathbb{R}^n$ that contains all the open subsets of $\mathbb{R}^n$ is called the Borel σ-algebra of $\mathbb{R}^n$.

2. A measure μ on Ω is called regular iff for each $A \subseteq \Omega$ there exists a μ-measurable set B such that $A \subset B$ and $\mu A = \mu B$.

3. A measure μ on $\mathbb{R}^n$ is called a Borel measure iff every Borel set is measurable.

4. A measure μ on $\mathbb{R}^n$ is Borel regular iff μ is Borel and for each $A \subseteq \mathbb{R}^n$ there exists a Borel set B with the property that $A \subseteq B$ and $\mu A = \mu B$.

5. A measure μ on $\mathbb{R}^n$ is called a Radon measure iff μ is Borel regular and for every compact set $K \subseteq \mathbb{R}^n$ $\mu K < \infty$.

Remark. The above definitions also hold with $\mathbb{R}^n$ replaced by any metrizable topological space.

Definition 1.23 A function $f : \Omega \to \mathbb{R}$ is said to be measurable (with respect to the measure μ) iff for every open set $G \subseteq \mathbb{R}$, $f^{-1}G \in \mathcal{F}$.

A function $F : \mathbb{R} \to \mathbb{R}$ is called a *distribution function* iff it is increasing, right-continuous, and satisfies $\lim_{x \to -\infty} F(x) = 0$ and $\lim_{x \to +\infty} F(x) = 1$.

The classical theory of the Lebesgue-Stieltjes integral gives the following result:

Proposition 1.11 *Each probability measure μ on $\mathcal{B}^1$ determines a distribution function F through the correspondance*

$$F(x) := \mu(-\infty, x], \tag{1.21}$$

for all $x \in \mathbb{R}$.
Conversely, every distribution function F determines a probability measure μ on $\mathcal{B}^1$ through (1.21). ∎

Definition 1.24 Let $(\Omega, \mathcal{F}, \pi)$ be a probability space. A random variable on $(\Omega, \mathcal{F}, \pi)$ is any π-measurable function $X : \Omega \to \overline{\mathbb{R}}$, i.e, X satisfies

$$\forall B \in \mathcal{B}^1 : \ X^{-1}B \in \mathcal{F}. \tag{1.22}$$

Any random variable on $(\Omega, \mathcal{F}, \pi)$ induces canonically a probability measure μ on $(\mathbb{R}, \mathcal{B}^1, \mu)$ via the correspondence

$$\mu B := \pi X^{-1}(B) = \pi X\{\omega \in B\}, \tag{1.23}$$

for all $B \in \mathcal{B}^1$, i.e., the following diagram commutes (a precise definition of *diagram* and *commute* will be given in the next section):

$$\begin{array}{ccc} \mathcal{F} & \xrightarrow{X} & \mathcal{B}^1 \\ \pi \downarrow & \swarrow \mu & \\ \overline{\mathbb{R}} & & \end{array} \tag{1.24}$$

or symbolically: $\mu = \pi \circ X^{-1}$.

The distribution function F is given by $F(x) = \mu(-\infty, x] = \pi\{X(\omega) \in (-\infty, x]\} =: \pi\{X \le x\}$.

Next it is briefly restated what it means for a sequence of probability measures $\{\mu_n\}$ on $\mathcal{B}^1$ to converge weak* to a probability measure μ on $\mathcal{B}^1$ (this type of convergence is sometimes also called *vague* convergence).

Proposition 1.12

$$\mu_n \xrightarrow{w^*} \mu \iff \forall f \in \mathcal{K}(K) : \int_{\mathbb{R}} f(x)\, d\mu_n \to \int_{\mathbb{R}} f(x)\, d\mu, \qquad (1.25)$$

where $\mathcal{K}(K) := \mathcal{K}_{\mathbb{R}}(\mathbb{R}|K)$. ∎

Remark. Proposition (1.12) also holds for all $f \in C_b(\mathbb{R})$, the set of all continuous real-valued functions f with domain $\mathbb{R}$ and $\lim_{|x| \to \infty} f(x) = 0$.

A sequence $\{X_n\}$ of random variables on $(\Omega, \mathcal{F}, \pi)$ is said to *converge in probability* to a random variable X on $(\Omega, \mathcal{F}, \pi)$, written $X_n \xrightarrow{p} X$, iff

$$\forall \varepsilon > 0 : \lim_{n \to \infty} \pi\{|X_n - X| > \varepsilon\} = 0, \qquad (1.26)$$

or equivalently,

$$\lim_{n \to \infty} \pi\{|X_n - X| \le \varepsilon\} = 1. \qquad (1.27)$$

This last statement can also, less rigorously of course, be rephrased as "$X_n \to X$ *with probability one.*"

A sequence of random variables $\{X_n\}$ on $(\Omega, \mathcal{F}, \pi)$ is said to *converge in distribution* to a distribution function F, written $F_n \xrightarrow{d} F$, iff the associated sequence $\{F_n\}$ of distribution functions converges weak* to F.

The following result is well-known.

Proposition 1.13 *Let* $\{X_n\}$ *be a sequence of random variables,* X *a random variable, and* $\{F_n\}$ *and* F *the associated distribution functions. If* $X_n \xrightarrow{p} X$ *then* $F_n \xrightarrow{d} F$. ∎

The next two theorems deal with the limit of partial sums of a sequence of random variables and are known as the Strong Law of Large Numbers and the Weak Law of Large Numbers, respectively.

Theorem 1.11 (Strong Law of Large Numbers) *Let* $\{X_n\}$ *be a sequence of random variables. Suppose that for all* $n \in \mathbb{N}$ *the expectation* $E(X_n) := \int_\Omega X_n(\omega) \, d\pi(\omega) < \infty$. *Let* $S_n := \sum_{j=1}^n X_n$. *Then*

$$\lim_{n \to \infty} \pi \left\{ \left| \frac{S_n - E(S_n)}{n} \right| \leq \varepsilon \right\} = 1. \tag{1.28}$$

∎

Theorem 1.12 (Weak Law of Large Numbers) *Let* $\{X_n\}$ *be a sequence of random variables with the property that for all* $n \in \mathbb{N}$ *the expectation* $E(X_n) < \infty$. *If* $S_n := \sum_{j=1}^n X_n$ *then*

$$\frac{S_n - E(S_n)}{n} \to 0, \tag{1.29}$$

for π-*a.e.* $\omega \in \Omega$.   ∎

Now the Lebesgue spaces are introduced and the Riesz Representation Theorem is presented.

***Definition* 1.25** Suppose $(\Omega, \mathcal{F}, \mu)$ is a measure space and $\mu B \geq 0$, for all $B \in \mathcal{F}$. Let $0 < p \leq \infty$. The Lebesgue space $L^p(\Omega, \mathcal{F}, \mathbb{K}, \mu)$ is the collection of all classes of μ-measurable functions $f : \Omega \to \mathbb{K}$ (one identifies functions that are equal μ-almost everywhere) with the property that $\int_\Omega |f(\omega)|^p \, d\mu < \infty$.

If the Borel field $\mathcal{F}$ is understood, $L^p(\Omega, \mathcal{F}, \mathbb{K}, \mu)$ is simply written as $L^p(\Omega, \mathbb{K}, \mu)$. For $p = \infty$ the above integral is replaced by ess sup $|f(\omega)| < \infty$. It is clear that the $L^p(\Omega, \mathbb{K}, \mu)$ form linear spaces. If

$$\|f\|_p := \left(\int_\Omega |f(\omega)|^p \, d\mu \right)^{1/p}, \tag{1.30}$$

for $0 < p < \infty$, and

$$\|f\|_\infty := \text{ess sup} \, |f(\omega)|, \tag{1.31}$$

for $p = \infty$, then $\| \cdot \|_p$ is a norm for $1 \leq p \leq \infty$. For this range of p-values the $L^p(\Omega, \mathbb{K}, \mu)$ are Banach spaces.

Sometimes the case $p = 0$ is also of interest. $L^0(\Omega, \mathbb{K}, \mu)$ is defined as the space of all classes of measurable, almost everywhere finite functions from Ω into $\mathbb{K}$ endowed with the topology of convergence in measure (convergence in measure is defined as in (1.26) with "probability" replaced by "measure").

The next theorem is one of several versions of the Riesz Representation Theorem and certainly not the most general one. However, for our purposes this particular version will suffice.

Theorem 1.13 (Riesz Representation Theorem) *Suppose that* $(\Omega, \mathcal{F}, \mathbb{K}, \mu)$ *is a σ-finite measure space. Let* $1 \le p \le \infty$ *and suppose that* q *is such that* $1/p + 1/q = 1$, *or* $q = \infty$ *if* $p = 1$. *Let* $\varphi \in (L^p(\Omega, \mathbb{K}, \mu))^*$. *Then there exists a unique function* $g \in L^q(\Omega, \mathbb{K}, \mu)$ *such that*

$$\varphi(f) = \int_\Omega f(\omega) g(\omega) \, d\mu, \qquad (1.32)$$

for all $f \in L^p(\Omega, \mathbb{K}, \mu)$. *Moreover,* $\|\varphi\| = \|g\|_q$. ∎

Definition 1.26 A Markov process is a quadruple $(\Omega, \mathcal{F}, \mu, P)$ where P is a non-expansive positive operator on $L^1(\Omega, \mathbb{R}, \mu)$.

Note that *non-expansive* refers to the L^1-norm of P, i.e.,

$$\|P\| = \sup\{|Pf| \, | \, \|f\|_{L^1} \le 1\} \le 1, \qquad (1.33)$$

and *positive* means that

$$0 \le f \in L^1(\Omega, \mathbb{R}, \mu) \implies Pf \ge 0. \qquad (1.34)$$

By Proposition 1.9, there exists a unique adjoint P^* of P acting on $(L^1(\Omega, \mathbb{R}, \mu))^* = L^\infty(\Omega, \mathbb{R}, \mu)$, the set of all functions $f : \Omega \to \mathbb{R}$ with ess sup $|f| < \infty$. By the Riesz Representation Theorem, the correspondence

$$L^1(\Omega, \mathbb{R}, \mu) \ni f \mapsto \int_\Omega f \cdot f^* \, d\mu, \qquad (1.35)$$

for a unique $f^* \in L^\infty(\Omega, \mathbb{R}, \mu)$, defines a linear functional, i.e., an element of $L^\infty(\Omega, \mathbb{R}, \mu)$. This duality will be denoted by

$$\langle f, f^* \rangle = \int_\Omega f \cdot f^* \, d\mu. \qquad (1.36)$$

Hence
$$\langle Pf, f^* \rangle = \langle f, P^* f^* \rangle,$$

for all $f \in L^1(\Omega, \mathbb{R}, \mu)$ and $f^* \in L^\infty(\Omega, \mathbb{R}, \mu)$.

Since P^* is also a non-expansive and positive operator (however, on $L^\infty(\Omega, \mathbb{R}, \mu)$), P or P^* is called a *Markov operator*.

Now let $\chi_A : A \to \mathbb{R}$ denote the *characteristic function* on $A \in \mathcal{F}$. Define

$$P(\omega, A) := (P^* \chi_A)(\omega). \qquad (1.37)$$

Then $P(\cdot, , A)$ is a function from Ω into $\mathbb{R}$. Furthermore,

1. range $P(\cdot, A) = [0, 1]$.

2. $P(\cdot, A)$ is $\mathcal{F}$-measurable for all fixed $A \in \mathcal{F}$.

3. If A_i is a countable collection of disjoint sets in $\mathcal{F}$, then

$$P(\cdot, \bigcup_i A_i) = \sum_i P(\cdot, A_i).$$

4. If $\mu A = 0$, then $P(\omega, A) = 0$.

A function $P(\cdot, A)$ that satisfies the above conditions (a) — (c) everywhere is called a *transition probability* for the Markov process. The value of $P(\omega, A)$ is the probability of transfer of $\omega \in \Omega$ into A.

Conversely, one can show that, given a σ-finite measure space $(\Omega, \mathcal{F}, \mu)$ and a function $P(\cdot, A)$, $A \in \mathcal{F}$ fixed, satisfying conditions 1 through 3 above, it is possible to obtain a Markov process by defining P^* on $L^\infty(\Omega, \mathbb{R}, \mu)$ by

$$(P^* f)(\omega) := \int_\Omega f(\omega') \, P(\omega, d\omega'). \qquad (1.38)$$

It is easy to see that the product of two Markov operators is again a Markov operator. In particular, $(P^*)^n$ is a Markov operator for any $n \in \mathbb{N}$, and

$$P^{n+1}(\omega, A) = \int_\Omega P^n(\omega', A) \, P(\omega, d\omega').$$

In later chapters mostly discrete Markov processes, called *Markov chains*, are considered. One way of obtaining these Markov chains is by setting $\Omega = \mathbb{N}$, $\mathcal{F} = 2^{\mathbb{N}}$, the power set on $\mathbb{N}$, and by letting μ be counting measure, i.e.,

$\mu\{n\} = 1$, for all $n \in \mathbb{N}$. In this case the operator P is a positive matrix $(P_{nk})_{n,k\in\mathbb{N}}$ satisfying $\sum_{n\in\mathbb{N}} P_{nk} \leq 1$. Also, $P(\omega, A) = \sum_{n\in A} P_{nk}$, $A \subseteq \mathbb{N}$, and

$$(Pf)(k) = \sum_{n\in\mathbb{N}} P_{nk} f(n), \tag{1.39}$$

$$(P^*f^*)(n) = \sum_{k\in\mathbb{N}} P_{nk} f^*(k). \tag{1.40}$$

This characterization of a Markov process follows directly from its description in terms of random variables.

A sequence $\{X_n\}_{n\in\mathbb{N}_0}$ of random variables on $(\Omega, \mathcal{F}, \mu)$ is called a Markov process iff

$$\pi\{X_{n+1} \in B | X_0, \ldots, X_n\} = \pi\{X_{n+1} | X_n\}, \tag{1.41}$$

for all $n \in \mathbb{N}_0$ and for all $B \in \mathcal{B}^1$ (here $\pi\{X|Y)\}$ denotes the conditional probability of X given Y).

In the following chapters a certain class of countable Markov processes is considered. To give an equivalent characterization of such Markov chains the concept of a *sequence space* has to be introduced. These sequence spaces will continue to play an important role in the theory of iterated function systems and fractal functions.

Let $S := \{1, \ldots, N\}$, $N \in \mathbb{N}$, $N > 1$, be a subset of $\mathbb{N}$, called the *state space*. Let $\Sigma := S^{\mathbb{N}}$ and let $\Sigma_n := S^{\mathbb{N}_n}$, where $\mathbb{N}_n := \{1, \ldots, n\}$ ($\mathbb{N}_n$ is sometimes called an *initial segment* of $\mathbb{N}$ of length n). The elements of Σ, respectively Σ_n, are infinite sequences, respectively n-tuples, of the form, $\mathbf{i} = (i_0 i_1 \ldots i_n \ldots)$ and $\mathbf{i}(n) = (i_0 i_1 \ldots i_n)$, respectively, where $i_j \in \{1, \ldots, N\}$. The set Σ_n is called a *finite tree* of length n with N branches at each branchpoint i_j. The set Σ is called a *sequence space*, its elements $\mathbf{i}$ *codes*, and i_n the *nth outcome* on $\mathbf{i}$. The nth outcome is said to occur at *time n*. A code in Σ_n is called a *finite code of length n*.

Let $S_0, \ldots, S_n$ be subsets of the state space S. Let $\mathcal{F}_n$ denote the collection of all unions of sets in Σ of the form

$$\{\mathbf{i} = (i_0 \, i_1 \, \ldots \, i_n \, \ldots) | \, i_0 \in S_0, \ldots, i_n \in S_n\}, \tag{1.42}$$

and let $\mathcal{F} := \bigcup_{n=0}^{\infty} \mathcal{F}_n$. Although each $\mathcal{F}_n$ is a Borel field, $\mathcal{F}$ is in general not. A measure will be defined on the smallest Borel field $\mathcal{G}$ containing $\mathcal{F}$, and then the completion of $\mathcal{G}$ is taken. In order to achieve this, a few more definitions and facts are needed.

A set of $\mathcal{F}$ is called a *cylinder set*. If C is a cylinder set of $\mathcal{F}_n$, then C can be written as

$$C = \bigcup_{k \in \mathbb{N}_0} B_k^n, \tag{1.43}$$

where the *basic cylinder sets* B_k^n are defined by

$$B_k^n := \{\mathbf{i} | \, i_0 = c_0, \ \ldots, \ i_n = c_n\}. \tag{1.44}$$

Now let

$$\nu B_k^n := \prod_{m=0}^{n} p_{i_m}, \tag{1.45}$$

where p_{i_m} denotes the probability that outcome i_m will occur. Thus, ν is a measure on $\mathcal{F}_n$ and $\nu \Sigma = \sum_{n \in \mathbb{N}} \prod_{m=0}^{n} p_{i_m} = (\sum_{j=1}^{N} p_j)^{\mathbb{N}} = 1$.

It is possible to extend ν to a probability measure μ on $\mathcal{G}$, and then to complete μ. The completion of μ is denoted by $(\Sigma, \mathcal{F}, \mu)$. (Here, in order to ease notation, the ^ is deleted from $\mathcal{F}$ and μ.)

It is possible to define a metric, the so-called *Fréchet metric*, on the set Σ: Let $d_F : \Sigma \times \Sigma :\to \mathbb{R}$ be given by

$$d_F(\mathbf{i}, \mathbf{j}) := \sum_{n \in \mathbb{N}} \frac{|i_n - j_n|}{(N+1)^n}, \tag{1.46}$$

for all $\mathbf{i} = (i_1 \ldots i_n \ldots), \mathbf{j} = (j_1 \ldots j_n \ldots) \in \Sigma$.

The metric space (Σ, d_F) is — by Tychonov's Theorem — compact, and it is easy to verify that Σ is homeomorphic to the classical Cantor set on N symbols. $(\Sigma, \mathcal{F}, \mu)$ or (Σ, d_F) is called a *code space*.

Now suppose S is a countably finite or infinite state space and $\{X_n\}_{n \in \mathbb{N}_0}$ is a sequence of random variables defined on a code space $(\Sigma, \mathcal{F}, \mu)$ with values in S, i.e., $X_n : \Sigma \to S$, for all $n \in \mathbb{N}$. The sequence $\{X_n\}_{n \in \mathbb{N}_0}$ is called a *countable Markov chain* if

$$\pi\{X_{n+1} = i_{n+1} | X_n = i_n, \ldots, X_0 = i_0\} = \pi\{X_{n+1} = i_{n+1} | X_n = i_n\}, \quad (1.47)$$

for all $i_0, \ldots, i_{n+1} \in S$.

The *starting or initial probability distribution* is a vector p whose components p_i are given by

$$p_i = \pi\{X_0 = i\},$$

and the transition matrix is defined by

$$P_{ij} = \pi\{X_{n+1} = j | X_n = i\},$$

provided $\pi\{X_n = i\} > 0$.

Assume that the matrix $P = (P_{ij})$ has a fixed point p^*, i.e., $Pp^* = p^*$. Then p^* is called a *limiting probability* or a *stationary (probability) distribution* (that $p^* = (p_i^*)$ is a probability vector, i.e., a vector satisfying $p_i^* \geq 0$ and $\sum_i p_i^* = 1$, follows directly from (1.39). If the cardinality of S is finite, say $\|S\|_c = N$, and if the $N \times N$ transition matrix P is positive, the following result holds.

Theorem 1.14 *Let $p_i(n)$ denote the probability that at time n, $X_n = i$, and let $p(n) := (p_1(n), \ldots, p_N(n))$ be the corresponding probability vector. Suppose that $p(n+1) = Pp(n)$. Then*

$$\lim_{n \to \infty} p(n) = p^*, \tag{1.48}$$

where p^* is a probability vector, independent of $p(0)$.
Furthermore, p^* is an eigenvector of P with associated eigenvalue 1. ∎

1.3 Algebra

In this rather brief section on abstract algebra, the reader is reminded of such rudimentary concepts as *diagram, semigroup, group, endomorphisms, free semigroups*, and *free groups*. These notions will be used in the next chapter when M. F. Dekking's approach to fractal sets is introduced, and also later in connection with dimension calculations for fractals. A rather short review of category theory and direct and inverse limits is also presented.

***Definition* 1.27** The quadruple (A, F, α, β) is called a diagram D of mappings if it consists of the following:

1. A system of sets $A = (A_i)_{i \in I}$, where I is an index set.

2. A system of mappings $F = (F_j)_{j \in J}$ indexed by a set J.

3. Mappings $\alpha : J \to I$ and $\beta : J \to I$, satisfying

   (a) dom $F_j = A_{\alpha(j)}$ and range $F_j = A_{\beta(j)}$, i.e., $F_j : A_{\alpha(j)} \to A_{\beta(j)}$, for all $j \in J$.

   (b) $\alpha(J) \cup \beta(J) = I$.

The diagram D is called finite iff J is finite.

The elements of F are called *arrows*, and an n-tuple $\tau := (F_{j_1}, \ldots, F_{j_n})$ of arrows a *path* in D, iff for all $\ell = 1, \ldots, n-1$, the arrow F_{j_ℓ} ends where the arrow $F_{j_{\ell+1}}$ begins. The arrow F_{j_1}, respectively F_{j_n}, is called the *initial point*, respectively *terminal point*, of the path τ. Along a path τ one can form the composition F_τ of the mappings $F_{j_n} \circ \cdots \circ F_{j_1}$.

Definition 1.28   1. A diagram D is called commutative iff for two paths τ and τ' having the same initial and terminal point in ϑ: $F_\tau = F_{\tau'}$.

2. A diagram D is called a sequence iff

   (a) at each A_i, $i \in I$, there begins or ends at most one arrow;

   (b) for any two distinct sets A_{i_1} and A_{i_2}, there exists exactly one path whose initial point is A_{i_1} and whose terminal point is A_{i_2}, for all $i_1, i_2 \in I$.

Next *semigroups* and *groups* are defined. Let X be a *set* of objects called the *elements* of the set. A *binary operation* in X is a function $\theta : X \times X \to X$, $(a,b) \mapsto \theta(a,b)$. Following the usual convention, one writes ab for $\theta(a,b)$, although ab may *not* mean ordinary multiplication. Moreover, ab is called the *product of a with b*.

A binary operation θ on X is called *associative* if

$$\forall a,b,c \in X : \theta(\theta(a,b),c) = \theta(a,\theta(b,c)), \tag{1.49}$$

or simply

$$\forall a,b,c \in X : (ab)c = a(bc). \tag{1.50}$$

An element e in X is called a *unit* or a *neutral element* if

$$\forall a \in X : ae = ea = a. \tag{1.51}$$

It is easy to establish that the unit in a set X — provided it exists — is unique.

Definition 1.29 A semigroup S is a pair (X,θ), where X is a set and θ an associative binary operation on X.

A subset of T of a semigroup S is called *stable (with respect to the binary operation θ in S)* provided that for all $a,b \in T$, $ab \in T$. If T is a stable subset of S, the restriction ρ of θ to $T \times T$ defines a binary operation on T. Together with this binary operation $\rho : T \times T \to T$, T becomes a semigroup. Since $T \subseteq S$, T is called a subsemigroup of S.

Now let X be an arbitrary subset of a semigroup S. If the smallest (with respect to set containment) subsemigroup of S that contains X is S itself, then X is said to be a *set of generators* for S, and that S is *generated by X*.

Now suppose S and S' are semigroups. A *homomorphism* of S into S' is a function $h : S \to S'$ such that

$$\forall a, b \in S : h(ab) = h(a)h(b). \tag{1.52}$$

An homomorphism is called a *monomorphism* iff it is injective, and an *epimorhism* iff it is surjective. A bijective homomorphism is called an *isomorphism*. If $S' = S$, then h is called an *endomorphism* of S. Isomorphic endomorphisms are called *automorphisms*.

Notation. For an injective mapping f from a set A into a set B, one writes $f : A \rightarrowtail \triangleright$ or $A \overset{f}{\rightarrowtail} B$, and for a surjective mapping g from a set A into a set B, $g : A \twoheadrightarrow \triangleright$ or $A \overset{g}{\twoheadrightarrow} B$.

Let X be an arbitrary set. A *free semigroup* on the set X is a semigroup S together with a function $f : X \to S$ such that, for every function $g : X \to S'$ from the set X into a semigroup S', there exists a unique homomorphism $h : S \to S'$ so that the following diagram commutes:

$$
\begin{array}{ccc}
X & \overset{f}{\to} & S \\
{\scriptstyle g}\downarrow & \swarrow{\scriptstyle h} & \\
S' & &
\end{array}
\tag{1.53}
$$

This definition expresses what is called the *universality* of a free semigroup.

The following facts about free semigroups are straightforward to prove.

Proposition 1.14 *Assume that the semigroup S together with the function $f : X \to S$ is a free semigroup on the set X. Then f is injective and its image fX generates S.* ■

Theorem 1.15 (Uniqueness Theorem) *Suppose that (S, f) and (S', f') are free semigroups on the same set X. Then there exists a unique isomorphism $j : S \to S'$ such that the following diagram commutes:*

$$
\begin{array}{ccc}
X & \overset{f}{\to} & S \\
{\scriptstyle f'}\downarrow & \swarrow{\scriptstyle j} & \\
S' & &
\end{array}
\tag{1.54}
$$

■

The proof of the following Existence Theorem for free semigroups is given because it introduces notation and terminology that will be used later.

Theorem 1.16 (Existence Theorem) *For any set X, there exists a free semigroup on X.*

Proof. Let $S := \{(x_1, \ldots, x_n) \mid x_1, \ldots, x_n \in X, n \in \mathbb{N}\}$. Define a binary operation on S as follows: If $\xi = (x_1, \ldots, x_n)$ and $\eta = (y_1, \ldots, y_m)$ are in S, define $\xi\eta$ to be the concatenated finite sequence

$$\xi\eta = (x_1, \ldots, x_n, y_1, \ldots, y_m).$$

It is easy to show that this binary operation is associative. Hence, S together with binary operation is a semigroup. Next define $f : X \to S$ by $f(x) = (x)$, where (x) is the finite sequence consisting of the single element x. To prove that the pair (S, f) is a free semigroup on X, let $g : X \to S'$ be a given arbitrary function from the set X into a semigroup S'. Define a function $h : S \to S'$ as follows:

$$h(\xi) = g(x_1) \cdots g(x_n),$$

for all $\xi = (x_1, \ldots, x_n) \in S$.

Since the binary operation on S' is associative, the function h is a homomorphism. Furthermore,

$$\forall x \in S : (h \circ f)(x) = h(f(x)) = h((x)) = g(x),$$

i.e., $h \circ f = g$.

To show that h is unique, let $h' : S \to S'$ be an arbitrary homomorphism satisfying $h' \circ f = g$. The following equalities hold for all $\xi = (x_1, \ldots, x_n) \in S$:

$$
\begin{aligned}
h'(\xi) &= h'((x_1) \cdots (x_n)) = h'((x_1)) \cdots h'((x_n)) \\
&= h'(f(x_1)) \cdots h'(f(x_n)) = g(x_1) \cdots g(x_n) \\
&= h(\xi),
\end{aligned}
$$

and therefore $h = h'$.   ∎

Remarks.

1. Every set X determines — up to isomorphisms — a unique free semi-group (S, f). By Proposition 1.14, X is identified with its image fX in S. Hence, X becomes a subset of S that generates S, and every function $g : X \to S'$ from X into an arbitrary semigroup S' extends to a unique homomorphism $h : S \to S'$. (S, f) is referred to as the *semigroup generated by* X.

2. The set X is also called the *alphabet,* and the elements of the free semi-group generated by X are called *words.*

The next result is an easy application of the Existence Theorem for free semi-groups.

Proposition 1.15 *Let X be a set of generators of a semigroup (S, f). Then every element of S can be written as the product of a finite sequence of elements in X.*

For our later purposes the concept of *group* is needed, and especially that of a *free group.* A *group* G is a semigroup that has the following two additional properties:

(a) $\exists e \in G \, \forall g \in G : ge = eg = g.$

(b) $\forall g \in G \, \exists g^{-1} \in G : gg^{-1} = g^{-1}g = e.$

For any two subsets H, K of a group G define the following new subsets:

$$H^{-1} := \{h^{-1} | h \in H\},$$

$$HK := \{hk | h \in H, \, k \in K\},$$

$$HK^{-1} := \{hk^{-1} | h \in H, \, k \in K\}.$$

Definition 1.30 A non-empty subset H of a group G is called a subgroup of G iff

$$HH^{-1} \subseteq H. \tag{1.55}$$

There exist special subgroups in a group G that can be used to *abelianize* it.

Definition 1.31 A group G is called abelian or commutative iff $gh = hg$, for all $g, h \in G$.

In order to justify the foregoing remark, a few more definitions and results from elementary group theory are needed.

Suppose H is an arbitrary subgroup of a group G. One can define a relation $\sim$ in the set G as follows: For any two elements $g, h \in G$, let $g \sim h$ iff $g^{-1}h \in H$. It is not at all difficult to show that the relation $\sim : G \times G \to G$ is reflexive, symmetric, and transitive, and hence an *equivalence relation* on the set G. This equivalence relation divides the set G into disjoint subsets whose union is G. These subsets are called the *equivalence classes modulo* $\sim$.

Denote by Q the totality of all equivalence classes modulo $\sim$. This set is called the *quotient set of the group G over its subgroup H*; symbolically,

$$Q = G/H.$$

An interesting question is under what conditions on the subgroup H the quotient set $Q = G/H$ becomes a *quotient group*. The following proposition gives the answer to this question. But first a definition.

Definition 1.32 A subgroup H of a group G is called normal iff $gH = Hg$, for all elements $g \in G$ (here $gH := \{gh|\, h \in H\}$).

Proposition 1.16 *Suppose H is a normal subgroup of a group G, and $Q = G/H$ the quotient set of G over H. The binary operation in Q defined by*

$$(gH)(g'H) := (gg')H,$$

for all $g, g' \in G$, makes Q into a group, the quotient group of G over H.

Furthermore, the natural projection $p : G \twoheadrightarrow H$, $p(g) := gH$, is an epimorphism whose kernel equals H. ∎

Now suppose X is a subset of a group G. Then there exists a smallest subgroup H (with respect to set containment) of G that contains X. This subgroup is called the *subgroup generated by X*. In case $H = G$, G is said to be *generated by X* and X to be *a set of generators for G*. The next result gives some information about the form of the elements of a group that is generated by a given set.

Proposition 1.17 *Let X be a set of generators for a group G. Then every element in G can be expressed as a product of a finite sequence of elements in $X \cup X^{-1}$.* ∎

Let $g, g' \in G$ be any two elements in a group G. The element $gg'g^{-1}g'^{-1}$ is called the *commutator of g and g'*. The subgroup $\Gamma(G)$ generated by all commutators in G is called the *commutator subgroup of G*. Clearly, $\Gamma(G)$ is normal, and thus $G/\Gamma(G)$ is a group; moreover, it is an abelian group. Therefore, $G/\Gamma(G)$ is called the *abelianization of G*.

Since every group is also a semigroup, the preceding definitions of homomorphism, monomorphism, endomorphism, isomorphism, and automorphism carry over to groups.

Proposition 1.18 *Suppose $h : G \to G'$ is a homomorphism of a group G into a group G'. Then*

1. *h maps the neutral element e_G of G into the neutral element $e_{G'}$ of G;*

2. *$h(x^{-1}) = (h(x))^{-1}$.*

∎

Next, the *direct product* of an arbitrary family of groups is introduced. For this purpose, let I be an arbitrary index set indexing a given family of groups $\{G_i | i \in I\}$. Denote by G the union of the sets G_i and by $\prod_{i \in I} G_i$ the cartesian product of the family of sets G_i. By the definition of cartesian product, each element of $\prod_{i \in I} G_i$ is a function $f : I \to G$ such that $f(i) \in G_i$ for every $i \in I$. One defines a binary operation θ on $\prod_{i \in I} G_i$ as follows: For any two elements $f, g \in \prod_{i \in I} G_i$, let $\theta(f, g)$ be the function given by

$$(fg)(i) := f(i)g(i) \in G_i, \qquad \text{for all } i \in I.$$

The neutral element of $\prod_{i \in I} G_i$ is the function $e : I \to G$ given by $e(i) = e_i \in G_i$, with e_i being the neutral element in G_i, and the inverse element is the function $f^{-1} : I \to G$ defined by $f^{-1}(i) = [f(i)]^{-1}$ for all $i \in I$. The pair $(\prod_{i \in I} G_i, \theta)$ is called the *direct product* of the given family of groups. In case the G_i are abelian groups, the direct product is also called the *direct sum* of the family $\{G_i | i \in I\}$ of abelian groups and written as $(\bigoplus_{i \in I} G_i, \theta)$.

Now suppose X is an arbitrarily given set. A *free group on the set X* is a group F together with a function $f : X \to F$ such that, for every function $g : X \to G$ from the set X into a group G, there exists a unique homomorphism $h : F \to G$ with the property

$$h \circ f = g.$$

The following two results are the analogues of Proposition 1.14 and Theorem 1.15.

Proposition 1.19 *Let (F, f) be a free group on the set X. Then the function $f : X \to F$ is injective and its image fX generates F.*   ∎

Theorem 1.17 (Uniqueness Theorem for Free Groups)   *Let (F, f) and (F', f') be free groups on the same set X. Then there exists a unique isomorphism $j : F \to F'$ such that $j \circ f = f'$.*   ∎

Now the Existence Theorem for Free Groups is established.

Theorem 1.18 (Existence Theorem for Free Groups)   *Let X be any set. Then there exists a free group on X.*

Proof. Given the set X, define a new set Y by

$$Y := X \times \{-1, 1\}.$$

Set $x^1 := (x, 1)$ and $x^{-1} := (x, -1)$. It is clear that the set Y generates a free semigroup S, whose words are finite formal products of elements of Y.

A word w is called *reduced* iff, for any $x \in X$, x^1 never stands next to x^{-1} in w. The symbol e will stand for the empty word. Now define F as the collection of all reduced words in S together with e.

To make F into a group, a binary relation has to be defined on it, and it must be shown that this binary relation satisfies the group axioms. So suppose that $u, v \in F$ are arbitrary. If $u = e$, define $uv = v$, and if $v = e$, define $uv = u$. If neither u nor v equals e, u and v are reduced words in S, and so uv is in S. Two cases are possible: Either $uv = e$ or $uv = w$, where w is a reduced word obtained by cancelling from uv all pairs of the form $x^1 x^{-1}$ and $x^{-1} x^1$. Hence, define a binary operation on F by

$$uv := \begin{cases} e \\ w \end{cases},$$

according to the two cases mentioned earlier.

It is straightforward to verify that F with this binary operation is a group with unit e.

Now define a function $f : X \to F$ by setting $f(x) := x^1 \in F$, for all $x \in X$. It remains to be shown that (F, f) is a free group on X.

For this purpose, suppose that G is an arbitrary group and that $g : X \to G$ is an arbitrary function from the set X into G. Let $w \in F$ be arbitrary. Then w is either the empty word e or w is of the form $w = x_1^{\varepsilon_1} x_2^{\varepsilon_2} \cdots x_n^{\varepsilon_n}$, where $\varepsilon_i := \pm 1$, $i = 1, \ldots, n$. Define a function $h : F \to G$ by

$$
h(w) := \begin{cases} e_G & \text{if } w = e \\ (g(x_1))^{\varepsilon_1} \cdots (g(x_n))^{\varepsilon_n} & \text{otherwise} \end{cases},
$$

(e_G is the unit in G). Obviously, h is a homomorphism satisfying $h \circ f = g$.

The uniqueness of h is established using an argument similar to the one given in Theorem 1.16. ∎

The uniqueness theorem 1.17 implies that every set determines essentially a unique free group (F, f). The injectivity of the function f allows us to identify the set X with its image fX is F. This identification is called the *embedding of X into F*; in symbols, $X \hookrightarrow F$. Hence, X is a set of generators for F. Furthermore, every function $g : X \to G$ from the set X into an arbitrary group G extends to a unique homomorphism $h : F \to G$. Therefore, F is called the *free group generated by the set X*.

It is sometimes useful to indicate the set X that generates a free semigroup or a free group. The notation $S[X]$, respectively $F[X]$, is used to express the fact that the free semigroup, respectively free group, is generated by the set X.

Before closing out this section, the relation between a free semigroup $S[X]$ and a free group $F[X]$ that are both generated by the same set X is investigated. By Theorem 1.16 and 1.18, and especially their respective proofs, it is not difficult to see that there is an obvious embedding $S[X] \hookrightarrow F[X]$ of a free semigroup $S[X]$ into a free group $F[X]$: The free group $F[X]$ is clearly a semigroup, and by the universality of $S[X]$ there exists a homomorphism h from $S[X]$ into the *semigroup* $F[X]$. The injectivity of h follows from the observation that the generators of $S[X]$ are mapped onto generators of $F[X]$, and that distinct words $x_1 \cdots x_n$ and $y_1 \cdots y_m$ in $S[X]$ can be identified with the distinct words $x_1^1 \cdots x_n^1$ and $y_1^1 \cdots y_m^1$ in $F[X]$.

Assume that the set X has finite cardinality, say $\|X\|_c = n$. Let $F[X]$ be the free group generated by this set X. The abelianization of $F[X]$ is the free abelian group $\mathbb{Z}^X$; the generators $\{x_1, \ldots, x_n\}$ of $F[X]$ correspond to the n-tuples $\{(1, 0, \ldots, 0), \ldots, (0, \ldots, 1)\}$ in $\mathbb{Z}^X$. Thinking of $\mathbb{Z}^X$ primarily as the set of all functions from X into $\mathbb{Z}$, the natural projection $p : F[X] \twoheadrightarrow \mathbb{Z}^X$ can be written as

$$
p(x)(y) = \delta_{xy}, \tag{1.56}
$$

for all $x, y \in X$. Here δ_{xy} denotes the *Kronecker delta*, defined by

$$\delta_{xy} := \left\{ \begin{array}{ll} 1 & x = y \\ 0 & \text{otherwise} \end{array} \right. ,$$

i.e., the generator x_i is mapped onto the n-tuple $(0, \ldots, 1, \ldots 0)$ in $\mathbb{Z}^X$, where 1 is in the ith position.

The notion of a *category* is one of the most important and far-reaching tools in modern mathematics: It is a unifying and clarifying device for apparently different and complicated concepts. It is therefore worthwhile to remind the reader of the definition of category and also a few related results from category theory. Some of these results will be used in later chapters, thus giving a more complete insight into nature of the subject.

***Definition* 1.33** A category $\mathfrak{K}$ consists of

1. a class $\text{obj}(\mathfrak{K})$ of objects,

2. disjoint sets $\mathfrak{K}(A, B)$ for each ordered pair (A, B) of elements of $\text{obj}(\mathfrak{K})$,

3. and compositions $\mathfrak{K}(B, C) \times \mathfrak{K}(A, B) \to \mathfrak{K}(A, C)$ (for short: $(f, g) \mapsto fg$, for $g \in \mathfrak{K}(A, B)$ and $f \in \mathfrak{K}(B, C)$) for each tripel (A, B, C) of elements of $\text{obj}(\mathfrak{K})$ such that

    (a) $\forall A \in \text{obj}(\mathfrak{K}) \ \exists I_A \in \mathfrak{K}(A, A)$: $I_A f = f$ and $g I_A = g$, for all $f \in \mathfrak{K}(B, A)$ and $g \in \mathfrak{K}(A, B)$.

    (b) $\forall f \in \mathfrak{K}(A, B) \ \forall g \in \mathfrak{K}(B, C) \ \forall h \in \mathfrak{K}(C, D) : h(gf) = (hg)f$.

The elements of $\mathfrak{K}(A, B)$ are called $\mathfrak{K}$-morphism from A to B and are denoted by $f : A \to B$ or $A \overset{f}{\to} B$. The collection of all $\mathfrak{K}$-morphisms is denoted by $\text{Mor}(\mathfrak{K})$. The elements I_A, $A \in \text{obj}(\mathfrak{K})$, are called identities. If $\text{obj}(\mathfrak{K})$ is a set, the category is called small.

Examples of categories are the following:

1. Let $\text{obj}(\mathfrak{S})$ be the class of all sets and let $\text{Mor}(A, B)$ be the set of all functions from A to B. This is the category $\mathfrak{S}$ of sets.

2. Let $\text{obj}(\mathfrak{G})$ be the class of all groups and let $\text{Mor}(A, B)$ be the set of all group homomorphisms. Then $\mathfrak{G}$ is the category of groups.

3. If obj($\mathfrak{T}$) is the class of all topological spaces and $\mathrm{Mor}(A, B)$ the set of all continuous functions from A to B, then $\mathfrak{T}$ is the category of topological spaces.

Definition 1.34 Suppose $\mathfrak{K}$ is a category. A $\mathfrak{K}$-morphism $f : A \to B$ is called a retraction iff there exists a $\mathfrak{K}$-morphism $g : B \to A$ such that $f \circ g = id_B$. B is called the retract of A under f.

Definition 1.35 A directed system $\{G_\alpha, r_\alpha^\beta\}$ of abelian groups over a directed set $(\mathbf{D}, \preceq)$ is a function that assigns to each element $\alpha \in \mathbf{D}$ an abelian group G_α and to each pair (α, β) of elements of $\mathbf{D}$ with $\alpha \preceq \beta$ a homomorphism $r_\alpha^\beta : G_\alpha \to G_\beta$ such that

(a) $\forall \alpha \in \mathbf{D} : r_\alpha^\alpha = id_{G_\alpha}$.   (b) $\forall \alpha \preceq \beta \preceq \gamma : r_\beta^\gamma r_\alpha^\beta = r_\alpha^\gamma$.

Remark. Analogously one may define directed systems of groups, rings, modules, homomorphisms, etc.

Suppose that $\{G_\alpha, r_\alpha^\beta\}$ is a directed system of abelian groups. Regard the G_α as pairwise disjoint sets and define the union $\bigcup_{\alpha \in \mathbf{D}} G_\alpha$. An equivalence relation on $\bigcup_{\alpha \in \mathbf{D}} G_\alpha$ can be defined as follows: For $g_\alpha \in G_\alpha$ and $g_\beta \in G_\beta$ let $g_\alpha \sim g_\beta \iff \exists \gamma \in \mathbf{D} : \alpha \preceq \gamma \wedge \beta \preceq \gamma \wedge r_\alpha^\gamma(g_\alpha) = r_\beta^\gamma(g_\beta)$. Denote the equivalence class of g_α by $[g_\alpha]$, and the set of all equivalence classes by G. The set G can be made into an abelian group by defining $+ : G \times G \to G$, $([g_\alpha], [g_\beta]) \mapsto [g_\alpha] + [g_\beta]$ as the equivalence class of $r_\alpha^\gamma(g_\alpha) + r_\beta^\gamma(g_\beta)$, and $-[g_\alpha]$ as $[-g_\alpha]$. The identity elements 0_α in G_α are to represent the identity element $0 \in G$. The so obtained abelian group is called the *direct* or *inductive limit* of the directed system $\{G_\alpha, r_\alpha^\beta\}$ and is written as

$$G = \varinjlim G_\alpha.$$

For a fixed $\alpha \in \mathbf{D}$, a homomorphism $r_\alpha : G_\alpha \to G$ can be defined by $G_\alpha \ni g_\alpha \mapsto [g_\alpha] \in G$. Clearly, if $\alpha \preceq \beta$ then $r_\alpha = r_\beta r_\alpha^\beta$.

In a similar fashion, one can make G for a directed system of rings or modules into a ring or module, respectively.

Example 1.1 Let G be a group and let G_α be the collection of all finitely generated subgroups of G. An ordering on these subgroups can be introduced by setting $\alpha \preceq \beta \implies G_\alpha \subseteq G_\beta$. The inclusion maps $G_\alpha \hookrightarrow G_\beta$ are taken as the r_α^β. Then $\{G_\alpha, r_\alpha^\beta\}$ is a directed system and its direct limit is G.

The following results is sometimes used to define the direct limit of a directed system:

Theorem 1.19 *Assume that $\{G_\alpha, r_\alpha^\beta\}$ is a directed system of abelian groups and $s_\alpha : G_\alpha \to G'$ a system of homomorphisms from G_α into an abelian group G' satisfying $s_\alpha = s_\beta r_\alpha^\beta$ for all $\alpha \preceq \beta$. Then there exists a unique homomorphism*

$$s : G = \varinjlim G_\alpha \to G_\alpha,$$

so that the following diagram commutes for all $\alpha \in \mathbf{D}$:

$$
\begin{array}{ccc}
G_\alpha & \overset{f}{\to} & G' \\
{\scriptstyle r_\alpha}\downarrow & \nearrow{\scriptstyle s} & \\
G & &
\end{array}
\qquad\qquad (1.57)
$$

■

One may define a *dual* object to directed system and direct limit. This dual object is called an *inverse system* and the corresponding limit the *inverse* or *projective limit*.

***Definition* 1.36** Let $(\mathbf{D}, \preceq)$ be a directed set and let $((S_\alpha, \mathcal{S}_\alpha)| \alpha \in \mathbf{D})$ be a collection of topological spaces. Suppose that for all $\alpha \preceq \beta$ there exist maps $p_\alpha^\beta : S_\beta \to S_\alpha$ such that

$$(a) \quad p_\alpha^\alpha = id_{S_\alpha}, \qquad (b) \quad \forall \alpha \preceq \beta \preceq \gamma : p_\alpha^\gamma = p_\alpha^\beta p_\beta^\gamma.$$

Then $\{S_\alpha, p_\alpha\}$ is called an inverse system of topological spaces over the directed system $(\mathbf{D}, \preceq)$.
The inverse limit S of $\{S_\alpha, p_\alpha\}$, written

$$S = \varprojlim S_\alpha,$$

is the subset of $\prod_{\alpha \in \mathbf{D}} S_\alpha$ consisting of mappings $s_\alpha : S \to S_\alpha$ with the property that $s_\alpha = p_\alpha^\beta s_\beta$ for all $\alpha \preceq \beta$. The topology of S is that induced by $\prod_{\alpha \in \mathbf{D}} S_\alpha$ and the map s_α is called the α-coordinate of $s \in S$.

***Example* 1.2** Consider the collection of all subsets S_α of a topological space $(S, \mathcal{S})$. These subspaces are ordered by inverse inclusion, i.e., $\alpha \preceq \beta \Longrightarrow S_\alpha \supset S_\beta$. If these inclusions are denoted by p_α^β, then $\{S_\alpha, p_\alpha\}$ is an inverse system and its inverse limit S is homeomorphic to $\bigcap_{\alpha \in \mathbf{D}} S_\alpha$.

The next two results will be used implicitly in later chapters.

Theorem 1.20 *The direct limit exists in the category of groups.*   ■

Theorem 1.21 *The inverse limit exists in the category of sets.*   ■

Chapter 2

Construction of Fractal Sets

The current chapter will focus on the construction of fractal sets, some of their properties, and the concept of *dimension* of a set. Over the last decades several methods for such constructions have emerged. They allow the generation and classification of a broad class of fractals which contains the classical Cantor sets, Weierstraß's nowhere differentiable function, the Sierpiński gasket, and Peano curves (the reason Peano curves are included in this list will become apparent when they are discussed later). Fractal sets are thought of as objects that are obtained by an *infinite* recursive or inductive process of successive microscopic refinements. This process is one of the reasons why the Hausdorff-Besicovitch dimension sometimes — but not always — exceeds the topological dimension of the fractal.

First, the above-mentioned classical examples of fractal sets are considered and some of their features discussed. By doing this, a new class of measures, so-called *Hausdorff measures*, is defined. These Hausdorff measures allow a more accurate description of fractals. Then the iterated function systems (IFSs) approach to fractals is introduced and discussed in detail. This approach contains an analytic as well as a probabalistic aspect. Both will be studied and related to each other by using the concept of **p**-*balanced measure*. Results concerning the uniqueness and existence of fractals are given. Next, moments of fractal sets are considered, and it is shown that there is a recursion relation that allows their unique and explicit calculation in terms of lower order moments and the parameters that define the IFS. A generalization of an IFS, called a *recurrent IFS*, is then presented. Properties of such recurrent IFSs are discussed. At the end of this section *iterated Riemann surfaces* are defined, after some basic complex-analytic definitions are given and results concerning Riemann surfaces are stated. The study of these iterated Riemann surfaces is closely related to the theory of *Julia sets*. This connection, however, will not be pursued.

The third section deals with M. F. Dekking's construction of fractal sets. In this approach a sequence of compact subsets $\{K_\nu\}$ of $\mathbb{R}^n$ is associated with the iterates of a given word w under a given endomorphism ϑ of a free semigroup. The limit — in the Hausdorff metric — $K_\vartheta(w)$ of this sequence is called a *recurrent set*. Properties of these recurrent sets are discussed and their relation to the fractals generated by IFSs given.

2.1   Classical Fractal Sets

The classical Cantor set provides one of the simplest examples of a fractal set. Its construction reflects the general principle upon which fractals are built: It is made up of infinitely many pieces, each of which is similar to the entire set. More precisely, let $C_0 := [0, 1]$, $C_1 := [0, 1/3] \cup [2/3, 1]$,..., and let C_{k+1} be obtained from C_k by removing the middle third of each of the intervals in C_k. Clearly, C_k consists of 2^k intervals each having length $(1/3)^k$. Note that the sets C_k can be obtained by successively applying the functions $f_i : [0, 1] \to [0, 1]$, $i = 1, 2$, given by

$$f_1(x) := \frac{1}{3}x, \tag{2.1}$$

and

$$f_2(x) := \frac{1}{3}x + \frac{2}{3}, \tag{2.2}$$

to each interval in the previous set C_{k-1}. The Cantor set C in then defined as $C := \bigcap_{k=0}^{\infty} C_k$. It is well-known that C is a *perfect* set, i.e., a set that is closed and dense in itself. The Cantor set C does not contain any intervals, and its one-dimensional Lebesgue measure is equal to zero. However, since C is uncountable, its zero-dimensional Lebesgue measure is infinite. These rather simple observations suggest that Lebesgue measure is too coarse to accurately measure the "size" of C (this indeed is a characteristic of fractal sets in general). However, Felix Hausdorff introduced a measure which can be used to associate a finite non-zero number with the set C.

2.1.1 Hausdorff measures and Hausdorff dimension

Let (X, d) be a metric space and let U be a non-empty subset of X. Define the *diameter* of U by $|U| := \sup\{d(x, y) \mid x, y \in U\}$. A countable ε-*cover* of a set $E \subseteq X$ is a collection of subsets $\{U_i\}$ of X such that $E \subseteq \bigcup_{i=0}^{\infty} U_i$ and $0 < |U_i| \leq \varepsilon$.

Definition 2.1 (Hausdorff measure) *Suppose E is a subset of a metric space (X, d). Let $s > 0$. For $\varepsilon > 0$, define*

$$\mathcal{H}_\varepsilon^s(E) := \inf \sum_{i=0}^{\infty} |U_i|^s,$$

where the infimum is taken over all countable ε-covers of E. Hausdorff s-dimensional outer measure of E is then given by

$$\mathcal{H}^s(E) := \lim_{\varepsilon \to \infty} \mathcal{H}_\varepsilon^s = \sup_{\varepsilon > 0} \mathcal{H}_\varepsilon^s. \tag{2.3}$$

A few remarks seem to be in order now.

Remarks.

1. Since $\mathcal{H}_\varepsilon^s$ increases as ε decreases, the limit in (2.3) exists, but may be infinite.

2. The restriction of $\mathcal{H}^s$ to the σ-algebra of $\mathcal{H}^s$-measurable sets is called *s-dimensional Hausdorff measure*.

3. Suppose that $X = \mathbb{R}^n$. If $s = 1$, then one-dimensional Hausdorff measure is indeed one-dimensional Lebesgue measure. For $s = n$, $n \in \mathbb{N}$, $n > 1$, n-dimensional Lebesgue measure is a constant multiple of n-dimensional Hausdorff measure (the reason for this lies in the fact that Lebesgue measure is defined in terms of the diameters of parallelepipeds). Therefore, s-dimensional Hausdorff measure generalizes (integral) Lebesgue measure.

Now suppose that $X = \mathbb{R}^n$. It is easy to see that for any set $E \subseteq \mathbb{R}^n$, $\mathcal{H}^s(E)$ is a non-increasing function of s. If $s' < s$ then

$$\mathcal{H}^s(E) \leq \varepsilon^{s-s'} \mathcal{H}^{s'}(E),$$

which implies that $\mathcal{H}^{s'}(E) = \infty$ if $\mathcal{H}^s(E) > 0$. Hence, there exists a unique number s^*, such that

$$\mathcal{H}^s(E) = \begin{cases} \infty & \text{if } 0 \leq s < s^* \\ 0 & \text{if } s^* < s < \infty. \end{cases}$$

This unique value s^* is called the *Hausdorff-(Besicovitch) dimension* of E. From now on $\dim_H E$ denotes the Hausdorff-Besicovitch dimension of a set E. It is important to note that $\mathcal{H}^s(E)$ at $s = \dim_H E$ may be zero, finite, or infinite. A subset E of $\mathbb{R}^n$ is called an *s-set* if E is $\mathcal{H}^s$-measurable and $0 < \mathcal{H}^s(E) < \infty$.

As n-dimensional Lebesgue measure of a set E scales by λ^n when E is magnified by a factor λ, one might expect that s-dimensional Hausdorff measure of a set scales by λ^s. The next proposition confirms this.

Proposition 2.1 *Let $E \subseteq \mathbb{R}^n$ and let $\lambda > 0$. Let λE denote the set $\{\lambda e : e \in E\}$. Then*

$$\mathcal{H}^s(\lambda E) = \lambda^s \mathcal{H}^s(E).$$

Proof. The proof is an easy exercise in applying the definition of Hausdorff measure. ∎

A more general result relating the s-dimensional Hausdorff measure of a set to its image under a certain type of transformation can be proven. But first the following definition is needed.

Definition 2.2 *A function $f : E \subseteq \mathbb{R}^n \to \mathbb{R}^m$ satisfies a Hölder condition on E with exponent $\alpha > 0$ iff there exists a positive constant c such that for all $x, x' \in E$,*

$$\|f(x) - f(x')\| \leq c\|x - x'\|^\alpha. \tag{2.4}$$

Note that $\alpha = 1$ corresponds to f being Lipschitz. A function $f : E \subseteq \mathbb{R}^n \to \mathbb{R}^m$ is said to belong to the class $\text{Lip}^\alpha(E, \mathbb{R}^m)$, $0 < \alpha \leq 1$, iff

$$\sup \left\{ \frac{\|f(x) - f(x')\|}{\|x - x'\|^\alpha} \mid x, x' \in E \right\} < \infty. \tag{2.5}$$

Proposition 2.2 *Suppose that $E \subseteq \mathbb{R}^n$ and that f satisfies a Hölder condition on E with exponent α. Then for all $s > 0$,*

$$\mathcal{H}^{s/\alpha}(fE) \leq c^{s/\alpha} \mathcal{H}^s(E).$$

Proof. Let $\{U_i\}$ be an ε-cover of E. Note that $\|f(E \cap U_i)\| \le c|U_i|$, and therefore $\{f(E \cap U_i)\}$ is a $c\varepsilon^\alpha$-cover of $f(E)$. Hence, $\sum_i \|f(E \cap U_i)\|^{s/\alpha} \le c^{s/\alpha} \sum_i |U_i|^s$. Taking the infimum over all ε-covers and then the supremum over all $\varepsilon > 0$ yields the result. ∎

Now consider the classical Cantor set C. As seen earlier, Lebesgue measure fails to describe the Cantor set. However, using Hausdorff measure the following rather well-known fact can be proven.

Proposition 2.3 *The Hausdorff-Besicovitch dimension of the Cantor set* C *is given by* $\log(2)/\log(3)$, *and its* $\log(2)/\log(3)$-*dimensional Hausdorff measure equals one.* ∎

Remark. This formula for the Hausdorff dimension of the Cantor set is derived later as a special case of a more general formula.

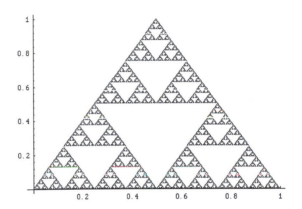

Figure 2.1: The Sierpiński triangle.

The *Sierpiński gasket* or *Sierpiński triangle* provides a two-dimensional example of a fractal set that was known long before the term "fractal" was introduced. This set had its origin in point-set topology. Its construction proceeds as follows: Let Δ be the triangle with vertices $(0,0)$, $(1,0)$, and $(1/2,1)$. Partition Δ into four congruent subtriangles $\Delta_1, \ldots, \Delta_4$ and delete the center subtriangle Δ_1. Applying the above procedure to each of the remaining three subtriangles yields nine triangles $\Delta_{12}, \Delta_{13}, \Delta_{14}, \ldots, \Delta_{42}, \Delta_{43}, \Delta_{44}$, together with the union $\Delta^{(2)} := \Delta_{11} \cup \Delta_{21} \cup \Delta_{31} \cup \Delta_{41}$ of their deleted centers. The indefinite continuation of this process then yields a sequence $\{\Delta^{(k)}\}$, where $\Delta^{(k)}$ denotes the union of the 3^{k-1} deleted centers after step k. The Sierpiński triangle is defined as $S := \Delta \setminus \bigcup_{k=1}^\infty \Delta^{(k)}$ (Fig. 2.1). Clearly, the one-dimensional

Lebesgue measure of S is infinite, and its two-dimensional Lebesgue measure is zero: The set S is a connected subset of $\mathbb{R}^2$ containing no area. It is an obvious consequence of the construction of S that any of its subparts is similar to the entire figure itself. Also, note that S can be obtained as the union of successive applications of the maps $f_i : \Delta \to \Delta$, $i = 1, 2, 3$,

$$f_1(x, y) := \left(\frac{1}{2}x, \frac{1}{2}y \right), \tag{2.6}$$

$$f_2(x, y) := \left(\frac{1}{2}x + \frac{1}{2}, \frac{1}{2}y \right), \tag{2.7}$$

$$f_3(x, y) := \left(\frac{1}{2}x + \frac{1}{4}, \frac{1}{2}y + \frac{1}{2} \right), \tag{2.8}$$

to each of their respective images.

Later it is shown that the Hausdorff-Besicovitch dimension of S is equal to $\log(3)/\log(2)$.

2.1.2   Weierstraß-like fractal functions

In connection with his work on infinite function series, Carl Weierstraß introduced in 1872 a function that is continuous but nowhere differentiable. Such functions are now called *fractal functions* since their graphs are in general fractal sets. The next chapter will focus entirely on these functions and their properties. Weierstraß's function is given by the infinite series $\sum_{i=1}^{\infty} \lambda^{s(i-1)} \sin(\lambda^i x)$, with parameters $1 < s < 2$ and $\lambda > 1$. In the 1930s Besicovitch and Ursell introduced a class of fractal functions $f : \mathbb{R} \to \mathbb{R}$ and calculated the Hausdorff-Besicovitch dimension of their graphs. Let us briefly summarize their results ([25]).

Theorem 2.1 *Suppose $f \in \mathrm{Lip}^\alpha(\mathbb{R})$ and let $G := graph(f)$. Then*

$$1 \le \dim_H G \le 2 - \alpha.$$

(*Here* $\mathrm{Lip}^\alpha(\mathbb{R}, \mathbb{R}) =: \mathrm{Lip}^\alpha(\mathbb{R})$.)   ■

Note that this result implies that if f has a finite derivative at all points $x \in \mathbb{R}$, then $\dim_H G = 1$.

Besicovitch and Ursell considered the following special class of functions in $\mathrm{Lip}^\alpha(\mathbb{R})$: Let $\phi : \mathbb{R} \to \mathbb{R}$,

$$\phi(x) := \begin{cases} 2x & \text{if } 0 \le x \le 1/2 \\ \phi(-x) = \phi(x+1) & \text{otherwise} \end{cases}.$$

Define functions $f_\phi : \mathbb{R} \to \mathbb{R}$ by

$$f_\phi(x) := \phi(x) + \sum_{i \in \mathbb{N}} b_i^{-\alpha} \phi(b_i x), \qquad (2.9)$$

for $0 < \alpha < 1$, $b_i > 0$, $i \in \mathbb{N}$. In their paper [25] they showed that if $b_{i+1} \leq B b_i$, for $B > 1$ and all $i \in \mathbb{N}$, then $f_\phi \in \mathrm{Lip}^\alpha(\mathbb{R})$, but in no higher Lipschitz class. The following theorem gives the equality between $2 - \alpha$ and dim_H under certain conditions on α and the b_i.

Theorem 2.2 (Besicovitch-Ursell) *Let f_ϕ be defined as in (2.9) and let G_ϕ be its graph. Assume that the sequence $\{\mu_i : i \in \mathbb{N}\}$ is such that $\mu_i \geq (1 - \alpha)/(\alpha) \cdot (2 - \dim_H G)/(\dim_H G - 1)$, for all $i \in \mathbb{N}$. Let $b_1 > 1$ and $b_{i+1} := b_i^{\mu_i}$, $i \in \mathbb{N}$. If $\lim_{i \to \infty} b_{i+1}/b_i = \infty$ and $\mu_i \to 1$ as $i \to \infty$, then $\dim_H G = 2 - \alpha$.* ∎

In Chapter 5 the functions f_ϕ will be constructed using IFSs.

2.2 Iterated Function Systems

In this section the concept of an iterated function system is introduced. In what follows these IFSs are used to construct special fractal sets, namely fractal functions (Chapter 5) and fractal surfaces (Chapter 8). IFSs are first defined via continuous operators on $C_\mathbb{R}(X)$, the Banach space of all continuous real-valued functions on a compact and metrizable topological space X endowed with the norm $\|f\|_\infty := \max\{|f(x)| \,|\, x \in X\}$, $f \in C_\mathbb{R}(X)$.

2.2.1 Definition and properties of iterated function systems

Let N be an integer greater than one, and let $\mathbf{w} := \{w_i : X \to X \,|\, i = 1, \ldots, N\}$ be a collection of Borel measurable functions on X. Let $\mathbf{p} := \{p_i \,|\, i = 1, \ldots, N\}$ denote a set of probabilities. Define an operator $T : C_\mathbb{R}(X) \to \mathbb{R}^X$ by

$$(Tf)(x) := \sum_{i=1}^{N} p_i (f \circ w_i)(x). \qquad (2.10)$$

The following definition of iterated function system is due to Barnsley and Demko, and most of the results presented in this section can also be found in [9].

Definition 2.3 The triple $(X, \mathbf{w}, \mathbf{p})$ is called an iterated function system (on X) with probabilities iff the associated set of probabilities $\mathbf{p}$ is such that the operator T defined in (2.10) maps $C_{\mathbb{R}}(X)$ into itself.

An IFS with probabilities $(X, \mathbf{w}, \mathbf{p})$ is called hyperbolic with contractivity s iff there exists a constant $0 \leq s < 1$ such that

$$\sup \left\{ \frac{d(w_i(x), w_i(x'))}{d(x, x')} \,\Big|\, x, x' \in X \right\} \leq s, \tag{2.11}$$

for $i = 1, \ldots, N$, where d denotes the metric on X.

Remarks.

1. If $w_i \in C(X, X)$, the linear space of all continuous functions from X into itself, for all $i = 1, \ldots, N$, then $(X, \mathbf{w}, \mathbf{p})$ is an IFS with probabilities for any set of associated probabilities $\mathbf{p}$. In this case the IFS is written as $(X, \mathbf{w})$ and referred to as simply an IFS.

2. The operator T clearly depends on $\mathbf{w}$ and $\mathbf{p}$ and should be more precisely denoted by $T(\mathbf{w}, \mathbf{p})$. However, in order to ease notation the less exact notation was chosen.

Let $\mathfrak{H}(X)$ denote the collection of all non-empty compact subsets of the set X. The function $\mathfrak{h} : \mathfrak{H}(X) \times \mathfrak{H}(X) \to \mathbb{R}$ defined by

$$\mathfrak{h}(A, B) := \max \left\{ \max_{a \in A} \min_{b \in B} d(a, b) + \max_{b \in B} \min_{a \in A} d(b, a) \right\}, \tag{2.12}$$

is easily seen to be a metric on $\mathfrak{H}(X)$. This metric is called the *Hausdorff metric*, and with it $(\mathfrak{H}(X), \mathfrak{h})$ becomes a complete metric space, called the *hyperspace of compact subsets (of the complete metric space X)*. The following elementary properties of $\mathfrak{h}$ will be used later:

(a) Suppose $f : X \to X$ is Lipschitz with Lipschitz constant s. Then for all $A, B \in \mathfrak{H}(X)$,

$$\mathfrak{h}(fA, fB) \leq s\, \mathfrak{h}(A, B). \tag{2.13}$$

(b) Let $\{A_i \,|\, i = 1, \ldots, n\}$ and $\{B_i \,|\, i = 1, \ldots, n\}$ be finite collections of sets in $\mathfrak{H}(X)$. Then

$$\mathfrak{h}\left(\bigcup_{i=1}^{n} A_i, \bigcup_{i=1}^{n} B_i \right) \leq \max\{\mathfrak{h}(A_i, B_i) \,|\, i = 1, \ldots, n\}. \tag{2.14}$$

Given an IFS with probabilities $(X, \mathbf{w}, \mathbf{p})$, a set-valued map $\mathfrak{w} : \mathfrak{H}(X) \to \mathfrak{H}(X)$ can be associated with $\mathbf{w}$ by setting

$$\mathfrak{w}(A) := \bigcup_{i=1}^{N} w_i(A). \tag{2.15}$$

In case $(X, \mathbf{w}, \mathbf{p})$ is a hyperbolic IFS with probabilities and contractivity s, $\mathfrak{w}$ is a contraction on the complete metric space $(\mathfrak{H}(X), \mathfrak{h})$ with the same contractivity s. (This follows immediately from the above-listed properties of $\mathfrak{h}$.) Next the class of fractal sets that will be used most frequently are defined.

Definition 2.4 Let $(X, \mathbf{w}, \mathbf{p})$ be a hyperbolic IFS with probabilities, let $\mathfrak{w}$ be the associated set-valued map, and let $E \in \mathfrak{H}(X)$. The attractor or the fractal (set) generated by the hyperbolic IFS $(X, \mathbf{w}, \mathbf{p})$ with probabilities is the unique compact set A defined by

$$A := \lim_{n \to \infty} \mathfrak{w}^n(E), \tag{2.16}$$

where the limit is taken in the Hausdorff metric.

Remark. Equation (2.16) has an algebraic interpretation as well: The attractor A is that set which is invariant under the semigroup $S[\mathfrak{w}]$ generated by the maps w_i, $i = 1, \ldots, N$.

That every hyperbolic IFS (with probabilities) has an attractor, and that this attractor is unique and independent of the set E, follows from the Banach Fixed-Point Theorem applied to the contraction $\mathfrak{w}$ and the complete metric space $(\mathfrak{H}(X), \mathfrak{h})$. Furthermore, the uniqueness part of the theorem does not exclude the possibility that there exist sets A *not* in $\mathfrak{H}(X)$ satisfying $\mathfrak{w}(A) = A$ (an obvious example is, for instance, the empty set). Note that one can also define an attractor for only an IFS with probabilities; however, this attractor does in general depend on the set E (the existence of an attractor in this case is guaranteed by the Schauder Fixed-Point Theorem).

Also note that the attractor depends in a subtle way on the set of associated probabilities, even in the case where the w_i's are continuous. This dependence will be studied shortly.

Two attractors of hyperbolic IFSs have already been encountered:

Example 2.1 The classical Cantor set C.

For this example $X = [0, 1]$ and the functions $w_i = f_i$, $i = 1, 2$ (2.1) and (2.2). Since $(X, \mathbf{w})$ is clearly a hyperbolic IFS (with $s = 1/3$), the attractor, the classical Cantor set in this case, is unique and uniquely determined by any initial interval $[a, b]$, $0 \le a < b \le 1$.

***Example* 2.2** The Sierpiński triangle S.

Here, for instance, $X = [0,1] \times [0,1]$ and $w_i = f_i$, $i = 1,2,3$, as defined in (2.6), (2.7), and (2.8). Again it is easy to verify that $(X, \mathbf{w})$ is a hyperbolic IFS with $s = 1/2$, and thus the attractor, the Sierpiński triangle, is uniquely and independently determined by any compact set $E \subseteq [0,1] \times [0,1]$.

Given an IFS $(X, \mathbf{w}, \mathbf{p})$ with probabilities on X, it is sometimes useful to work with a compact subset K of X instead of using X itself. The following proposition shows that this can be easily achieved.

Proposition 2.4 *Suppose that $K \subseteq X$ is compact. Then there exists a $\hat{K} \in \mathfrak{H}(X)$ such that $\hat{K} \supseteq K$ and $\mathfrak{w}(\hat{K}) \subseteq \hat{K}$.*

Proof. Set $\hat{K} := \overline{\cup_{i=0}^{\infty} \mathfrak{w}^i(K)}^{\mathfrak{H}(X)}$. Then $\hat{K}$ is compact and by its definition $\mathfrak{w}(\hat{K}) \subseteq \hat{K}$. ∎

Now let $(X, \mathbf{w}, \mathbf{p})$ be an IFS with probabilities. Consider the following discrete time Markov process $(X, \mathcal{B}(X), \mu, P)$, where $\mathcal{B}(X)$ denotes the algebra of Borel subsets of X, and where

$$P(x, B) := \sum_{i=1}^{N} p_i \, \chi_{w_i(x)}(B). \tag{2.17}$$

Recall that the dual of $C_{\mathbb{R}}(X)$ is the set of all (finite) regular Borel measures $\mathcal{M}_{\mathbb{R}}(X)$. The operator T as defined in (2.10) maps the space $C_{\mathbb{R}}(X)$ continuously into itself, and thus the adjoint operator T^* (more precisely, $T^*(\mathbf{w}, \mathbf{p})$) maps the set $\mathcal{P}(X)$ of probability measures on X weak* continuously into itself. By the Schauder Fixed-Point Theorem T^* has a fixed point $\mu \in \mathcal{P}(X)$. Furthermore,

$$(T^*\nu)B = \sum_{i=1}^{N} p_i \, (w_i^{\#} \circ \nu)B = \sum_{i=1}^{N} p_i \int_X \chi_{w_i(x)} d\nu(x), \tag{2.18}$$

where $(w_i^{\#} \circ \nu)B := \nu(w_i^{-1}B)$. Therefore, the fixed point μ satisfies

$$\mu B = \sum_{i=1}^{N} p_i \int_X \chi_{w_i(x)} d\mu(x) = \int_X P(x, B) d\mu(x), \tag{2.19}$$

i.e., μ is the stationary probability measure for the Markov process defined above. This measure has a special name.

Definition 2.5 The measure μ defined in (2.19) is called the **p**-balanced measure of the IFS $(X, \mathbf{w}, \mathbf{p})$ with probabilities.

The **p**-balanced measure of a hyperbolic IFS $(X, \mathbf{w}, \mathbf{p})$ with probabilities is also *attractive* for probability measures $\nu \in \mathcal{M}_{\mathbb{R}}(E)$, where $E \subseteq X$. More precisely,

$$T^{*n}\nu \xrightarrow{w^*} \mu, \text{ as } n \to \infty, \tag{2.20}$$

for all probability measures ν with support in $E \subseteq X$.

To show this, the so-called *Hutchinson metric* $\mathfrak{d}$ on the space $\mathcal{P}(X) \subseteq \mathcal{M}_{\mathbb{R}}(X)$ of probability measures on X must be introduced.

Definition 2.6 *Let* $\mu, \nu \in \mathcal{P}(X)$ *and denote by* $\mathrm{Lip}_{\mathbb{R}}^{(\leq 1)}(X)$ *the set of all Lipschitz functions* $\phi : X \to \mathbb{R}$ *with Lipschitz constant less than or equal to one. Define*

$$\mathfrak{d}(\mu, \nu) := \sup\{\mu(\phi) - \nu(\phi) | \, \phi \in \mathrm{Lip}_{\mathbb{R}}^{(\leq 1)}(X)\}. \tag{2.21}$$

Here the notation $\mu(\phi) = \int_X \phi(x)\, d\mu(x)$ *was used.*

That $\mathfrak{d}$ is indeed a metric is an easy exercise left for the reader. It should be remarked that Lipschitz functions map bounded sets to bounded sets, and thus $\phi \in \mathrm{Lip}_{\mathbb{R}}^{(\leq 1)}(X)$ is also an element of $BC_{\mathbb{R}}(X)$, the collection of all continuous functions $f : X \to \mathbb{R}$ that are bounded on bounded subsets of X. Moreover, the convergence in the topology induced by the $\mathfrak{d}$-metric implies convergence in the topology of $\mathcal{M}_{\mathbb{R}}(X)$.

It is left to the reader to establish the next proposition.

Proposition 2.5 *The linear space* $(\mathcal{P}(X), \mathfrak{d})$ *is a complete metric space.* ∎

The Hutchinson metric was designed to obtain the following result:

Proposition 2.6 *The operator* $T^* : \mathcal{P}(X) \to \mathcal{P}(X)$, *given by*

$$(T^*\nu)B = \sum_{i=1}^{N} p_i\, (w_i^{\#} \circ \nu)B, \tag{2.22}$$

is a contraction with contractivity s.

Proof. Let $\mu, \nu \in \mathcal{P}(X)$, and let $\phi \in \mathrm{Lip}_{\mathbb{R}}^{(\leq 1)}(X)$. Then

$$\mathfrak{d}(T^*\mu, T^*\nu) = \sum_{i=1}^{N} p_i\, (w_i^{\#} \circ \mu)(\phi) - \sum_{i=1}^{N} p_i\, (w_i^{\#} \circ \nu)(\phi)$$

$$= \sum_{i=1}^{N} p_i(\mu(\phi \circ w_i) - \nu(\phi \circ w_i)).$$

Now, recall that $\sup\{d(w_i(x), w_i(x'))/d(x, x') \mid x, x' \in X\} \le s$, for $i = 1, \ldots, N$. Hence,

$$
\begin{aligned}
\eth(T^*\mu, T^*\nu) &= \sum_{i=1}^{N} p_i s(\mu(s^{-1}\phi \circ w_i) - \nu(s^{-1}\phi \circ w_i)) \\
&\le \sum_{i=1}^{N} p_i s\, L(\mu, \nu) = s L(\mu, \nu).
\end{aligned}
$$

Here the fact that the Lipschitz constant of $s^{-1}\phi \circ w_i$ is less than or equal to $s^{-1} \cdot 1 \cdot s = 1$ was used. ∎

These results are now summarized in a theorem.

Theorem 2.3 *Let* $(X, \mathbf{w}, \mathbf{p})$ *be a hyperbolic IFS with probabilities, and let* A *be its attractor. Then there exists a unique probability measure* μ *on* $\mathcal{P}(A)$ *such that* $T^*\mu = \mu$, *where* T^* *is given in (2.22). Moreover, every probability measure* $\nu \in \mathcal{P}(X)$ *is attracted to* μ *in the sense that* $T^{*n}\nu \to \mu$, *where the convergence is in the Hutchinson metric* $\eth$. *Furthermore,* μ *is also the stationary probability measure for the discrete time Markov process (2.17).* ∎

At this point it is natural to ask what the support of the measure μ is, and how it is related to the attractor A. The setting in which these questions can be answered most easily involves the concept of a *code space* associated with an IFS with probabilities. Recall from Chapter 1, in particular Section 1.2., the definition of state space, code, cylinder set and the Fréchet metric (cf. (1.42), (1.43), and (1.46)). Here the same notation and terminology as in Section 1.2 is used, i.e., the code space Σ is defined by

$$\Sigma := \{1, \ldots, N\}^{\mathbb{N}}, \tag{2.23}$$

and the *codes* are given by

$$\mathbf{i} := (i_1\, i_2\, \ldots\, i_n\, \ldots), \tag{2.24}$$

where each $i_n \in \{1, \ldots, N\}$.

Furthermore, recall that (Σ, d_F) is a compact metric space. If *left-shift* maps $\tau_i : \Sigma \to \Sigma$

$$\tau_i(i_1\, i_2\, \ldots\, i_n\, \ldots) := (i\, i_1\, i_2\, \ldots\, i_n\, \ldots), \tag{2.25}$$

for all $i \in \{1, \ldots, N\}$, are defined, then the pair $(\Sigma, \boldsymbol{\tau})$ is a hyperbolic IFS (for any choice of probabilities $\mathbf{p}$). To see this, let us show that each map τ_i is Lipschitz with Lipschitz constant less than one. Let $\mathbf{i}, \mathbf{i}' \in \Sigma$. Then

$$d_F(\tau_i(\mathbf{i}), \tau_i(\mathbf{i}')) = \sum_{j=1}^{\infty} \frac{|(\tau_i(\mathbf{i}))_j - (\tau_i(\mathbf{i}'))_j|}{(N+1)^j}$$

$$= \sum_{j=2}^{\infty} \frac{|i_j - i'_j|}{(N+1)^j} < \frac{d_F(\mathbf{i}, \mathbf{i}')}{N+1}.$$

Applying Theorem 2.3 to the hyperbolic IFS $(\Sigma, \boldsymbol{\tau})$ yields the existence of a unique probability measure ρ, the $\mathbf{p}$-balanced measure for this IFS.

For the measure space $(\Sigma, \mathcal{F}, \rho)$ ((1.42), (1.43), and (1.44)) the analogues of the operators T and T^* can be defined. These are denoted by Θ and Θ^*, respectively:

$$(\Theta f)(\mathbf{i}) := \sum_{i=1}^{N} p_i \, f(\tau_i(\mathbf{i})), \qquad (2.26)$$

for all $f \in C_{\mathbb{R}}(\Sigma)$, and

$$(\Theta^* \nu) E := \sum_{i=1}^{N} p_i \, (\tau_i^{\#} \circ \nu) E, \qquad (2.27)$$

for all $\nu \in \mathcal{M}_{\mathbb{R}}(\Sigma)$ and for all $E \in \mathcal{F}$.

Observe that for $\mathbf{i} \in \Sigma$,

$$\tau_i^{-1}(\mathbf{i}) = \begin{cases} \emptyset & \text{for } i_1 \neq i \\ \mathbf{j} & \text{for } i_1 = i, \end{cases}$$

where $j_n = i_{n+1}$, and that, as usual, $\tau_i^{-1} E := \{\tau_i^{-1}(\mathbf{i}) \,|\, \mathbf{i} \in E \subseteq \Sigma\}$.

It is an easy exercise to show that the attractor of the hyperbolic IFS $(\Sigma, \boldsymbol{\tau})$ is Σ itself. The next theorem relates the attractor A of the hyperbolic IFS $(X, \mathbf{w}, \mathbf{p})$ to Σ, the attractor of $(\Sigma, \boldsymbol{\tau})$.

Theorem 2.4 *Suppose that $(X, \mathbf{w}, \mathbf{p})$ is a hyperbolic IFS with probabilities and that A is its unique attractor. Then there exists a continuous surjective mapping $\gamma : \Sigma \twoheadrightarrow \beth$,*

$$\gamma(\mathbf{i}) = \lim_{n \to \infty} w_{i_1} \circ w_{i_2} \circ \ldots \circ w_{i_n}(x), \qquad (2.28)$$

where the limit exists and is independent of $x \in X$. In other words, $A = \gamma \Sigma$.

Proof. Let us denote $w_{i_1} \circ \ldots \circ w_{i_n}(x)$ by $w_{\mathbf{i}(n)}(x)$. Notice that

$$|w_{\mathbf{i}(n)}(x) - w_{\mathbf{i}(n')}(x')| < s^{\min\{n,n'\}}|X|,$$

for all $x, x' \in X$. Hence, the limit in (2.28) exists *and* is independent of $x \in X$. Now set $A(x) := \lim_{n \to \infty} \mathfrak{w}^n(x)$. Then the limit in (2.28) clearly belongs to $A(x)$.

To show the continuity of γ, let $\varepsilon > 0$ be given. Choose an integer n so large that $s^n|K| < \varepsilon$. Now if $\mathbf{i}, \mathbf{i}' \in \Sigma$ are such that they agree through n terms, i.e.,

$$d_F(\mathbf{i}, \mathbf{i}') = \sum_{j=n}^{\infty} \frac{|i_j - i_j'|}{(N+1)^j} < \sum_{j=n}^{\infty} \frac{N}{(N+1)^j} = (N+1)^{1-n},$$

then

$$|\gamma(\mathbf{i}) - \gamma(\mathbf{i}')| < \varepsilon.$$

To prove surjectivity of γ, suppose that $x' \in A(x)$. By the definition of $A(x)$ there exists a sequence $\{\mathbf{i}_\nu\} \subseteq \Sigma$ such that

$$\lim_{n \to \infty} w_{\mathbf{i}_\nu(n)}(x) = x'.$$

The compactness of Σ now implies that the sequence $\{\mathbf{i}_\nu\}$ has at least one limit point, say, $\mathbf{i}$. Since $d_F(\mathbf{i}_n, \mathbf{i}) \to 0$ as $n \to \infty$, the cardinality $c(n)$ of the set

$$\{k \in \mathbb{N}|\ (\mathbf{i}_\nu)_j = i_j \text{ for } 1 \leq j \leq k\}$$

goes to ∞. Therefore,

$$|w_{\mathbf{i}_\nu(n)}(x) - w_{\mathbf{i}(n)}(x)| < s^{c(n)}|X|,$$

and hence $\gamma(\mathbf{i}) = x'$.

This argument also shows that $A(x)$ is independent of $x \in X$. ∎

Remarks.

1. Let $(i_1\, i_2\, \ldots i_n \ldots) \in \Sigma$ and let $A_{i_1\, i_2 \ldots i_n} := w_{i_1} \circ w_{i_2} \circ \ldots \circ w_{i_n} A$. Then Theorem 2.4 shows that

$$\gamma(i_1\, i_2 \ldots i_n \ldots) = \bigcap_{n=1}^{\infty} A_{i_1\, i_2 \ldots i_n}. \tag{2.29}$$

2. If $\{A_1, \ldots, A_N\}$ is disjoint, then γ is also injective.

Next let us consider a hyperbolic IFS with probabilities $\mathbf{p}$ and associated attractor A. For any $\nu \in \mathcal{P}(X)$, consider the sequence $\{T^{*n}\nu\}$ of the iterates of the adjoint to (2.10). The above arguments show that for any $f \in C_{\mathbb{R}}(X)$ the sequence $\{\int T^n f \, d\nu\}$ is Cauchy; therefore, the sequence $\{T^{*n}\nu\}$ converges weak* to some fixed point κ of T^*. Now suppose that there was a closed set E such that $\kappa E > 0$ and $E \cap A = \emptyset$. Then there exists a function $g \in C_{\mathbb{R}}(X)$ with range equal to $[0,1]$ and the property that

$$g \equiv \begin{cases} 1 & \text{on } E \\ 0 & \text{on an open set } O \supset A \end{cases}.$$

Hence,

$$\kappa E \leq \int f \, d\kappa = \int f \, d(T^{*n}\kappa) = \int T^n f \, d\kappa$$

$$= \sum p_{i_1} p_{i_2} \cdots p_{i_n} \int f(w_{\mathbf{i}(n)}(x)) \, d\kappa(x),$$

where the sum is taken over all $(i_1 \, i_2 \ldots i_n)$. Theorem 2.4 implies that $w_{\mathbf{i}(n)}(x) \to A$ uniformly in $x \in X$ as $n \to \infty$. Thus, $f(w_{\mathbf{i}(n)}(x)) \to 0$, and therefore $\int f \, d\kappa = 0$. This contradiction now shows that if $\kappa E > 0$, then $E \subseteq A$.

Our next goal is to describe the relationship between the $\mathbf{p}$-balanced measure μ of a hyperbolic IFS with probabilities $(X, \mathbf{w}, \mathbf{p})$ and the $\mathbf{p}$-balanced measure ρ of the hyperbolic IFS (Σ, τ). This will be done by "lifting" the surjection γ to the function space of continuous mappings. More precisely, let $\gamma : \Sigma \twoheadrightarrow \sqsupset$ be the surjection defined in Theorem 2.4. The *contravariant* linear operator $\Gamma : C_{\mathbb{R}}(A) \to C_{\mathbb{R}}(\Sigma)$ defined by

$$(\Gamma f)(\mathbf{i}) := f(\gamma(\mathbf{i})), \tag{2.30}$$

$f \in C_{\mathbb{R}}(A)$, is clearly injective and has norm one. Also, $\Gamma 1 = 1$. Hence, the mapping

$$\Gamma f \mapsto f \tag{2.31}$$

is well-defined and a continuous linear surjective mapping of $\Gamma C_{\mathbb{R}}(A)$ onto $C_{\mathbb{R}}(A)$.

Now suppose $\nu \in \mathcal{M}_{\mathbb{R}}(A)$. Define $\kappa_0 : C_{\mathbb{R}}(\Sigma) \to \mathbb{R}$ by $\kappa_0(\Gamma f) := \nu(f)$, that is, κ_0 is the composition of the mapping defined in (2.31) with ν. Thus,

κ_0 is a continuous linear functional on $\Gamma(C_{\mathbb{R}}(A))$, and by the Hahn-Banach Extension Theorem there exists an extension $\kappa \in \mathcal{M}_{\mathbb{R}}(\Sigma)$. Therefore,

$$\nu = \kappa_0 \circ \Gamma = \kappa \circ \Gamma = \Gamma^* \kappa,$$

which shows that the adjoint Γ^* of Γ is a surjection from $\mathcal{M}_{\mathbb{R}}(\Sigma)$ onto $\mathcal{M}_{\mathbb{R}}(A)$. Using the order-theoretic formulation of the Hahn-Banach Extension Theorem ([22], Section 32), more can be shown: namely, that Γ^* maps $\mathcal{P}(\Sigma)$ onto $\mathcal{P}(A)$.

Now let $\tilde{T}$ be the restriction of T to $C_{\mathbb{R}}(A)$. Theorem 2.4 implies the commutativity of the following diagram:

$$
\begin{array}{ccc}
\Sigma & \overset{\gamma}{\twoheadrightarrow} & A \\
\tau_i \downarrow & & \downarrow w_i \\
\Sigma & \overset{\gamma}{\twoheadrightarrow} & A
\end{array}
\qquad (2.32)
$$

Recalling Definitions 2.10, 2.26, and 2.30, one thus obtains

$$(\Gamma T)(f)(\mathbf{i}) = (Tf)\gamma(\mathbf{i}) = \sum_{i=1}^{N} p_i \, (f \circ w_i)\gamma(\mathbf{i}) = \sum_{i=1}^{N} p_i \, (f \circ \gamma)\tau_i(\mathbf{i})$$

$$= \Theta f(\gamma(\mathbf{i})) = \Theta(\Gamma f)(\mathbf{i}) = (\Theta\Gamma)(f)(\mathbf{i}).$$

Hence, for any integer n greater than one

$$\Gamma T^n = \Theta^n \Gamma,$$

and also,

$$T^{*n}\Gamma^* = \Gamma^* \Theta^{*n}.$$

Then, given any $\kappa \in \mathcal{P}(A)$, there exists a $\nu \in \mathcal{P}(\Sigma)$ such that $\kappa = \Gamma^* \nu$, and

$$T^{*n}\kappa = T^{*n}\Gamma^* \nu = \Gamma^* \Theta^{*n} \nu \to \Gamma^* \rho,$$

the **p**-balanced measure of the hyperbolic IFS $(\Sigma, \boldsymbol{\tau})$. Thus, $\Gamma^* \rho$ is the unique fixed point if T^* in $\mathcal{P}(A)$, and by our earlier results also in $\mathcal{P}(X)$.

The objective to characterize the relation between the **p**-balanced measure μ of a hyperbolic IFS with probabilities $(X, \mathbf{w}, \mathbf{p})$ and the **p**-balanced measure ρ of the hyperbolic IFS $(\Sigma, \boldsymbol{\tau})$ has now been achieved. The above arguments yield the following theorem.

Theorem 2.5 *Suppose $(X, \mathbf{w}, \mathbf{p})$ is a hyperbolic IFS with probabilities. Then the **p**-balanced measure μ is given by $\mu E = (\gamma^{\#} \circ \rho)E$, for all $E \in \mathcal{F}$. Furthermore, the support of μ is the attractor A, independent of $\mathbf{p}$.* ∎

2.2.2 Moment theory and iterated function systems

Theorem 2.3 may be used to find a recursion relation between the moments of a certain class of IFSs. For illustrative purposes it is assumed that X is a compact subset of $(\mathbb{K}^n, |\cdot|)$, where $|\cdot|$ is the metric given by $|z| := \sqrt{\sum_{j=1}^n z_i \bar{z}_i}$.

Let N be an integer greater than one. Define *affine* maps $w_i : X \to X$ by

$$w_i(z) := A_i z + \zeta_i, \text{ for all } i = 1, \ldots, N, \tag{2.33}$$

where A_i is the diagonal matrix

$$A_i := \begin{pmatrix} a_{i1} & 0 & \cdots & 0 \\ 0 & a_{i2} & \cdots & 0 \\ 0 & 0 & \ddots & 0 \\ 0 & 0 & \cdots & a_{in} \end{pmatrix},$$

$a_{ij} \in \mathbb{K}$ with $|a_{ij}| < 1$, for all $i = 1, \ldots N; j = 1, \ldots n$, and $\zeta_i \in \mathbb{K}^n$. It follows immediately from Eq. (2.33) that w_i is a contraction with constant of contractivity $c \leq \max\{\max\{|a_{ij}| \, | \, i = 1, \ldots, N\} : j = 1, \ldots, n\} < 1$. Hence, $(X, \mathbf{w})$ is a hyperbolic IFS for any choice of probabilities $\mathbf{p}$. Its $\mathbf{p}$-balanced measure is denoted by μ. By Theorem 2.3,

$$\int_X f \, d\mu = \sum_{i=1}^N p_i \int_X f \circ w_i \, d\mu \tag{2.34}$$

for all $f \in C_{\mathbb{K}}(X)$. (That functions in $C_{\mathbb{K}}(X)$ as well as functions in $C_{\mathbb{R}}(X)$ can be used should be clear from previous results.)

In order to proceed the following definition is needed.

Definition 2.7 Let $z \in \mathbb{K}^n$ and let $\alpha \in \mathbb{N}_0^n := (\mathbb{N} \cup \{0\})^n$. For any probability measure $\nu \in \mathcal{P}(X)$, the αth moment of ν is defined by

$$\mathbf{M}(\nu, \alpha) := \int_X z^\alpha \, d\nu(z). \tag{2.35}$$

Remark. In Definition 2.7 multi-index notation has been used: For $z = (z_1, \ldots, z_n)^t \in \mathbb{K}^n$ and $\alpha = (\alpha_1, \ldots, \alpha_n)^t \in \mathbb{N}_0^n$, set

$$z^\alpha := \prod_{j=1}^n z_j^{\alpha_j}.$$

If $\alpha, \beta \in \mathbb{N}_0^n$, define

$$\alpha \leq \beta \iff \forall j \in \mathbb{N} : \alpha_j \leq \beta_j.$$

Furthermore, let

$$\alpha - \beta := (\alpha_1 - \beta_1, \ldots, \alpha_n - \beta_n)^t.$$

Since $z^\alpha \in C_{\mathbb{K}}(X)$, Eq. (2.34) can be used with $f(z) := z^\alpha$. Choosing $\alpha \in \mathbb{N}^n$, the αth moment of the **p**-balanced measure μ of the above-defined hyperbolic IFS is then given by

$$\mathbf{M}(\mu, \alpha) := \int_X z^\alpha \, d\mu(z) = \int_X \prod_{j=1}^n z_j^{\alpha_j} \, d\mu(z)$$

$$= \sum_{i=1}^N p_i \int_X \prod_{j=1}^n [a_{ij} z_i + \zeta_{ij}]^{\alpha_j} \, d\mu(z).$$

Each factor in the preceding equation reduces to

$$[a_{ij} z_i + \zeta_{ij}]^{\alpha_j} = \sum_{\beta_j=1}^{\alpha_j} \binom{\alpha_j}{\beta_j} a_{ij}^{\beta_j} \zeta_{ij}^{\alpha_j - \beta_j} z_j^{\beta_j},$$

and thus, using the definition of moments, one obtains

$$\mathbf{M}(\mu, \alpha) = \sum_{i=1}^N p_i \, a_i^\alpha \cdot \mathbf{M}(\mu, \alpha) + \sum_{0 \leq \beta < \alpha} \binom{\alpha}{\beta} a_i^\beta \zeta_i^{\alpha - \beta} \cdot \mathbf{M}(\mu, \beta),$$

where $\beta := (\beta_1, \ldots, \beta_n)^t$ and $a_i := (a_{i1}, \ldots, a_{in})^t$.. Since $c < 1$, this equation can be rewritten as

$$(1 - \sum_{i=1}^N p_i \, a_i^\alpha) \mathbf{M}(\mu, \alpha) = \sum_{0 \leq \beta < \alpha} \binom{\alpha}{\beta} a_i^\beta \zeta_i^{\alpha - \beta} \mathbf{M}(\mu, \beta). \qquad (2.36)$$

Note that $\mathbf{M}(\mu, 0) = \int_X d\mu = 1$. Therefore, by Eq. (2.36), the moment $\mathbf{M}(\mu, \alpha)$ can be explicitly and recursively calculated in terms of the lower order moments $\mathbf{M}(\mu, \beta)$. These arguments now give the next theorem.

Theorem 2.6 *Denote by μ the p-balanced measure of the hyperbolic IFS $(X, \mathbf{w})$ defined above. Then the moments $\{\mathbf{M}(\mu, \alpha); \alpha \in \mathbb{N}_0^n\}$ can be recursively calculated via Eq. (2.36). This calculation involves only the parameters which define the IFS.* ■

2.2.3 Recurrent iterated function systems

Next a generalization of an IFS, called a *recurrent IFS*, is presented. To see how such a generalization is obtained, recall the interpretation of an IFS in terms of a discrete time Markov process (Eq. (2.17) and (2.19)). The IFSs considered so far have the property that each state in the Markov process is allowed to *communicate* with any other state, including itself, i.e., given a set of non-zero probabilities **p** and that the process at time n is in state $i \in \{1,\dots,N\}$, where N is the number of maps in the IFS, the probability of transfer in one time unit into any other state $j \in \{1,\dots,N\}$ is equal to p_j, *independent* of i. One may want to relax this condition by only requiring that given the process is in state i at time n, the probability of tranfer into state j is a prescribed *conditional* probability $p_{ij} \geq 0$. This weaker condition is basically what defines a recurrent IFS.

Before the precise definition of a recurrent IFS can be given, the following notions need to be introduced.

Definition 2.8 Let $A = (a_{ij})$ be an $N \times N$ matrix with coefficients in $\mathbb{K}$.

1. A linear subspace V of the linear space $\mathbb{K}$ is said to be invariant under A iff $AV := \{Av \,|\, v \in V\} = V$.

2. The matrix A is called irreducible iff $\{0\}$ and $\mathbb{K}$ are the only subspaces left invariant under A.

3. The matrix A is called row-stochastic iff

   (a) $\forall i \in \{1,\dots N\} : \sum_{j=1}^{N} a_{ij} = 1.$
   (b) $\forall i,j \in \{1,\dots N\} \exists i_1,\dots i_n \in \{1,\dots N\} : \quad i = i_1, \; j = i_n \wedge$
   $a_{i_1 i_2} \cdots a_{i_{n-1} i_n} > 0.$

Now suppose (X,d) is a complete, separable, and locally compact metric space. (Note that the setting here is slightly more general than in Section 2.2.) Let N be a fixed integer greater than one. The linear space of all Lipschitz functions $\phi : X \to Y$ with Lipschitz constant less than one is denoted by $\mathrm{Lip}^{(<1)}(X,Y)$. The linear space $\mathrm{Lip}^{(<1)}(X,X)$ is written as $\mathrm{Lip}^{(<1)}(X)$.

Definition 2.9 (Recurrent IFS) Suppose that $P := (p_{ij})$ is an irreducible row-stochastic $N \times N$ matrix. Let $\mathbf{w} := \{w_i : X \to X \,|\, i = 1,\dots N\}$ be a given collection of Lipschitz continuous functions. The triple $(X, \mathbf{w}, P)$ is called a recurrent IFS (on X). The recurrent IFS $(X, \mathbf{w}, P)$ is called hyperbolic iff $w_i \in \mathrm{Lip}^{(<1)}(X)$ for all $i = 1,\dots N$.

The above-mentioned interpretation of a recurrent IFS in terms of a Markov process is now studied further.

Let $(X, \mathbf{w}, P)$ be a recurrent IFS, let $S := \{1, \ldots N\}$, and let $(\Sigma, \mathcal{F}, \mu)$ be the associated code space (Section 1.2). Recall that $\Sigma = S^{\mathbb{N}}$. A Markov chain on S is obtained by defining a countable sequence of random variables $\{Y_n : \Sigma \to S \mid n \in \mathbb{N}_0\}$,

$$Y_n(\mathbf{i}) := i_n, \tag{2.37}$$

for all $n \in \mathbb{N}_0$ and for all $\mathbf{i} \in \Sigma$, and a transition matrix

$$\pi\{Y_{n+1}(\mathbf{i}) = i_{n+1} | Y_n(\mathbf{i}) = i_n\} := p_{ij}. \tag{2.38}$$

To generate a Markov process on X using the Markov chain defined above, one proceeds as follows: Let $\tilde{X} := X \times S$. Let the random variables $Z_n : X^{\mathbb{N}} \to X$ be given by

$$Z_n := w_{Y_n(\mathbf{i})} Z_{n-1}, \tag{2.39}$$

for all $n \in \mathbb{N}_0$, and for some initially chosen $x_0 =: Z_0(\mathbf{x}) \in X$. In order to satisfy the Markov property (1.41), $\{Z_n | n \in \mathbb{N}_0\}$ needs to be extended to $\tilde{X}$. This is done by pairing up Z_n and $i_n = Y_n(\mathbf{i})$. More precisely, a countable sequence of random variables $\{\tilde{Z}_n | n \in \mathbb{N}_0\}$ is defined by

$$\tilde{Z}_n := (Z_n, Y_n(\mathbf{i})), \tag{2.40}$$

for all $n \in \mathbb{N}_0$, and a transition probability by ((1.37))

$$\tilde{P}((x, i), \tilde{B}) := \sum_{j=1}^{N} p_{ij} \, \chi_{\tilde{B}}(w_j(x), j), \tag{2.41}$$

for all $(x, i) \in \tilde{X}$ and all Borel sets $\tilde{B} \in \tilde{\mathcal{B}}(\tilde{X})$.

Now let π^* be the stationary probability distribution for the Markov chain $\{Y_n | n \in \mathbb{N}_0\}$, that is, $P\pi^* = \pi^*$. By previous results it is known that, if $(X, \mathbf{w}, P)$ is hyperbolic, then there exists a stationary distribution $\tilde{\pi}^*$ for the Markov process $\{\tilde{Z}_n | n \in \mathbb{N}_0\}$. Denote by μ the X-projection of $\tilde{\pi}^*$, i.e.,

$$\mu = \text{proj}_X \tilde{\pi}^*, \tag{2.42}$$

where $\text{proj}_X : \tilde{X} \twoheadrightarrow X$, $(x, \mathbf{i}) \mapsto x$. Since the mapping γ defined in (2.28) is continuous and surjective, the measure μ is unique and attractive. To obtain its "lift" $\tilde{\mu}$, the *inverse* Markov chain to $\{Y_n | n \in \mathbb{N}_0\}$ has to be introduced.

Define a transition probability matrix by

$$q_{ij} := \frac{\pi_j^*}{\pi_i^*} p_{ji}. \tag{2.43}$$

The matrix $Q = (q_{ij})$ is easily seen to be row-stochastic and irreducible. Also, $Q\pi^* = \pi^*$.

Suppose the probability that $Y(\mathbf{i}_n) = Y(\mathbf{j}_n)$, for $\mathbf{i}, \mathbf{j} \in \Sigma$, needs to be calculated. This probability is given by

$$\sum_{j_0=1}^{N} \pi_{j_0}^* p_{j_0 j_1} \cdots p_{j_{n-1} j_n}.$$

Using the definition of Q it is seen that the above expression is equal to

$$\sum_{j_0=1}^{N} \pi_{j_0}^* \frac{\pi_{j_1}^*}{\pi_{j_0}^*} q_{j_1 j_0} \cdots \frac{\pi_{j_n}^*}{\pi_{j_{n-1}}^*} q_{j_n j_{n-1}} = \pi_{j_n}^* q_{j_n j_{n-1}} \cdots q_{j_1 j_0}.$$

This last equality, however, is equal to the probability that a Markov chain $\{\check{Y}_n \,|\, n \in \mathbb{N}_0\}$ with probability transition matrix Q and stationary distribution π^* will have for its first n values $(j_n j_{n-1} \dots j_1)$. Hence, the chain $\{\check{Y}_n \,|\, n \in \mathbb{N}_0\}$ is "time-reversed." The stationary distribution of the time-reversed Markov chain $\{\check{Y}_n \,|\, n \in \mathbb{N}_0\}$ is denoted by $\check{\pi}^*$. Now it should be clear that if we let

$$\tilde{\mu}\tilde{B} := \check{\pi}^* \{\mathbf{i} \in \Sigma \,|\, (\gamma(\mathbf{i}), i_1) \in \tilde{B}\}, \tag{2.44}$$

then $\tilde{\mu}$ is the unique stationary distribution for the Markov process $\{\tilde{Z}_n \,|\, n \in \mathbb{N}_0\}$. It can also be shown ([10]) that if $\tilde{\nu}$ is a probability measure on $\tilde{X}$ with the property that $\tilde{\nu}(X \times i) = \pi_i^*$, for $i \in \{1, \dots, N\}$, the Markov process $\{\tilde{Z}_n^{\tilde{\nu}} \,|\, n \in \mathbb{N}_0\}$ whose stationary distribution is $\tilde{\nu}$ converges in distribution to $\tilde{\mu}$. Note that this is in sharp contrast to the *pointwise* convergence in the previous subsection.

As in Section 2.2, the support of the measure μ is called the attractor of the recurrent IFS. The support of μ is an element of $\mathfrak{H}(X)$.

The action of the Markov process $\{Y_n \,|\, n \in \mathbb{N}_0\}$ associated with a recurrent IFS on its state space $S = \{1, \dots, N\}$ can also be described in terms of a *labelled directed graph*. For this purpose some basic definitions are recalled.

Definition 2.10 1. A (finite) directed graph G is a set consisting of (i) a finite set V of objects called vertices, and (ii) a set $\mathsf{E} \subseteq \mathsf{V} \times \mathsf{V}$ of ordered pairs of vertices called (directed) edges.

2. An edge $(\mathsf{v}, \mathsf{v}) \in \mathsf{E}$ from a vertex to itself is called a loop.

3. A directed graph G is called complete iff $\mathsf{E} = \mathsf{V} \times \mathsf{V}$.

4. A directed graph G is called labelled iff there exists a mapping $\ell : \mathsf{E} \to \mathbb{R}$.

Now let $\mathsf{V} := S$ and let $\mathsf{E} := \{(\mathsf{v}, \mathsf{v}') \in \mathsf{V} \times \mathsf{V} \,|\, p_{\mathsf{v},\mathsf{v}'} > 0\}$. This clearly defines a directed graph G_Y associated with the Markov chain $\{Y_n \,|\, n \in \mathbb{N}_0\}$. If $\ell : \mathsf{E} \to \mathbb{R}$ is defined by

$$\ell(\mathsf{v}, \mathsf{v}') := p_{\mathsf{v}, \mathsf{v}'},$$

then G_Y becomes a labelled directed graph. (The labels on G_Y are the non-zero probabilities of transfer from one state to another. One may also refer to this labelling as the *communication relation* between states in a Markov chain.)

It is worthwhile mentioning that the labelled directed graph associated with an IFS — recall that in this case $p_{ij} = p_j$, $i, j \in \{1, \ldots, N\}$ — is very special: It is *complete*.

Next, the recurrent IFS structure on several compact metrizable topological spaces is investigated. The issues that will be addressed deal only with the point-set topological setting.

Let N be an integer greater than one, and suppose that $(X, \mathbf{w}, P)$ is a hyperbolic recurrent IFS. For each $j = 1, \ldots, N$, let $(X_j, d_j) := (X, d)$. Also, for each $(i, j) \in \mathbf{I}$ and for $E \in \mathfrak{H}(X)$, define set maps $\mathfrak{w}_{ij} : \mathfrak{H}(X_j) \to \mathfrak{H}(X_i)$ by

$$\mathfrak{w}_{ij}(E) := w_i(E).$$

Let $\mathbf{I}(i) := \{j \in \{1, \ldots, N\} \,|\, p_{ji} > 0\}$ and let s_{ij} denote the contractivity constant of $\mathfrak{w}_{ij}$. The set map $\mathfrak{w} : \mathfrak{H}(X) \to \mathfrak{H}(X)$ associated with $\{w_i \,|\, i = 1, \ldots, N\}$ extends to a mapping $\mathsf{X}\mathfrak{w} : \mathsf{X}\mathfrak{H}(X) \to \mathsf{X}\mathfrak{H}(X)$ by setting

$$\mathsf{X}\mathfrak{w}(E_1, \ldots, E_N) := \left(\bigcup_{j \in \mathbf{I}(1)} \mathfrak{w}_{1j}(E_j), \ldots, \bigcup_{j \in \mathbf{I}(N)} \mathfrak{w}_{Nj}(E_j) \right), \tag{2.45}$$

where $\mathsf{X}\mathfrak{H}(X) := \times_{j=1}^{N} \mathfrak{H}(X_j)$. To make $\mathsf{X}\mathfrak{H}(X)$ into a compact metric space, an extension $\mathsf{X}\mathfrak{h}$ of the Hausdorff metric is defined by

$$\mathsf{X}\mathfrak{h}((E_1, \ldots, E_N), (E_1', \ldots, E_N')) := \max_{1 \le j \le N} \{\mathfrak{h}_j(E_j, E_j')\}, \tag{2.46}$$

with $E_j, E'_j \in \mathfrak{H}(X_j)$, $j = 1, \ldots, N$. (That $\mathsf{X}\mathfrak{h}$ is a metric on $\mathsf{X}\mathfrak{H}(X)$ is readily shown; the compactness follows immediately from Tychonov's Theorem. The fact that the maps w_{ij} are in $\mathrm{Lip}^{(<1)}(\mathfrak{H}(X_j), \mathfrak{H}(X_i))$ gives the following proposition, whose easy though messy proof is left for the diligent reader:

Proposition 2.7 *The set map* $\mathsf{X}\mathsf{w} : \mathsf{X}\mathfrak{H}(X) \to \mathsf{X}\mathfrak{H}(X)$ *defined in Eq. (2.45) satisfies*

$$\mathsf{X}\mathfrak{h}(\mathsf{X}\mathsf{w}(\mathsf{X}E), \mathsf{X}\mathsf{w}(\mathsf{X}E') \le s\,\mathsf{X}\mathfrak{h}(\mathsf{X}E, \mathsf{X}E')),$$

for all $\mathsf{X}E$, $\mathsf{X}E' \in \mathsf{X}\mathfrak{H}(X)$, *where* $s := \max\{s_{ij}\,|\,(i,j) \in \mathbf{I}\}$. *Thus,* $\mathsf{X}\mathsf{w} \in \mathrm{Lip}^{(<1)}(\mathsf{X}\mathfrak{H}(X))$. ∎

The next result is then an immediate consequence of Proposition 2.7.

Corollary 2.1 *There exists a unique compact set* $\mathsf{X}A \in \mathsf{X}\mathfrak{H}(X)$ *such that*

$$\mathsf{X}\mathsf{w}(\mathsf{X}A) = \mathsf{X}A,$$

or equivalently,

$$A_i = \bigcup_{j \in \mathbf{I}(i)} \mathsf{w}_{ij}(A_j), \tag{2.47}$$

for all $i = 1, \ldots, N$. ∎

The connection between the previous results, concerning a single hyperbolic recurrent IFS and its probabilistic interpretation, and the present setting will now be established. To this end, note that if the index j is restricted to only a single value, then w_{ij} reduces to w. This gives the next result.

Corollary 2.2 *Let* $(X, \mathbf{w}, P)$ *be a hyperbolic recurrent IFS. Let* A *denote the support of the unique stationary probability measure* μ *given by (2.42). Then there exist unique sets* $A_i \subseteq A \in \mathfrak{H}(X)$, $i = 1, \ldots, N$, *such that*

$$A = \bigcup_{i=1}^{N} A_i,$$

and

$$A_i = \bigcup_{j \in \mathbf{I}(i)} w_i(A_j),$$

for $i = 1, \ldots, N$. ∎

By Corollary 2.2 it is therefore justified to call the set $\mathsf{X}A$ — whose existence is guaranteed by Corollary 2.1 — the attractor of the hyperbolic recurrent IFS $(X, \mathbf{w}, P)$. Note that, unlike the case of a hyperbolic IFS, the subsets A_i of the attractor A are not copies of A itself but are related to each other via the maps w_i. In the next section this characterization is discussed further and related to M. F. Dekking's construction of fractal sets.

2.2.4   Iterated Riemann surfaces

Now a class of IFSs in a complex-analytic setting is considered. It is assumed that the reader is familiar with the rudimentary concepts of complex function theory and some of the more advanced topcis such as holomorphy and Riemann surfaces. Nevertheless, for the purpose of completeness and coherence, as well as notation and terminology, a few definitions are given and some relevant results quoted. This subsection is independent of the general approach, and the reader who is not interested in this rather special setting may want to skip it.

The one-point compactification of the complex plane $\mathbb{C}$ is denoted by $\mathbb{C}^+$, i.e., $\mathbb{C}^+ := \mathbb{C} \cup \{\infty\}$. The set $\mathbb{C}^+$ may be identified with the unit sphere $\{(x, y, z) \mid x^2 + y^2 + z^2 = 1\}$ in $\mathbb{R}^3$, and is therefore commonly called the *Riemann sphere*. By a *region* $G \subseteq \mathbb{C}^+$ is meant an open and connected subset of $\mathbb{C}^+$.

***Definition* 2.11** A function $f : G \to \mathbb{C}^+$ is called holomorphic in G iff f is (complex) differentiable at every point of G; it is called holomorphic at a point $z_0 \in G$ iff f is holomorphic in an entire ε-neighborhood $N_\varepsilon(z_0) \subseteq \mathbb{C}^+$ of z_0.

Remarks.

1. Some authors call *analytic* what was called holomorphic above, and reserve holomorphy for analytic mappings between abstract Riemann surfaces.

2. Holomorphy of $f(z)$ at ∞ is always understood as holomorphy of $f(1/z)$ at 0.

Next the concept of a *surface* is introduced. A Hausdorff topological space M with a countable base for its topology is called a surface iff every point $x \in M$ has an open neighborhood U homeomorphic to some open subset of

$\mathbb{C}$, or equivalently, $\mathbb{R}^2$. Note that a surface is a special case of what is called a *manifold*.

Using Proposition 1.5, it is seen that any surface M can be covered by a countable collection $\{U_i \mid i \in \mathbb{N}\}$ of open sets, such that for each U_i there exists a homeomorphism $\varphi_i : U_i \to O_i$, where O_i is an open subset of $\mathbb{C}$.

Definition 2.12 Let M be a surface. The set of pairs $\mathcal{A} := \{(U_i, \varphi_i) \mid i \in \mathbb{N}\}$ is called an atlas for M. If $x \in U_i$, then (U_i, φ_i) is called a chart at x and $z := \varphi_i(x)$ a local coordinate for x.

Suppose (U_i, φ_i) and (U_j, φ_j) are charts at $x \in M$ giving local coordinates z_i and z_j, respectively. Then $z_i \mapsto \varphi_i \circ \varphi_j^{-1}(z_i) = z_j$ describes the change in local coordinates for x. If $U_i \cap U_j \neq \emptyset$, then the functions

$$\varphi_i \circ \varphi_j^{-1} : \varphi_j(U_i \cap U_j) \to \varphi_i(U_i \cap U_j) \qquad (2.48)$$

are called the *coordinate transition functions*.

An atlas $\mathcal{A}$ is called holomorphic if all its coordinate transition functions are holomorphic. In order to define a *complex structure* on a surface M, one has to introduce the relation of *compatibility* between holomorphic atlases for M.

Two holomorphic atlases $\mathcal{A} := \{(U_i, \varphi_i)\}$ and $\mathcal{B} := \{(V_j, \psi_j)\}$ are called *compatible*, in symbols, $\mathcal{A} \sim \mathcal{B}$, iff for all $(U_i, \varphi_i) \in \mathcal{A}$ and $(V_j, \psi_j) \in \mathcal{B}$ satisfying $U_i \cap V_j \neq \emptyset$, the functions

$$\varphi_i \circ \psi_j^{-1} : \psi_j(U_i \cap V_j) \to \varphi_j(U_i \cap V_j)$$

are holomorphic. It is easily verified that $\sim$ is an equivalence relation on the class of all holomorphic atlases for M. A complex structure on M is the pair $(M, \mathfrak{A})$ where $\mathfrak{A}$ is an equivalence class of compatible holomorphic atlases for M.

Definition 2.13 An (abstract) Riemann surface is a surface M with a complex structure $(M, \mathfrak{A})$.

It can be shown that the Riemann surface constructed for functions such as $f(z) = z^{1/2}$ or $f(z) = \log(z)$ are also abstract Riemann surfaces in the above sense (cf. [101]). The next result gives an example of a class of abstract Riemann surfaces.

Proposition 2.8 *Let Ω be a lattice in $\mathbb{C}$. Then $\mathbb{C}/\Omega$ is an abstract Riemann surface.* ∎

By a *lattice* in $\mathbb{C}$ is meant a discrete subgroup of the additive group $\mathbb{C}$ that is of the form $\{n_1\omega_1 + n_2\omega_2 | n_1, n_2 \in \mathbb{Z}\}$, for some fixed $\mathbb{R}$-independent $\omega_1, \omega_2 \in \mathbb{C}$, i.e., $\omega_1 \neq 0 \neq \omega_2$ and $\omega_1/\omega_2 \notin \mathbb{R}$. It is our next goal to briefly look at the Riemann surface of an *algebraic function*. w is called an algebraic function of z iff $A(z, w) = 0$, for some polynomial $A(z, w)$. In the case that $A(z, w)$ is irreducible, this can be expressed as

$$A(z, w) = a_n(z)w^n + a_{n-1}(z)w^{n-1} + \ldots + a_0(z),$$

where $a_i(z)$ is a polynomial in z, $i = 1, \ldots, n$, and $a_n(z) \not\equiv 0$. The integer n is called the degree of $A(z, w)$ in w. If $A(z, w)$ is reducible, then the degree n of $A(z, w)$ in w is defined as $\sum n_i$, where n_i is the degree in w of its irreducible factor $A_i(z, w)$. From now on let us assume that $A(z, w)$ is irreducible. For a fixed value of $z \in \mathbb{C}^+$, $A(z, w)$ is a polynomial of degree n having n distinct roots $w_1, \ldots, w_n$. Such values of z are called the *regular points* for $A(z, w)$. The *critical points* for $A(z, w)$ are those values in $\mathbb{C}^+$ that satisfy at least one of the following:

1. $z = \infty$;

2. $a_0(z) = 0$;

3. $A(z, w) = 0$ has a repeated root w.

The set of all critical points for $A(z, w)$ is denoted by C_A. For points $z_0 \in \mathbb{C}^+/C_A$ the following holds.

Proposition 2.9 *Suppose that $z_0 \in \mathbb{C}^+/C_A$. Then there exists an open disk $\overset{\circ}{D}(z_0)$ centered at z_0 and a set of single-valued holomorphic functions $f_i : \overset{\circ}{D}(z_0) \to \mathbb{C}^+$, $i = 1, \ldots, n$, such that*

1. *$f_i(z_0) = w_i$, $i = 1, \ldots, n$;*

2. *for every $z \in \overset{\circ}{D}(z_0)$, the roots w_i of $A(z, w) = 0$ are all distinct, simple, and of the form $w = f_i(z)$.*

∎

The Riemann surface for $A(z, w)$ will be denoted by M_A.

Theorem 2.7   *1. If $A(z, w)$ is an irreducible polynomial, then M_A is connected; otherwise, M_A consists of finitely many connected components M_{A_i}, one for each irreducible factor $A_i(z, w)$ of $A(z, w)$.*

2. If $A(z,w)$ is a polynomial, then M_A is compact.

3. Any compact abstract Riemann surface can be identified with a Riemann surface M_A for some algebraic function $A(z,w) = 0$.

∎

After this rather short and compact digression, a special class of IFSs is defined. Let $X := \mathbb{C}^+$ and let $A(z,w)$ be an algebraic function of degree N in w. Define maps $w_i : X \to X$, $i = 1, \ldots, N$, to be the solutions of

$$A(z,w) = 0, \text{ for } z \in \mathbb{C}^+,$$

with possible repeated roots separated. It is easy to see that choosing a set of probabilities according to $p_i := 1/N$, $i = 1, \ldots, N$, the triple $(X, \mathbf{w}, \mathbf{p})$ becomes an IFS on X. This IFS is called the *iterated Riemann surface* generated by the algebraic function $A(z,w)$. The remaining terms such as hyperbolicity and attractor are defined just as before.

***Example* 2.3** Let $\{\zeta_1, \zeta_2, \zeta_3\} \subseteq \mathbb{C}$ be a given non-collinear collection of points. Let $A(z,w) := (2w - z - \zeta_1)(2w - z - \zeta_2)(2w - z - \zeta_3)$. Then the maps $w_i(z)$, $i = 1, 2, 3$, are given by

$$w_i(z) = 1/2(z - \zeta_i), \text{ for } i = 1, 2, 3.$$

Choosing $p_i := 1/3$, $i = 1, 2, 3$, it is seen that the attractor of the iterated Riemann surface generated by $A(z,w) = (2w - z - \zeta_1)(2w - z - \zeta_2)(2w - z - \zeta_3)$ is a Sierpiński triangle with vertices at ζ_1, ζ_2, and ζ_3.

Attractors for iterated Riemann surfaces are closely related to *Julia sets*. However, the present scope of this book does not allow us to investigate this relation any further. The interested reader is referred to [9] and the references given therein.

2.3 Recurrent Sets

In 1982 M. F. Dekking ([45]) introduced a construction of fractal sets that is primarily based on an interplay between algebra and analysis; in his setting fractals — Dekking calls them recurrent sets — are limits of a sequence of compact subsets of $\mathbb{R}^n$ associated with iterates of a given word under an appropriate free semigroup endomorphism. This construction was generalized by T. Bedford in [15] and is closely related to recurrent IFSs and the *mixed self-similar sets* introduced by Ch. Bandt in [5].

2.3.1   The construction of recurrent sets

Let X be a finite set whose cardinality will be denoted by $\|X\|_c$. Denote by $S[X]$ and $F[X]$ the free semigroup, respectively free group, generated by the set X (Section 1.3). Recall that the abelianization of a free group $F[X]$ with generators $\{x_1, \ldots, x_{\|X\|_c}\}$ is the free abelian group $\mathbb{Z}^X$ whose generators are the $\|X\|_c$-tuples $\{(1, 0, \ldots, 0), \ldots, (0, \ldots, 1)\}$, and that the generator x_i is mapped onto the generator $(0, \ldots, 1, 0, \ldots, 0)$ in $\mathbb{Z}^X$, where 1 is in the ith position.

Every endomorphism $\vartheta : S[X] \to S[X]$ can be extended to a free group endomorphism by setting $\vartheta(x^{-1}) = \vartheta(x)^{-1}$. This extension will also be denoted by ϑ. The main idea of Dekking's construction is to associate compact subsets in $\mathbb{R}^n$ with words in $S[X]$, and the iterates of ϑ with the iterates of a representation L_ϑ of ϑ in $\mathbb{R}^n$. The following diagram shows the steps that are involved in obtaining this goal:

$$
\begin{array}{ccccccccc}
S[X] & \overset{i}{\hookrightarrow} & F[X] & \overset{p}{\twoheadrightarrow} & \mathbb{Z}^X & \overset{j}{\hookrightarrow} & \mathbb{R}^X & \overset{g}{\twoheadrightarrow} & \mathbb{R}^n \\
\downarrow{\vartheta} & & \downarrow{\vartheta} & & \downarrow{\vartheta^*} & & \downarrow{\vartheta^*} & & \downarrow{L_\vartheta} \\
S[X] & \overset{i}{\hookrightarrow} & F[X] & \overset{p}{\twoheadrightarrow} & \mathbb{Z}^X & \overset{j}{\hookrightarrow} & \mathbb{R}^X & \overset{g}{\twoheadrightarrow} & \mathbb{R}^n
\end{array} \; ,
$$

where $i : S[X] \hookrightarrow F[X]$ denotes the embedding of the free semigroup $S[X]$ into $F[X]$, and $p : F[X] \twoheadrightarrow \mathbb{Z}^X$ is the projection of the free group $F[X]$ onto its abelianization ((1.56)). The map $j : \mathbb{Z}^X \hookrightarrow \mathbb{R}^X$ is the canonical embedding of $\mathbb{Z}^X$ into $\mathbb{R}^X$, and $g : \mathbb{R}^X \twoheadrightarrow \mathbb{R}^n$ is a linear surjective map. Finally, ϑ^* denotes the induced endomorphism on $\mathbb{Z}^X$ and $\mathbb{R}^X$, respectively. (To simplify notation the same symbol was used for ϑ^* and its lift on $\mathbb{R}^X$.)

Now let $f := g \circ j \circ p \circ i$. Since g is linear, $f : S[X] \to \mathbb{R}^n$ is a homomorphism, i.e.,

$$f(ww') = f(w) + f(w'), \tag{2.49}$$

for all words $w, w' \in S[X]$. Furthermore, it is seen that

$$f \circ \vartheta = L_\vartheta \circ f. \tag{2.50}$$

The map $L_\vartheta : \mathbb{R}^n \to \mathbb{R}^n$ is called a *representation* of ϑ. Since g is a linear surjection, L_ϑ is uniquely defined up to a linear map in $\ker g$. Since the kernel of g is a ϑ^*-invariant subspace of $\mathbb{R}^X$, L_ϑ is determined — in the above sense — by this subspace. Two representations of ϑ deserve to be mentioned explicitly ([45])

***Example* 2.4** Suppose that $\ker g = \{0\}$. Then $n = \|X\|_c$, and choosing $g := id_{\mathbb{R}^X}$ gives $L_\vartheta = \vartheta^*$. This representation is referred to as the *full representation* of ϑ.

***Example* 2.5** Suppose ϑ^* has a real eigenvalue λ. Let v be the associated eigenvector. The set $\{\xi \in \mathbb{R}^X \mid \langle \xi, v \rangle = 0\}$ is clearly a ϑ^*-invariant subspace of $\mathbb{R}^X$. Define the linear surjection g by

$$g := \langle \cdot, v \rangle.$$

Then, for $w \in S[X]$,

$$
\begin{aligned}
f(\vartheta(w)) &= g((j \circ p \circ i)\vartheta(w)) &= \langle (j \circ p \circ i)\vartheta(w), v \rangle \\
&= \langle \vartheta^*(j \circ p \circ i)(w), v \rangle &= \langle j \circ p \circ i(w), \lambda v \rangle \\
&= \lambda g(j \circ p \circ i(w)) &= \lambda f(w).
\end{aligned}
$$

Hence, $L_\vartheta = \lambda(\cdot)$. If ϑ is a non-trivial endomorphism of $S[X]$, then the Perron-Frobenius Theorem (1.8) implies the existence of a positive eigenvalue λ larger than or equal to the absolute value of all other eigenvalues of ϑ^*. This particular representation is called the *Frobenius representation* of ϑ.

The mapping f as defined earlier only associates points in $\mathbb{R}^n$ with words in $S[X]$. For the generation of fractal sets more flexibility is needed; words should be associated with more general compact sets, such as polygonal lines or squares. This can be achieved by defining a map $K : S[X] \to \mathfrak{H}(\mathbb{R}^n)$, satisfying

$$K(ww') := K(w) \cup (K(w') + f(w)). \tag{2.51}$$

Here the usual notation $A + v := \{a + v \mid a \in A\}$ for a set $A \subseteq \mathbb{R}^n$ and a vector $v \in \mathbb{R}^n$ is used.

The foregoing requirement can also be expressed in the following way. Let $w := x_1 x_2 \ldots x_m \in S[X]$. Then to w corresponds the compact set

$$
\begin{aligned}
K(w) = \; & K(x_1) \cup (f(x_1) + K(x_2)) \cup (f(x_1 x_2) + K(x_3)) \cup \ldots \\
& \cup (f(x_1 \ldots x_{m-1}) + K(x_m)).
\end{aligned}
$$

If, for instance, K is defined as the map

$$K(x) = [0, f(x)], \tag{2.52}$$

for $x \in X$ and $[a, b] = \{a + t(b - a) \mid 0 \le t \le 1\}$, then for a word $w = x_1 x_2 \ldots x_m \in S[X]$ the set $K(w)$ is the polygonal line with vertices at 0, $f(x_1)$,

$f(x_1)+f(x_2), \ldots, f(x_1)+\ldots+f(s_m) = f(s_1 s_2 \ldots s_m)$. Furthermore, $K(\vartheta(w))$ is a polygonal line from 0 to $f(\vartheta(w))$, and $K(\vartheta^\nu(w))$, $\nu \in \mathbb{N}$, connects the points 0 and $f(\vartheta^\nu(w))$.

Of interest is the behavior of the sets $K(\vartheta^\nu(w))$ as $\nu \to \infty$, for a fixed word $w \in S[X]$. To obtain convergence (in the Hausdorff metric), the sets $K(\vartheta^\nu(w))$ need to be rescaled. This is possible if it is assume that L_ϑ is an expansive map, i.e., if all its eigenvalues have modulus greater than one. This assumption then allows the definition of contraction mappings $K_\nu(\vartheta)$: $S[X] \to \mathfrak{H}(\mathbb{R}^n)$:

$$K_\nu(\vartheta)(w) := L_\vartheta^{-\nu} K(\vartheta^\nu(w)). \tag{2.53}$$

In addition to the preceding assumption, *virtual* letters need to be introduced; these are those elements $x \in X$ that satisfy $K(x) = \emptyset$. The set of all virtual letters will be denoted by Q. The empty word will be called a virtual letter. Hence, $Q \neq \emptyset$.

The next definition introduces the concept of Y-stability, $Y \subseteq X$, that is needed in passing from $K(\vartheta^\nu(w))$ to $K(\vartheta^{\nu+1}(w))$, $w \in S[X]$.

Definition 2.14 Suppose that Y is a subset of X. A free semigroup endomorphism ϑ of $S[X]$ is called Y-stable iff there exists a $\nu_0 \in \mathbb{N}$ so that for all $x \in X$ exactly one of the following occurs:

(a) $\vartheta^\nu(x) \in S[Y]$, for all $\nu \geq \nu_0$;

(b) $\vartheta^\nu(x) \notin S[Y]$, for all $\nu \geq \nu_0$.

If $Y = Q$, the set of virtual letters, then the letters that satisfy (b) are called essential. The set of essential letters is denoted by E.

For the proof of the main result the following lemma is needed.

Lemma 2.1 *Let L be an endomorphism of $\mathbb{R}^n$, i.e., a bounded linear operator from the normed linear space $(\mathbb{R}^n, d_E)$ onto itself (here d_E denotes the Euclidean norm). Suppose that the eigenvalues $\lambda_1, \ldots, \lambda_n$ of L satisfy $|\lambda_1| \leq |\lambda_2| \leq \cdots \leq |\lambda_n|$. Then for any $\lambda > |\lambda_n|$ there exists a positive constant c such that*

$$\|L^\nu v\| \leq c \lambda^\nu \|v\|,$$

for all $v \in \mathbb{R}^n$ and $\nu \in \mathbb{N}$.

Proof. The result is an easy consequence of the Spectral Radius Theorem (1.10). ∎

Now the main theorem can be stated (also [45]).

Theorem 2.8 (Recurrent Sets) *Suppose that ϑ is a free semigroup endomorphism on $S[X]$ and that L_ϑ is an expansive representation of ϑ. Furthermore, assume that the maps $f : S[X] \to \mathbb{R}^n$, $K : S[X] \to \mathfrak{H}(\mathbb{R}^n)$, and $K_\nu(\vartheta) : S[X] \to \mathfrak{H}(\mathbb{R}^n)$ are defined as in (2.49), (2.51) and (2.53), respectively. Let $Q = \{x \in X \,|\, K(x) = \emptyset\}$ and assume that ϑ is Q-stable. Then there exists a unique non-empty compact set $K_\vartheta(w)$ that is the limit (in the Hausdorff metric $\mathfrak{h}$) of the sequence $\{K_\nu(\vartheta)(w)\}$, where $w \in S[X]$ is any word containing at least one essential letter.*

Remark. The set $K_\vartheta(w)$ in Theorem 2.8 is called a *recurrent set*.

Proof. Let ν_0 be that natural number whose existence is asserted by the Q-stability of ϑ. For any $\nu \geq \nu_0$, let

$$\vartheta^\nu(w) = \vartheta^{\nu_0}\vartheta^{\nu-\nu_0}(w) =: \vartheta^{\nu_0}(w_{\nu 1} w_{\nu 2} \ldots w_{\nu\mu(\nu)}).$$

Let

$$w'_{\nu\kappa} := f(\vartheta^{\nu_0}(w_{\nu 1} \ldots w_{\nu\kappa-1})).$$

Then, for all $\nu \geq \nu_0$,

$$
\begin{aligned}
\mathfrak{h}(K_\nu(\vartheta), K_{\nu+1}(\vartheta)) &= \mathfrak{h}\Big(\bigcup_{\kappa=1}^{\mu(\nu)} L_\vartheta^{-\nu} K(\vartheta^{\nu_0}(w_{\nu\kappa})) \\
&\quad + w'_{\nu\kappa}, \bigcup_{\kappa=1}^{\mu(\nu)} L_\vartheta^{-\nu-1} K(\vartheta^{\nu_0+1}(w_{\nu\kappa})) + w'_{\nu\kappa}\Big) \\
&= \mathfrak{h}\Big(\bigcup_{w_{\nu\kappa} \in E} L_\vartheta^{-\nu} K(\vartheta^{\nu_0}(w_{\nu\kappa})) \\
&\quad + w'_{\nu\kappa}, \bigcup_{w_{\nu\kappa} \in E} L_\vartheta^{-\nu-1} K(\vartheta^{\nu_0+1}(w_{\nu\kappa})) + w'_{\nu\kappa}\Big) \\
&\leq \max_{e \in E}\{\mathfrak{h}(L_\vartheta^\nu K(\vartheta^{\nu_0}(e)), L_\vartheta^{-\nu} K(\vartheta^{\nu_0}(e)))\} \\
&=: \delta
\end{aligned}
$$

In the preceding estimate the simple fact that $\mathfrak{h}(A + v, B + v) = \mathfrak{h}(A, B)$, for $A, B \in \mathfrak{H}(\mathbb{R}^n)$ and $v \in \mathbb{R}^n$, was used.

Since L_ϑ is assumed to be expansive, there exists a $\lambda > 1$ satisfying $\lambda < \min\{|\lambda_i| \,|\, \lambda_i \in \sigma(L_\vartheta)\}$.

By Lemma 2.1 there exists a positive constant c such that

$$\mathfrak{h}(K_\nu(\vartheta), K_{\nu+1}(\vartheta)) \leq c \lambda^{-\nu} \delta.$$

Thus, by the Banach Fixed-Point Theorem, there exists a $K_\vartheta(w) \in \mathfrak{H}(\mathbb{R}^n)$ such that $K_\nu(\vartheta)(w) \overset{\nu \to \infty}{\longrightarrow} K_\vartheta(w)$.   ∎

Remark. The fixed point or invariance condition of $K_\vartheta(\cdot)$ can be expressed as

$$L_\vartheta K_\vartheta(x) = \bigcup_{i=1}^{N} (K_\vartheta(y_j) + f(y_1 \ldots y_{i-1})), \tag{2.54}$$

or equivalently, as

$$K_\vartheta(x) = \bigcup_{i=1}^{N} \left(L_\vartheta^{-1} K_\vartheta(y_i) + \sum_{j<i} L_\vartheta^{-1} f(y_j) \right), \tag{2.55}$$

where $\vartheta(x) = y_1 \ldots y_N$, and $x \in X$.

The following two examples will illustrate the results just obtained.

Example 2.6 Let $X := \{x, \overline{x}\}$, and let ϑ be defined by

$$x \overset{\vartheta}{\mapsto} x\,x\,x, \quad \overline{x} \overset{\vartheta}{\mapsto} \overline{x}\,x\,\overline{x}.$$

Let f be given by $f(x) = f(\overline{x}) := 1$. Then, since $f(\vartheta(x)) = f(x) + f(\overline{x}) + f(x) = 3 = f(\vartheta(\overline{x}))$, ϑ can be represented by its Frobenius representation:

$$L_\vartheta f = 3 \cdot f.$$

Define the map $K : S[X] \to \mathfrak{H}(\mathbb{R})$ by

$$K(\overline{x}) := [0,1], \text{ and } K(x) := \emptyset.$$

The set of virtual letters is then $Q = \{x\}$. It is easy to see that the endomorphism ϑ leaves $S[Q]$ invariant and is Q-stable. Therefore, by Theorem 2.8, the sequence $K_\nu(\vartheta)(\overline{x})$ converges in the Hausdorff metric to a unique compact subset C of $\mathbb{R}$. Note that

$$K_1(\vartheta)(\overline{x}) = 1/3(K(\overline{x}) \cup (f(\overline{x}) + K(x)) \cup (f(\overline{x}x) + K(\overline{x})))$$

$$= [0, 1/3] \cup [2/3, 1],$$

$$K_2(\vartheta)(\overline{x}) = [0, 1/9] \cup [2/9, 1/3] \cup [2/3, 5/9] \cup [8/9, 1],$$

$$\vdots$$

$$K_\nu(\vartheta)(\overline{x}) \;=\; [0,(1/3)^\nu] \cup [2(1/3)^\nu,(1/3)^{\nu-1}] \cup \ldots \cup [1-(1/3)^\nu,1],$$

$$\vdots$$

Hence, C is the classical Cantor set.

Example 2.7 Let $X := \{x_i, \overline{x}_i \; : \; i = 1,2,3\}$, and let $\vartheta : S[X] \to S[X]$ be given by

$$x_1 \overset{\vartheta}{\mapsto} \overline{x}_1\, x_2\, \overline{x}_1, \quad \overline{x}_1 \overset{\vartheta}{\mapsto} x_1\, \overline{x}_2\, x_1,$$

$$x_2 \overset{\vartheta}{\mapsto} \overline{x}_2\, x_3\, \overline{x}_2, \quad \overline{x}_2 \overset{\vartheta}{\mapsto} x_2\, \overline{x}_3\, x_2,$$

$$x_3 \overset{\vartheta}{\mapsto} \overline{x}_3\, x_1\, \overline{x}_3, \quad \overline{x}_3 \overset{\vartheta}{\mapsto} x_3\, \overline{x}_1\, x_3.$$

Define $f : S[X] \to \mathbb{R}^2$ by

$$f(x_1) = f(\overline{x}_1) := (1,0),$$

$$f(x_2) = f(\overline{x}_2) := (-1/2, 1/2\sqrt{3}),$$

$$f(x_3) = f(\overline{x}_3) := (-1/2, -1/2\sqrt{3}),$$

and the mapping $K : S[X] \to \mathfrak{H}(\mathbb{R}^2)$ as follows:

$$K(x_i) := \{\alpha f(x_i)\,|\,\alpha \in [0,1]\}, \quad K(\overline{x}_i) := \emptyset,$$

for $i = 1,2,3$. It is easy to verify that ϑ is Q-stable ($\nu = 1$). Furthermore, it is easy to see that all letters are essential, i.e., $E = X$.

A representation of ϑ is given by

$$L_\vartheta f = \begin{pmatrix} \sqrt{3}\cos(\pi/6) & \sqrt{3}\sin(\pi/6) \\ -\sqrt{3}\sin(\pi/6) & \sqrt{3}\cos(\pi/6) \end{pmatrix} f.$$

The recurrent set $K_\vartheta(x_1)$ is one of the *dragon fractals* (see Fig. 2.2).

Notice that in the above example ϑ leaves $S[Q]$ invariant and is Q-stable. That the former statement always implies the latter will be proven next. But first a definition.

Definition 2.15 A free semigroup endomorphism ϑ on $S[X]$ is called mixing iff there exists a $\nu_0 \in \mathbb{N}$ such that all letters in X appear in each of the words $\vartheta^{\nu_0}(x)$, for $x \in X$.

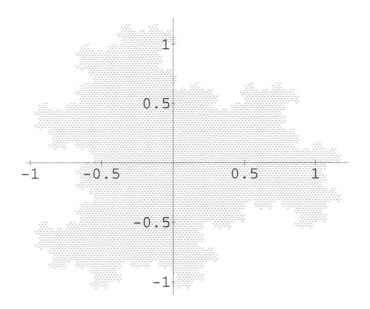

Figure 2.2: A dragon fractal.

The next proposition proves useful.

Proposition 2.10 *Let $S[X]$ be a free semigroup, and let $Y \subset X$. The following are necessary conditions for a free semigroup endomorphism ϑ to be Y-stable:*

    (a) $S[X]$ *is ϑ-invariant.*

    (b) ϑ *is mixing.*

Proof. (a) Let $Z := \{x \in X \setminus Y \mid \exists \nu \in \mathbb{N} \text{ such that } \vartheta^\nu(x) \in S[Y]\}$. The cardinality of Z is denoted by ν_0. Since Y is ϑ- invariant, it follows that for all $\nu \geq \nu_0$ and all $x \in Z \cup Y$, $\vartheta^\nu(x) \in S[Y]$. Now, if $x \notin Z \cup Y$ then $\vartheta^\nu(x) \notin S[Y]$, for all $\nu \in \mathbb{N}_0$.
(b) Two cases need to be considered: If $Y = X$, then $\vartheta^\nu(x) \in S[X]$; if $Y \neq X$, then, since ϑ is mixing, $\vartheta^\nu(x) \notin S[Y]$, for all $\nu \geq \nu_0$. ∎

It is worthwhile mentioning that virtual letters can be essential. The extreme case is given by the following example, which is a slight modification of Example 2.6.

Example 2.8 Let $X := \{x, \bar{x}, \bar{y}\}$, and let ϑ be given by

$$x \overset{\vartheta}{\mapsto} \bar{y}\,\bar{y}\,\bar{y}, \qquad \bar{x} \overset{\vartheta}{\mapsto} \bar{x}\,x\,\bar{y}, \qquad \bar{y} \overset{\vartheta}{\mapsto} \bar{y}\,\bar{y}\,\bar{y}.$$

Define $f : S[X] \to \mathbb{R}$ by

$$f(x) = f(\overline{x}) = f(\overline{y}) := 1,$$

and $K : S[X] \to \mathfrak{H}(\mathbb{R})$ by

$$K(x) = [0, 1], \quad K(\overline{x}) = K(\overline{y}) := \emptyset.$$

Clearly, $Q = \{\overline{x}, \overline{y}\}$ and $E = \{\overline{x}\} \subset Q$. It is not difficult to check that the recurrent set $K_{\vartheta}(\overline{x})$ is again the classical Cantor set C.

As the preceding example shows, some virtual letters are superfluous, i.e., their inclusion does not change the limit set $K_{\nu}(\vartheta)(\overline{x})$. It would therefore be useful if Q was equal to $X \setminus E$, as was the case in Example 2.6. The following proposition shows that this assumption can always be made.

Proposition 2.11 *Suppose $K_{\vartheta}(w)$ is a recurrent set as generated in Theorem 2.8. Then there exists a map $\tilde{K} : S[X] \to \mathbb{R}^n$ such that the associated set of virtual letters is equal to $X \setminus E$. Furthermore, $\tilde{K}_{\vartheta}(w) = K_{\vartheta}(w)$, if $w \in S[X]$ is any word containing at least one virtual letter.*

Proof. Let $w \in S[X]$ be a word containing at least one virtual letter, and let $K_{\vartheta}(w) = \lim_{\nu \to \infty} L_{\vartheta}^{-\nu} K(\vartheta^n(w))$ be the associated recurrent set. Let ν_0 be that integer such that for all $\nu \geq \nu_0$ either $\vartheta^{\nu_0}(w) \in S[X]$, i.e., $w \notin E$, or $\vartheta^{\nu_0}(w) \notin S[X]$, i.e., $w \in E$. Note that if L_{ϑ} is expansive, every iterate of ϑ is also expansive. It follows now from the Banach Fixed Point Theorem that $K_{\vartheta}(w) = K_{\vartheta^{\nu_0}}(w)$. Thus, without loss of generality it may be assumed that $\nu_0 = 1$.

Now define a mapping $\tilde{K} : S[X] \to \mathbb{R}^n$ by

$$\tilde{K}(x) := L_{\vartheta}^{-1} K(\vartheta(x)), \qquad (2.56)$$

for all $x \in X$.

Suppose then $w \in S[X]$. The iterates of $w = x_1 x_2 \ldots x_{\nu}$ under ϑ can be written in the following form:

$$\vartheta^{\nu}(w) = x_{1\nu} x_{2\nu} \ldots x_{\kappa(\nu)\nu},$$

$\nu \in \mathbb{N}_0$. Equations (2.49) and (2.51) then imply

$$
\begin{aligned}
K(\vartheta^{\nu+1})(w) &= \bigcup_{\mu=1}^{\kappa(\nu)} (K(\vartheta(x_{\mu\nu})) + f(\vartheta(x_{1\nu} x_{2\nu} \ldots x_{\mu-1\nu}))) \\
&= \bigcup_{\mu=1}^{\kappa(\nu)} L_{\vartheta}(\tilde{K}(x_{\mu\nu}) + f(x_{1\nu} x_{2\nu} \ldots x_{\mu-1\nu})) \qquad (2.57) \\
&= L_{\vartheta} \tilde{K}(\vartheta^{\nu}(w)).
\end{aligned}
$$

The set of essential letters is thus given by $E = \{x \in X : \tilde{K}(x) \neq \emptyset\}$. Therefore, one may define

$$\tilde{Q} := \{x \in X \mid \tilde{K}(x) = \emptyset\}. \tag{2.58}$$

Hence, $\tilde{Q} = X \setminus E$.

Furthermore, if $\tilde{E} := E$ and w in Eq. (2.57) is chosen to be $x \in X$, then it is easy to see that ϑ is $\tilde{Q}$-stable. Thus, $\tilde{E} = X \setminus \tilde{Q}$.

Finally, applying $L_\vartheta^{-(\nu+1)}$ to both sides of Eq. (2.57) yields

$$\tilde{K}_\nu(\vartheta)(w) := L_\vartheta^{-\nu} \tilde{K}(\vartheta^\nu(w)) = L_\vartheta^{-(\nu+1)} K(\vartheta^{\nu+1}(w)),$$

and hence $K_\vartheta(w) = \tilde{K}_\vartheta(w)$.   ∎

2.3.2 Subshifts of finite type and the connection to recurrent IFSs

In Section 2.2 it was seen that the code space (Σ, d_F) associated with an IFS $(X, \mathbf{w}, \mathbf{p})$ provides a way of describing the attractor and the dynamics of the maps in the IFS (Theorem 2.4 and Diagram (2.32)). In the theory of *symbolic dynamics*, one studies the structure of an attractor and the dynamics of the maps generating it via the *right-shift operator* $\sigma : \Sigma \twoheadrightarrow \Diamond$, defined by $\sigma(i_1\, i_2 \ldots i_n \ldots) := (i_2 \ldots i_n \ldots)$, and certain compact subsets of Σ, called *subshifts of finite type* or *topological Markov chains*. These subshifts then allow us to relate recurrent sets to attractors of recurrent IFSs.

Definition 2.16   1. A compact subset $\Upsilon \subseteq \Sigma$ is called a subshift iff $\sigma\Upsilon = \Upsilon$.

2. A subshift Υ is of finite type iff there exists an $N \times N$- matrix $M = (m_{ij})$ whose entries are zeros and ones such that

$$\Upsilon = \Upsilon(M) = \{(i_1\, i_2 \ldots i_n \ldots) \in \Sigma \mid m_{i_k i_{k+1}} = 1, \text{ for } k \in \mathbb{N}\}. \tag{2.59}$$

Remark. The reader may recall the graph-theoretic interpretation of a Markov process associated with a recurrent IFS (Definition 2.10 and arguments thereafter). A similar interpretation can be applied to a subshift of finite type: Let G_Υ be the graph whose vertex set V equals the state space $S = \{1, \ldots, N\}$, and whose edge set E equals $\{(i,j) : m_{ij} = 1\}$. A subshift Υ of finite type then consists of all sequences of vertices that form an infinite (directed) path in G_Υ. (Recall that a *(directed) path of length n* in a directed graph is a finite sequence of vertices $\{v_0, v_1, \ldots, v_n\}$ such that $v_{j-1} = v_j$, for all $j \in \{1, \ldots, n\}$.)

The next theorem states necessary and sufficient conditions for a compact subset of $\mathbb{R}^n$ to be the image of a subshift $\Upsilon \subseteq \Sigma$ under the continuous surjection γ defined in Eq. (2.28) (also [5]).

Theorem 2.9 *Assume that $w_1, \ldots, w_N$ are contractive on $(\mathbb{R}^n, d_E)$. Let A be the attractor of the hyperbolic IFS $(\mathbb{R}^n, \mathbf{w})$, and let γ be defined as in (2.28). Let $K \in \mathfrak{H}(\mathbb{R}^n)$. Then*

(a) *There exists a compact subset Υ of Σ with $\sigma\Upsilon \subseteq \Upsilon$ and $\gamma\Upsilon = K$ iff*

$$K \subseteq \bigcup_{i=1}^{N} w_i(K). \tag{2.60}$$

(b) *The set $\Upsilon := \gamma^{-1}K$ statisfies $\sigma\Upsilon \supseteq \Upsilon$ iff*

$$K \subseteq \bigcup_{i=1}^{N} w_i^{-1}K. \tag{2.61}$$

Hence, a compact subset K of $\mathbb{R}^n$ is the image of a subshift Υ of Σ under γ iff K satisfies Eqs. (2.60) and (2.61).

Proof. (a) NECESSITY: Let $K \in \mathfrak{H}(\mathbb{R}^n)$ be given. Suppose $\Upsilon \subseteq \Sigma$ is such that $\gamma\Upsilon = K$ and $\sigma\Upsilon \subseteq \Upsilon$. Note that for all $x \in K$, there exists a code $\mathbf{i} = \mathbf{i}(x) := (i_1 i_2 \ldots)$ in $\gamma^{-1}(x) \cap \Upsilon$. Thus, $\sigma(\mathbf{i}) = (i_2 i_3 \ldots) \in \Upsilon$. Furthermore, $(w_{i_1} \circ \gamma)(i_2 i_3 \ldots) = (\gamma \circ \tau_{i_1})(i_2 i_3 \ldots) = x$, where τ_i denotes the left shift map as defined by Eq. (2.25). Hence, $x \in w_{i_1}(K)$.
SUFFICIENCY: Now suppose that Eq. (2.60) holds, and $x \in K$ and a positive integer n are given. Then, an $x_1 \in K$ and an $i_1 \in S$ can be found such that $w_{i_1}(x_1) = x$, and an $x_2 \in K$ and an $i_2 \in S$ such that $w_{i_2}(x_2) = x_1$, and finally an $x_n \in K$ and an $i_n \in S$ with $w_{i_n}(x_n) = x_{n-1}$. By the surjectivity of γ there exists a code $\mathbf{j} \in \Sigma$ with $\gamma(\mathbf{j}) = x_n$. Define a code $\mathbf{i}$ by setting $\mathbf{i} := (i_1 i_2 \ldots i_n \mathbf{j})$. Then, for all $n \in S$, $\gamma(\sigma^n(\mathbf{i})) \in \Upsilon$. If sets $\{\Upsilon_n\}_{n \in \mathbb{N}}$ are defined by

$$\Upsilon_0 := \gamma^{-1}K,$$

$$\Upsilon_n := \{\mathbf{i} \in \Sigma : \sigma^m(\mathbf{i}) \in K_0, \text{ for } m = 0, 1, \ldots n\}, \; n \in \mathbb{N},$$

then $K_n \in \mathfrak{H}(\mathbb{R}^n)$ and $\gamma\Upsilon_n = K$. Now let

$$\Upsilon := \bigcap_{n \in \mathbb{N}} \Upsilon_n.$$

Then Υ is also compact and satisfies $\gamma(\Upsilon) = K$. It is clear from the definition of Υ that $\sigma\Upsilon \subseteq \Upsilon$.

(b) NECESSITY: Suppose $\gamma\Upsilon = K$ and $\sigma\Upsilon \supseteq \Upsilon$. Choose an $x \in K$ and a code $\mathbf{i} = (i_1 \, i_2 \ldots) \in \Upsilon$ with $\gamma(\mathbf{i}) = x$. Since Υ is a subset of $\sigma\Upsilon$ we can find an $i_0 \in S$ such that $(i_0 \, i_1 \, i_2 \ldots) \in \Upsilon$. Diagram (2.32) now implies $w_{i_0}(x) = \gamma(i_0 \, i_1 \, i_2 \ldots) \in K$. Hence, Eq. (2.61) follows.

SUFFICIENCY: Now assume that Eq. (2.61) holds. Let $\mathbf{j} \in \Upsilon$ be such that $\gamma(\mathbf{j}) = x$, for $x \in K$. Then an $i_0 \in S$ can be found so that $w_{i_0}(x) \in K$. Thus, $(i_0 \, \mathbf{j}) \in \Upsilon$. Hence, $\sigma\Upsilon \supseteq \Upsilon$.

To show the last statement, notice that if $\sigma\Upsilon \supseteq \Upsilon$ then for each $\mathbf{i} \in \Upsilon$ there exists a $\mathbf{j} \in \Upsilon$ such that $\sigma(\mathbf{j}) = \mathbf{i}$. However, $\mathbf{i} \in \Upsilon$ implies $\mathbf{j} \in \Upsilon$; hence, $\sigma\Upsilon \subseteq \Upsilon$.   ■

Now suppose a subshift Υ of finite type is given. Denote by B the set $\gamma\Upsilon$. For each $i \in S$ consider the sets

$$B_i := B \cap w_i(A) = \{\gamma(i_0 \, i_1 \ldots)| \, (i_0 \, i_1 \ldots) \in \Upsilon \wedge i_0 = i\}.$$

It is not difficult to see that

$$B_i = \bigcup_{\{j_1: \, m_{ij_1}=1\}} w_i(B_{j_1}). \tag{2.62}$$

Writing B_{j_1} as a union of its parts $w_{j_1}(B_{j_2})$ and proceeding in this manner yields the following equation:

$$\begin{aligned} B_i &= \bigcup w_i \circ w_{j_1}(B_{j_2}) \\ &= \bigcup w_i \circ w_{j_1} \circ \ldots \circ w_{j_m}(B_{j_m}), \end{aligned} \tag{2.63}$$

where the union is taken over all indices $j_1, j_2, \ldots \in S$ such that $m_{ij_1} = 1$, $m_{j_1 j_2} = 1, \ldots$. Hence, the matrix M defining the subshift of finite type Υ, or equivalently, the directed graph G_Υ, describes which function compositions are allowed.

At this point the reader is reminded of Corollary 2.2. The transition probability matrix $P = (p_{ij})$ for a hyperbolic recurrent IFS defines a *connection matrix* $C = (c_{ij})$ by setting $c_{ij} := 1$ if $p_{ij} > 0$ and $c_{ij} := 0$, otherwise. Then C can be used to define a subshift of finite type $\Upsilon = \Upsilon(C)$. If $p_{ij} = p_i$ for all $j \in S$, then C is the matrix whose entries are $c_{ij} = 1$ for all $i, j \in S$. The induced subshift of finite type Υ is then equal to Σ and G_Υ is complete.

The next objective is a generalization of Dekking's construction of recurrent sets. A slight generalization was first introduced by Dekking in [47], a more general approach was undertaken by T. Bedford ([15]). Here, Bedford's approach is presented. The main idea is to replace Eq. (2.49), or equivalently, $f = L_\vartheta^{-1} f \vartheta$, by a weaker condition. More precisely, suppose that for each letter $x \in X$ a homeomorphism $\phi_x : \mathbb{R}^n \to \mathbb{R}^n$ is given with the properties that $\phi_x(0) = 0$ and $\phi_x^{-1} \in \mathrm{Lip}_{\mathbb{R}}^{(<s)}(X)$, for all $x \in X$ and some $0 < s < 1$. Suppose that $\vartheta : S[X] \to S[X]$ is a free semigroup endomorphism, and that $f : S[X] \to \mathbb{R}^n$ satisfies

$$f(x) = \sum_{i=1}^m \phi_{y_i}^{-1} f(y_i), \tag{2.64}$$

where $\vartheta(x) = y_1 y_2 \ldots y_m$. Before the more general Eq. (2.64) can be used to construct recurrent sets, a few preliminaries are needed.

One starts with a subshift of finite type $\Upsilon \subseteq \Sigma$, i.e., a matrix M whose entries are zeros and ones, and then defines a surjection from this subshift onto a recurrent set. Given a free semigroup $S[X]$ and a free semigroup endomorphism $\vartheta : S[X] \to S[X]$, a natural way of defining a matrix associated with ϑ is the following: Let E be the set of essential letters, and let $x_i, x_j \in E$. Define m_{ij} as the number of appearences of x_j in the word $\vartheta(x_i)$, and set $M_\vartheta := (m_{ij})$. It is clear that M_ϑ is a non-negative $\|E\|_c \times \|E\|_c$-matrix whose entries may be greater than one. In order to obtain a subshift of finite type, the alphabet X has to be enlarged and a new free semigroup endomorphism defined.

Let R denote the set of all letters that appear more than once in a word $\vartheta(x)$, $x \in X$. The idea is to introduce a spurious letter for each letter that is repeated in a word. This method is illustrated in the case of a letter that appears twice. The general procedure is then immediate.

Suppose then, given $x \in R$, that $\vartheta(x) = w_1 y w_2 y w_3$, for words $w_1, w_2, w_3 \in S[X]$. Let $\tilde{X} := X \cup \{\tilde{y}\}$, and if $y \in E$, let $\tilde{E} := E \cup \{\tilde{y}\}$. Next, $\vartheta : S[X] \to S[X]$ is extended to a free semigroup endomorphism on $S[\tilde{X}]$. This extension is denoted by $\tilde{\vartheta}$ and defined by

$$\tilde{\vartheta}(z) := \vartheta(z), \quad \text{if } z \in X \text{ and } z \neq x, \tag{2.65}$$

$$\tilde{\vartheta}(x) := w_1 y w_2 \tilde{y} w_3, \tag{2.66}$$

$$\tilde{\vartheta}(\tilde{y}) := \vartheta(y). \tag{2.67}$$

Obviously, $\tilde{\vartheta} : S[\tilde{X}] \to S[\tilde{X}]$ and is a free semigroup endomorphism extending ϑ. If $\phi_{\tilde{y}} := \phi_y$ and $f(\tilde{y}) := f(y)$, for all $y, \tilde{y} \in \tilde{X}$, then $K_{\tilde{\vartheta}}(x) = K_\vartheta(x)$. Proceeding in a similar fashion for all letters in R, an extended alphabet is obtained as well as an extension of ϑ without changing the recurrent set. The matrix M_ϑ, as constructed above, then has the required properties.

From now on, unless specified otherwise, it is assumed that ϑ is a free semigroup endomorphism with the property that each letter $x \in X$ appears at most once in any word $\vartheta(y)$, $y \in X$.

The subshift of finite type defined via the matrix M_ϑ given earlier is denoted by

$$\Upsilon_\vartheta := \Upsilon(M_\vartheta) = \{(i_1 \, i_2 \, \ldots) \in \Sigma \, | \, m_{i_k i_{k+1}} = 1, \text{ for } k \in \mathbb{N}\},$$

where $\Sigma := E^{\mathbb{N}}$.

Now let $\mathsf{X}\mathfrak{H}(\mathbb{R}^n) := \times_{x \in E}\mathfrak{H}(\mathbb{R}^n)$, and let $\mathsf{X}\mathfrak{h}$ be defined as in Eq. (2.46), with $N = \|E\|_c$. The next theorem, which is due to T. Bedford ([15]), gives the construction of a recurrent set in $\mathsf{X}\mathfrak{H}(\mathbb{R}^n)$.

Theorem 2.10 *Let $x \in X$ and let $\vartheta(x) = y_1 y_2 \ldots y_m$. Then the mapping $\Xi_\vartheta : \mathsf{X}\mathfrak{H}(\mathbb{R}^n) \to \mathsf{X}\mathfrak{H}(\mathbb{R}^n)$, given on coordinates $x \in E$ by*

$$A_x \longmapsto \bigcup_{y_j \in \vartheta(x) \cap E} \left(\phi_{y_j}^{-1} A_{y_j} + \sum_{k<j} \phi_{y_k}^{-1} f(y_k) \right), \tag{2.68}$$

is contractive on the complete metric space $(\mathsf{X}\mathfrak{H}(\mathbb{R}^n), \mathsf{X}\mathfrak{h})$. If $E \neq \emptyset$ then the unique fixed point $\mathsf{X}K_\vartheta := (K_\vartheta(x_1), \ldots, K_\vartheta(x_{\|E\|_c}))$ of Ξ_ϑ is an element of $\mathsf{X}\mathfrak{H}(\mathbb{R}^n)$. Furthermore, there exists a continuous surjection $\gamma : \Upsilon_\vartheta \twoheadrightarrow \mathsf{X}K_\vartheta \subseteq \mathsf{X}_{x \in E}\mathbb{R}^n$.
If $\phi_x = L$, for all $x \in X$, where L is an expansive linear map from $\mathbb{R}^n$ onto itself, then $K_\vartheta(x) = \lim_{\nu \to \infty} L^{-\nu} K(\vartheta^\nu(x))$.

Proof. Suppose an element $(i_1 \, i_2 \, \ldots) \in \Upsilon_\vartheta$ is given. Then words $w_j, w_j' \in S[X]$ can be found such that $\vartheta(i_j) = w_j \, i_{j+1} \, w_j'$. Define $k_j := \sum_{x \in w_j} \phi_x^{-1} f(x)$. Let $\gamma : \Upsilon_\vartheta \twoheadrightarrow \mathsf{X}_{x \in E}\mathbb{R}^n$ be that surjection whose xth coordinate γ_x, $x \in E$, is defined by

$$\gamma_x(i_1 \, i_2 \, \ldots) := \begin{cases} k_2 + \sum_{m=3}^\infty (\phi_{i_{m-1}} \circ \cdots \circ \phi_{i_1})^{-1}(k_m), & \text{for } x = i_1 \\ 0, & \text{for } x \neq i_1. \end{cases}$$

The range of k_j is finite; hence, there exists a positive constant c such that $|k_j| \leq c$, independent of $(i_1 \, i_2 \, \ldots) \in \Upsilon_\vartheta$ and $j \in \mathbb{N}$. This then implies that if $\mathbf{i}' \in C_j(\mathbf{i}) := \{\mathbf{i}'' \in \Upsilon_\vartheta \, | \, \mathbf{i} = (i_1 \, i_2 \, \ldots), \, i_m = i_m'', \text{ for } m = 1, \ldots, j\}$,

$$|\gamma(\mathbf{i}, \mathbf{i}')| \leq \sum_{m=j+1}^\infty (\phi_{i_{m-1}} \circ \cdots \circ \phi_{i_1})^{-1}(k_m) \leq c \sum_{m=j+1}^\infty s^m \leq c_1 s^j,$$

for some positive constant c_1. Hence, γ is continuous (in the product topology). Now define $K_\vartheta(x) := \gamma_x \Upsilon_\vartheta \subseteq \mathbb{R}^n$, $x \in E$, and thus $\mathsf{X} K_\vartheta$ as $(K_\vartheta(x_1), \ldots, K_\vartheta(x_{\|E\|_c}))$.

Next it is shown that $\mathsf{X} K_\vartheta$ is invariant under the mapping Ξ_ϑ. It clearly suffices to prove invariance under each of the mappings $\Xi_{\vartheta,x}$, $x \in E$. It is then enough to establish the following equality:

$$\gamma_x(\tau_x C_0(y)) = \phi_y^{-1} \gamma_y C_0(y) + \sum_{z \in w} \phi_z^{-1} f(z),$$

with $\vartheta(x) = w\, y\, w'$. For, if $y \in \vartheta(x) \cap E$, then

$$C_0(x) = \bigcup_{y \in \vartheta(x) \cap E} \tau_x C_0(y).$$

(Here τ_x is the left shift map as defined in Eq. (2.25).) However, it is already known that

$$\gamma_x(x\, y \ldots) = \phi_y^{-1} \gamma_y(y \ldots) + \sum_{z \in w} \phi_z^{-1} f(z).$$

Hence, this coordinatewise invariance implies the invariance of $\mathsf{X} K_\vartheta$ under Ξ_ϑ.

The proof that Ξ_ϑ is a contraction on $(\mathsf{X}\mathfrak{H}(\mathbb{R}^n), \mathsf{X}\mathfrak{h})$ follows immediately from Proposition 2.7. Thus, by the Banach Fixed-Point Theorem, $\mathsf{X} K_\vartheta \in \mathsf{X}\mathfrak{H}(\mathbb{R}^n)$.

It remains to show the last statement. So suppose that L is an expansive linear map from $\mathbb{R}^n$ onto itself. It follows from Theorem 2.8 that $K_\vartheta(x) = \lim_{\nu \to \infty} L^{-\nu}(\vartheta^\nu(x))$, for $x \in E$. Assume that $(i_1\, i_2\, i_3 \ldots) \in \Upsilon_\vartheta$. By the definition of M_ϑ there exist words $w_j, w_j' \in S[X]$ such that $\vartheta(i_{j-1}) = w_j\, i_j\, w_j'$, for all $j \in \mathbb{N} \setminus \{1\}$. Thus, proceeding inductively, there is a word $w'' \in S[X]$ so that

$$\vartheta^\nu(i_1) = (\vartheta^{\nu-1} w_1)(\vartheta^{\nu-2} w_2) \cdots (\vartheta w_{\nu-1})(w_\nu\, i_\nu\, w'').$$

Then

$$L^{-\nu} f((\vartheta^{\nu-1} w_1) \cdots (\vartheta w_{\nu-1})(w_\nu)) = L^{-1} f(w_1) + L^{-2} f(w_2) + \ldots$$
$$+ L^{-\nu} f(w_\nu)$$

$$= k_2 + \sum_{\mu=3}^{\mu=\nu} L^{-\nu}(k_\mu),$$

with $k_\mu = k_\mu(\mathbf{i})$ defined as before. Defining the map $K : S[X] \to \mathfrak{H}(\mathbb{R}^n)$ by

$$K(x) = \{0\}, \text{ for all } x \in E,$$

then

$$L^{-\nu}K(\vartheta^\nu x) = \left\{ k_2 + \sum_{\mu=3}^{\nu} L^{-\mu} k_\mu(\mathbf{i}) \,|\, \mathbf{i} \in \Upsilon_\vartheta, \, i_1 = x \right\},$$

and therefore, $\lim_{\nu \to \infty} L^{-\nu} K(\vartheta^\nu(x)) = K_\vartheta(x)$. ∎

Remark. In this more general setting the invariance condition now reads

$$K_\vartheta(x) = \bigcup_{i=1}^{N} \left(\phi_{y_i}^{-1} K_\vartheta(y_i) + \sum_{j<i} \phi_{y_j}^{-1} f(y_j) \right), \tag{2.69}$$

where $\vartheta(x) = y_1 \ldots y_N$, for $x \in X$.

Now it is shown how IFSs fit into this framework. Suppose that F is a closed subset of $\mathbb{R}^n$, and that $(F, \mathbf{w})$ is a hyperbolic IFS on F with attractor A. It is clear from Theorem 2.4 that A is the closure of the set

$$\{x \in F \,|\, (w_{i_1} \circ \ldots \circ w_{i_\nu})(x) = x \wedge i_j \in \{1, \ldots N\} \wedge N \in \mathbb{N}_0\}.$$

Let $\mathbf{i}$ be a *periodic code* in Υ, i.e., an element of Σ of the form $\mathbf{i} = (i_1 \, i_2 \ldots i_\nu \ldots) \in \Upsilon$ with $i_{N+\nu} = i_\nu$, for all $\nu \in \mathbb{N}$. Then define maps $\Phi : \mathbb{R}^n \to \mathbb{R}^n$ by

$$\Phi_i(x) := \phi_{i_j}^{-1}(x) + \sum_{y \in w_j} \phi_y^{-1} f(y), \tag{2.70}$$

where $\vartheta(i_j) = w_j i_{j+1} w'_j$.

Proposition 2.12 *Let Φ be defined as in Eq. (2.70), and set $\Phi_{\mathbf{i}} := \Phi_1 \circ \ldots \circ \Phi_N$. Denote by E the set of all points $x \in F$ for which $\Phi_{\mathbf{i}}(x) = x$, for some $\Phi_{\mathbf{i}}$. Then $\overline{E} = K_\vartheta(i_1)$.*

Proof. Use the definition of the projection γ and the fact that periodic sequences are dense in Υ. ∎

Next it is shown how a hyperbolic IFS is obtained from a given free semigroup $S[X]$, a given free semigroup endomorphism ϑ, and mappings f and ϕ as defined earlier. Only the case $\phi_x = \phi$, for all $x \in X$, will be considered. (The more general setting follows easily from this special case but is notationally more challenging.)

Define the maps $K : S[X] \to \mathbb{R}^n$ by

$$K(x) := \{0\}, \text{ for all } x \in X, \tag{2.71}$$

and mappings $w_{x,i} : \mathbb{R}^n \to \mathbb{R}^n$ by

$$w_{x,i}(y) := \phi^{-1}(y + f(x_1 x_2 \dots x_N)), \tag{2.72}$$

for a fixed $x \in X$ with $\vartheta(x) = x_1 x_2 \dots x_N$ and $i = 1, \dots, N = N(x)$. It is clear that the maps $w_{x,i}$ are contractions on $\mathbb{R}^n$.

Theorem 2.11 *Let $K_0(x) := K(x) = \{0\}$ for all $x \in X$. Then $K_\nu(\vartheta(x)) \subseteq K_{\nu+1}(\vartheta)(x)$, for all $\nu \in \mathbb{N}_0$, and*

$$K_\vartheta(x) = \overline{\bigcup_{\nu \in \mathbb{N}_0} K_\nu(\vartheta)(x)}.$$

Furthermore, if $\vartheta(x) = x_1 x_2 \dots x_N$ then

$$K_\vartheta(x) = \bigcup_{i=1}^{N} w_{x,i}(K_\vartheta(x_i)). \tag{2.73}$$

Proof. Let $\vartheta(x) = x_1 x_2 \dots x_N$. By the definition of K,

$$K(\vartheta^{\nu+1}(x)) = K(\vartheta^\nu(x_1)\vartheta^\nu(x_2)\dots\vartheta^\nu(x_N))$$

$$= K(\vartheta^\nu(x_1) \cup [f(\vartheta^\nu(x_1)) + K(\vartheta^\nu(x_2))] \cup \dots$$
$$\cup [f(\vartheta^\nu(x_1 x_2 \dots x_{N-1})) + K(\vartheta^\nu(x_N))]),$$

and thus

$$K_{\nu+1}(x) = \phi^{-1}[K_\nu(x_1) \cup (f(x_1) + K_\nu(x_2) \cup \dots$$
$$\cup (f(x_1 x_2 \dots x_{N-1}) + K_\nu(x_N))]$$
$$= \bigcup_{i=1}^{N} w_{x,i}(K_\nu(x_i)). \tag{2.74}$$

Note that, since $K_0(\vartheta)(x) = \{0\} \subset K_1(\vartheta)(x)$, for all $x \in X$, $K_{\nu+1}(\vartheta)(x) \subseteq K_\nu(\vartheta)(x)$, for all $x \in X$ and all $\nu \in \mathbb{N}$. Letting $\nu \to \infty$ in Eq. (2.74) gives the last statement. ■

Now it should be clear how to interpret $K_\nu(\vartheta)(\,\cdot\,)$ as a recurrent IFS: For $x \in X$, let $\vartheta(x) = x_1(x)x_2(x)\dots x_N(x)$, and let $w_{x,i}(y) = \phi^{-1}(y +$

$f(x_1(x)x_2(x)\ldots x_N(x)))$, for $i = 1,\ldots N(x)$. Since $w_{x,i}$ maps $K_\vartheta(x_i)$ into $K_\vartheta(x)$, the directed graph G has as its edge set E those compositions $w_{x,i}\circ w_{x',i'}$ for which $x' = x_i(x)$. Assigning positive probability to these edges defines the recurrent IFS. The resulting attractor A and its subparts $A_{x,i}$ then satisfy

$$K_\vartheta(x) = \bigcup_{i=1}^{N(x)} A_{x,i}. \tag{2.75}$$

(The reader may compare this last equality to Eq. (2.47).)

2.4 Graph Directed Fractal Constructions

A construction of fractal sets related to IFSs and recurrent sets is due to Mauldin and Williams ([134, 135]). In this section their approach is discussed and it is shown how it relates to the constructions given earlier. The presentation given here follows the papers by Mauldin and Williams as well as Edgar's approach ([61]).

***Definition* 2.17** Let (X,d) and (X',d') be metric spaces. A mapping S : $X \to X'$ is called a similitude iff there exists a positive number s such that

$$\forall\, \xi, \xi' \in X : \; d'(S(\xi), S(\xi')) = s\, d(\xi, \xi'). \tag{2.76}$$

The number s is called the similarity constant (associated with S).

***Definition* 2.18** A directed multigraph $\mathsf{G} = (\mathsf{V}, \mathsf{E})$ is a directed graph with the additional property that there may be more than one edge $e \in \mathsf{E}$ connecting a given pair (u, v) of vertices in V. A Mauldin-Williams graph is a pair (G, s) where G is a directed multigraph and $s : \mathsf{E} \to \mathbb{R}^+$ a function. If $0 < s(e) < 1$ for all $e \in \mathsf{E}$, then the Mauldin-Williams graph is called strictly contracting.

Let $u, v \in \mathsf{V}$. The set of all edges $e = (u, v) \in \mathsf{E}$ is denoted by E_{uv}. With each vertex $v \in \mathsf{V}$ one associates a non-empty complete metric space X_v, and with each edge $e \in \mathsf{E}$ a similitude S_e such that $S_e : X_v \to X_u$ if $e \in E_{uv}$, and $s(e)$ is its similarity constant. The collection $\{S_e \,|\, e \in \mathsf{E}\}$ is called a *realization* of the Mauldin-Williams graph (G, s). The following theorem is proven in [134] and also in [61].

Theorem 2.12 *Let $\{S_e | e \in \mathsf{E}\}$ be a realization of the contracting Mauldin-Williams graph (G, s). Then there exists a unique vector element $(X_v)_{v \in \mathsf{V}}$ in $\prod_{v \in \mathsf{V}} \mathfrak{H}(X_v)$ such that*

$$X_u = \bigcup_{\substack{v \in \mathsf{V} \\ e \in E_{uv}}} S_e X_v. \tag{2.77}$$

Proof. The proof is straightforward: Since $(\mathfrak{H}(X_v), \rho_v)$ is complete, so is $(\prod_{v \in \mathsf{V}} \mathfrak{H}(X_v), \varrho)$ with $\varrho := \bigvee\{\rho_v | v \in \mathsf{V}\}$. (Here, $f \vee g := \max\{f(x), g(x) | x \in X\}$, for arbitrary functions f and g defined on a set X.) Let $(Y_v)_{v \in \mathsf{V}} \in \prod_{v \in \mathsf{V}} \mathfrak{H}(X_v)$. The mapping $\Psi : \prod_{v \in \mathsf{V}} \mathfrak{H}(X_v) \to \prod_{v \in \mathsf{V}} \mathfrak{H}(X_v)$ given by

$$\Psi(Y_v)_{v \in \mathsf{V}} := \left(\bigcup_{\substack{v \in \mathsf{V} \\ e \in E_{uv}}} S_e Y_v \right)_{u \in \mathsf{V}}$$

is a contraction. The result now follows from the Banach Fixed-Point Theorem. ∎

To establish the connection to recurrent IFSs, it suffices to notice that, if G is a directed *graph* and $S_e := \mathfrak{w}_{ij}$, for $e \in E_{ij}$, then Theorem 2.12 is essentially Corollary 2.1.

A slightly more general fractal set can be constructed by using the following approach: Let $\mathsf{G} = (\mathsf{V}, \mathsf{E})$ be a directed multigraph. A finite concatenation of edges $\mathbf{e} = e_1 \cdots e_k$ such that if $e_i = (u_i, v_i)$ and $e_{i+1} = (u_{i+1}, v_{i+1})$, $v_i = u_{i+1}$, is called a *path* of length k in G. The set of all paths of length k beginning at $v \in \mathsf{V}$ is denoted by $E_v^{(k)}$, and the set of all paths of length k by $E^{(k)}$. Two assumption on G are made:

(a) Given two arbitrary — not necessarily distinct — vertices u and v in V, there exists a path $\mathbf{e}$ along the edges of G connecting u and v (such a graph is called *strongly connected*).

(b) There are at least two edges leaving each vertex $v \in \mathsf{V}$.

As above, a function $s : \mathsf{E} \to \mathbb{R}^+$ is associated with each edge. It is also assumed that the resulting Mauldin-Williams graph is strictly contracting. The function s is extended to values on a path $\mathbf{e} = e_1 e_2 \cdots e_k$ by setting

$$s(\mathbf{e}) := s(e_1) s(e_2) \cdots s(e_k).$$

(To ease notation, this extension is again denoted by s.) Let Y be a complete metric space. With each vertex $v \in \mathsf{V}$ a non-empty compact set $X_v \subseteq Y$

is associated. Assume also that $X = \overline{\overset{\circ}{X}}$ and that $|X| = 1$. For each path $\mathbf{e} \in E^{(k)}$, sets $X_{\mathbf{e}}$ are chosen recursively as follows:

(i) If $\mathbf{0}$ is the empty path from v to v, let $X(\mathbf{0}) := X_v$.

(ii) For a path $\mathbf{e} \in E^{(k)}$ with terminal vertex v, the set $X_{\mathbf{e}}$ is geometrically similar to X_v with reduction value $s(\mathbf{e})$.

(iii) For a path $\mathbf{e}$ with terminal vertex v, the sets $X_{\mathbf{e}e}$, $e \in E_v$, are such that

  (a) $X_{\mathbf{e}e} \subset X_{\mathbf{e}}$.

  (b) $\bigcap_{e \in E_v} \overset{\circ}{X}_{\mathbf{e}e} = \emptyset$.

The set

$$X_v = \bigcap_{k \in \mathbb{N}_0} \bigcup_{\mathbf{e} \in E_v^{(k)}} X_{\mathbf{e}} \tag{2.78}$$

is called a *Mauldin-Williams fractal* or a *digraph recursive fractal* based on $(X_v)_{v \in \mathsf{V}}$ and ratios $(s(e))_{e \in \mathsf{E}}$. Note that the preceding conditions are equivalent to requiring OSC (Definition 3.10).

By choosing contractive similitudes S_e, $e \in \mathsf{E}$, and defining

$$X_{\mathbf{e}} := S_{e_1} S_{e_2} \cdots S_{e_k} X_v,$$

where $\mathbf{e} = e_1 e_2 \cdots e_k \in E_{uv}^{(k)}$, the previous construction is obtained. Finally, it is worthwhile mentioning that one can also place the subsets $X_{\mathbf{e}e}$ randomly into $X_{\mathbf{e}}$ ([134]). The resulting X_v are *random* Mauldin-Williams fractals.

Chapter 3

Dimension Theory

In this chapter the concept of *dimension* of a topological space $(X, \mathcal{T})$ is introduced. Since dimension theory is a branch of topology, it will not be possible to give an in-depth presentation of the subject. Therefore, only the most basic and for our purposes most important issues are discussed. The concentration will be on *topological, metric,* and *measure-theoretic and probabilistic* dimensions. Because of the limited scope of this monograph it is impossible to provide the proofs of all stated theorems. The reader who is interested in this fascinating topic may want to consult some of the references given in the bibliography ([32, 63, 97, 141, 142, 149]).

3.1 Topological Dimensions

Recall that a topological space $(X, \mathcal{T})$ satisfies the *separation axiom T_1* iff two distinct points in X have neighborhoods that do not contain the other point. I.e., if $x \neq y \in X$, then there exists a neigborhood U_x of x and U_y of y such that $x \notin U_y$ and $y \notin U_x$. The topological space $(X, \mathcal{T})$ is called *regular* iff it satisfies Axiom T_1 and for each closed set F and each point $x \notin F$, there exists an open set G containing F and a neighborhood U_x of x so that $G \cap U_x = \emptyset$. The following proposition characterizes regular spaces.

Proposition 3.1 *Suppose that the topological space $(X, \mathcal{T})$ satisfies Axiom T_1. Then $(X, \mathcal{T})$ is regular iff for every neighborhood U_x of any point $x \in X$ there exists an open neighborhood V_x of x such that $\overline{V_x} \subseteq U_x$.* ■

Now suppose that $(X, \mathcal{T})$ is a regular space. The *small inductive* or *Menger-Urysohn dimension* of $(X, \mathcal{T})$, denoted by ind X, is inductively defined as

87

follows:

Definition 3.1   1. ind $X := -1$, iff $X = \emptyset$.

2. ind $X \leq n$, for $n \in \mathbb{N}_0$, iff for every point $x \in X$ and each neighborhood U_x of x there exists an open set G such that

$$x \in U_x \subseteq G, \text{ and } \operatorname{ind} \partial X \leq n - 1.$$

3. ind $X = n$ iff the inequality ind $X < n$ does not hold.

4. ind $X := \infty$ iff ind $X \geq n$ for all $n \in \mathbb{N}_0 \cup \{-1\}$.
(Here ∂X denotes the boundary of X, i.e., $\partial X := \overline{X} - \overset{\circ}{X}$.)

For separable metrizable topological spaces the following results hold.

Theorem 3.1 (Addition Theorem) *Suppose Y and Z are separable subspaces of a metrizable topological space $(X, \mathcal{T})$. Then*

$$\operatorname{ind}(Y \cup Z) \leq \operatorname{ind} Y + \operatorname{ind} Z + 1.$$

∎

Theorem 3.2 (Cartesian Product Theorem) *Suppose X_1 and X_2 are separable metrizable topological spaces with at least one of them non-empty. Then*

$$\operatorname{ind}(X_1 \times X_2) \leq \operatorname{ind} X_1 + \operatorname{ind} X_2.$$

∎

Remark.   The dimension function ind : $2^X \to \{-1\} \cup \mathbb{N} \cup \{\infty\}$ is monotone, i.e., if A and B are elements of the *power set* 2^X of X with $A \subseteq B$, then ind $A \leq$ ind B.

A topological space satisfying Axiom T_1 is called *normal* iff for each pair of disjoint closed sets F_1 and F_2 there exist open sets $G_1 \supseteq F_1$ and $G_2 \supseteq F_2$ such that $G_1 \cap G_2 = \emptyset$.

Assume that $(X, \mathcal{T})$ is a normal topological space. The *large inductive* or *Brouwer-Čech dimension* of X, written Ind X, is inductively defined as follows:

Definition 3.2   1. Ind $X := -1$, iff $X = \emptyset$.

2. Ind $X \leq n$, with $n \in \mathbb{N}_0$, iff for every closed set $F \subseteq X$ and each open set $G \supseteq F$ there exists an open set $O \subseteq X$ such that

$$F \subseteq O \subseteq G, \text{ and } \operatorname{Ind} \partial X \leq n - 1.$$

3. $\operatorname{Ind} X := n$ iff the inequality $\operatorname{Ind} X < n$ does not hold.

4. $\operatorname{Ind} X := \infty$ iff $\operatorname{Ind} X > n$ for all $n \in \mathbb{N}_0 \cup \{-1\}$.

Both the small and the large inductive dimension is a topological invariant. More precisely, suppose that X and Y are homeomorphic regular, respectively normal, topological spaces. Then $\operatorname{ind} X = \operatorname{ind} Y$, respectively, $\operatorname{Ind} X = \operatorname{Ind} Y$.

The next result justifies the terminology "small inductive" and "large inductive."

Theorem 3.3   1. *If $(X, \mathcal{T})$ is a normal topological space, then $\operatorname{ind} X \leq \operatorname{Ind} X$.*

2. *If $(X, \mathcal{T})$ is a separable metrizable topological space, then $\operatorname{ind} X = \operatorname{Ind} X$.* ∎

Any finite set, the space of rational numbers, the space of irrational numbers, and the Cantor set are examples of topological spaces with small inductive and large inductive dimension equal to zero.

The Cartesian Product Theorem also holds for the large inductive dimension. However, the Sum Theorem reads differently.

Theorem 3.4 *Suppose that the topological space $(X, \mathcal{T})$ is the topological sum of the spaces $(X_1, \mathcal{T}_1)$ and $(X_2, \mathcal{T}_2)$, i.e., $X = X_1 \cup X_2$ and $\mathcal{T} = \{T \mid T \in \mathcal{T}_1 \cup \mathcal{T}_2 \wedge T \cap \mathcal{T}_1 \in \mathcal{T}_1 \wedge T \cap \mathcal{T}_2 \in \mathcal{T}_2\}$. If $\operatorname{Ind} X_i \leq n$, for $i = 1, 2$, then $\operatorname{Ind} X \leq n$.* ∎

The question arises under what condition(s) on the topological space $(X, \mathcal{T})$ the small inductive dimension is equal to the large inductive dimension. The proposition below provides the answer.

Proposition 3.2 *Assume that $(X, \mathcal{T})$ is a separable metrizable topological space. Then $\operatorname{ind} X = \operatorname{Ind} X$.* ∎

A third dimension function plays an important role. Before defining it, the *order* of a cover of a topological space has to be introduced.

Suppose that $(X, \mathcal{T})$ is a topological space and that $\mathcal{U}$ is a class of subsets of the point set X. Let $x \in X$. The *order of $\mathcal{U}$ at x*, written $\operatorname{ord}_x \mathcal{U}$, is the number of members of $\mathcal{U}$ that contain x. If this number is infinite, then $\operatorname{ord}_x \mathcal{U} := \infty$. The order of $\mathcal{U}$ is then defined as

$$\operatorname{ord} \mathcal{U} := \sup\{\operatorname{ord}_x \mathcal{U} \,|\, x \in X\}.$$

Definition 3.3 Suppose that for every finite open cover $\mathcal{U}$ of a normal topological space there exists an open cover $\mathcal{V}$ such that

$$\mathcal{V} < \mathcal{U}, \quad \text{and} \quad \operatorname{ord} \mathcal{V} \leq n + 1.$$

Then X is said to have *covering* or *Čech-Lebesgue dimension* $\leq n$. We write $\dim X \leq n$.

The space $(X, \mathcal{T})$ has covering dimension n, if $\dim X < n$ does not hold. If, for all $n \in \mathbb{N}$, the inequality $\dim X \leq n$ is false, then $\dim X := \infty$. As usual, $\dim \emptyset := -1$.

Remark. The notation $\mathcal{V} < \mathcal{U}$ expresses the fact that the cover $\mathcal{V}$ is a *refinement* of the cover $\mathcal{U}$. In other words, $\mathcal{V} < \mathcal{U}$ iff $\forall U \in \mathcal{U} \; \exists V \in \mathcal{V} : V \subseteq U$.

The following lemma is needed to prove the *Coincidence Theorem*.

Lemma 3.1 *Let $(X, \mathcal{T})$ be a separable metrizable topological space. Then* $\dim X \leq \operatorname{ind} X$. ∎

Theorem 3.5 (Coincidence Theorem) *Each separable metrizable topological space $(X, \mathcal{T})$ satisfies*

$$\dim X = \operatorname{ind} X = \operatorname{Ind} X.$$

∎

The next theorem, whose proof is rather deep, states that the dimension of Euclidean space $\mathbb{E}^n := (\mathbb{R}^n, \langle \cdot, \cdot \rangle_E)$ is n.

Theorem 3.6 (Fundamental Theorem of Dimension Theory)   *For every natural number n*

$$\dim \mathbb{E}^n = \operatorname{ind} \mathbb{E}^n = \operatorname{Ind} \mathbb{E}^n = n. \tag{3.1}$$

∎

Let $I^n := \times_{i=1}^n [0,1]$ be the *unit n-cube* and let $\mathbb{S}^n := \{x \in \mathbb{R}^{n+1} \mid \|x\| = 1\}$ be the *unit n-sphere*. Define the *Hilbert cube* $I^{\aleph_0}$ by

$$I^{\aleph_0} := \left\{ (x_1 \, x_2 \, \ldots) \in \mathbb{R}^{[0,1]} \mid \forall\, i \in \mathbb{N} : |x_i| \le 1/i \right\}.$$

This implies the following corollary to Theorem 3.6.

Corollary 3.1 *Let* $n \in \mathbb{N}$*. Then*

1. $\dim I^n = \operatorname{ind} I^n = \operatorname{Ind} I^n = n = \dim \mathbb{S}^n = \operatorname{ind} \mathbb{S}^n = \operatorname{Ind} \mathbb{S}^n$.

2. $\dim I^{\aleph_0} = \operatorname{ind} I^{\aleph_0} = \operatorname{Ind} I^{\aleph_0} = \infty$.

∎

Before closing this subsection a few results are stated that will be implicitly used in later developments and which provide some examples ([109]).

Theorem 3.7 *1. Every separable metrizable topological space of dimension 0 is homeomorphic to a subset of the Cantor set.*

2. *Each 0-dimensional separable metrizable topological space* $(X, \mathcal{T})$ *can be decomposed into disjoint closed sets of diameter less than* ε *(*ε *arbitray positive number). I.e.,*

$$X = \bigcup_{i=1}^N F_i, \quad F_i \cap F_j = \emptyset, \; i \ne j, \; |F_i| < \varepsilon.$$

∎

Theorem 3.8 *1. Every separable metrizable topological space of dimension n has a base consisting of open sets with boundary of dimension at most* $n - 1$.

2. *Every n-dimensional separable metrizable topological space is homeomorphic to a subset of the cube* I^{2n+1}.

3. *Each n-dimensional separable metrizable topological space* $(X, \mathcal{T})$ *can be decomposed into closed sets of diameter less than* ε*,* ε *arbitrary positive number, in such a way that no point belongs simultaneously to* $n + 2$ *of these sets:*

$$X = \bigcup_{i=1}^N F_i, \quad |F_i| < \varepsilon,$$

$$F_{i_0} \cap F_{i_1} \cap \ldots \cap F_{i_{N+1}} = \emptyset, \quad if \; i_0 < i_1 < \ldots < i_{N+1}.$$

∎

3.2   Metric Dimensions

In Section 2.1 an example of a metric dimension, namely the Hausdorff dimension, was already encountered. Here it is shown that Hausdorff dimension can be defined in a more general setting and in more general terms. Other metric dimensions are also discussed and related to Hausdorff dimension. Throughout this subsection $(X, \mathcal{T}_d)$ or simply X denotes a separable (semi-)metrizable topological space whose topology $\mathcal{T}_d$ is induced by a (semi-)metric d.

Let (X, d) be a separable (semi-)metrizable topological space, and let $h : \mathbb{R}_0^+ \to \mathbb{R}_0^+$ be an increasing continuous function from the non-negative reals into themselves that is right-continuous at 0. This function h is referred to as a *dimension function*. Suppose that E is a subset of X and that $\mathcal{C}_\varepsilon := \{C_{i,\varepsilon}\}_{i\in\mathbb{N}}$ is a countable ε-cover of E (Section 2.1.1). Define

$$\mathcal{H}^h_\varepsilon(E) := \inf \left\{ \sum_{C_{i,\varepsilon} \in \mathcal{C}_\varepsilon} h(|U_{i,\varepsilon}|) \,|\, \mathcal{C}_\varepsilon \text{ is an } \varepsilon\text{-cover of E} \right\}. \qquad (3.2)$$

Note that $|\cdot|$ denotes the diameter of a set $U \subseteq X$ with respect to the (semi-)metric d on X.

One easily verifies that $\mathcal{H}^h_\varepsilon(\cdot)$ is an outer measure on X. Furthermore, by the properties of h, $\mathcal{H}^h_\varepsilon$ increases as ε decreases. Therefore, define

$$\mathcal{H}^h(E) := \lim_{\varepsilon \to 0+} \mathcal{H}^h_\varepsilon(E). \qquad (3.3)$$

Restricting $\mathcal{H}^h(\cdot)$ to the σ-algebra of $\mathcal{H}^h$- measurable sets yields a measure on X, called *h-Hausdorff measure*. (The class of $\mathcal{H}^h$-measurable sets is non-empty; it includes the Borel sets.) Clearly, if for an arbitrary positive real number s, $h(x) := x^s$, then s-dimensional Hausdorff measure as introduced in Definition 2.1 is obtained.

If g and h are two dimension functions with $h(t)/g(t) \to 0$ as $t \to 0+$ then, by arguments similar to those given in Section 2.1.1, it is easily seen that $\mathcal{H}^h(E) = 0$ whenever $\mathcal{H}^g(E) < \infty$.

Assume that a set X and two semi(metrics) d_1 and d_2 are given. The Hausdorff dimension induced by d_1 and d_2 is denoted by $\dim_H^{(1)}$ and $\dim_H^{(2)}$, respectively. For $x \in X$ and $r > 0$, let $B_i(x, r) := \{y \in X \,|\, d_i(x, y) < r\}$, $i = 1, 2$. Then the following result can be proven.

Proposition 3.3 *Suppose that $Y \subseteq X$ and that for all $x \in Y$ the inequality*

$$\liminf_{r \to 0+} \frac{\log d_2(B_1(x,r))}{\log d_1(B_1(x,r))} \geq \delta \tag{3.4}$$

holds. Then $\dim_H^{(1)} Y \geq \delta \, \dim_H^{(2)} Y$.

Proof. Apply the definition of Hausdorff measure and dimension. ∎

Sometimes it is advantageous to consider measures that are *comparable* to Hausdorff measure. One such class consists of so-called *net measures*. These measures are constructed similarly to Hausdorff measure, but in their construction a smaller class of sets is used.

Suppose that $\mathcal{N}$ is a *net of sets*, i.e., a collection of subsets of X with the property that if $N_1, N_2 \in \mathcal{N}$, then $N_1 \cap N_2 = \emptyset$, or $N_1 \subseteq N_2$, or $N_1 \supseteq N_2$. It is furthermore assumed that each set in $\mathcal{N}$ is contained in finitely many others. An example of such a net of sets in $X := \mathbb{R}^n$ consists of the following collection:

$$\mathcal{N} := \left\{ \times_{i=1}^n [2^{-k} m_i, 2^{-k}(m_i + 1)) \,|\, k \in \mathbb{N}_0 \wedge m_i \in \mathbb{Z} \right\}. \tag{3.5}$$

The elements in this collection are called *intervals*. Let h be a dimension function and let $\varepsilon > 0$. Define the h-net measure $\mathcal{M}^h(\cdot)$ on $\mathbb{R}^n$ by

$$\mathcal{M}^h(E) := \lim_{\varepsilon \to 0+} \inf \sum_{i \in \mathbb{N}} h(|N_{i,\varepsilon}|), \tag{3.6}$$

where the infimum is taken over all countable ε-covers of $E \subseteq \mathbb{R}^n$ consisting of sets from $\mathcal{N}$. As before, it is easy to see that $\mathcal{M}^h$ is an outer measure. Also notice that $\mathcal{M}^h(E)$ is finite if E is a bounded subset of $\mathbb{R}^n$.

Now suppose $h(x) := x^s$, for some positive real number s.

Proposition 3.4 *The measures $\mathcal{H}^s$ and $\mathcal{M}^s$ are comparable on subsets of $\mathbb{R}^n$, in other words, there exists a constant $c = c(n)$ such that for all $E \subseteq \mathbb{R}^n$,*

$$\mathcal{H}_\varepsilon^s(E) \leq \mathcal{M}_\varepsilon^s(E) \leq c(n)\mathcal{H}_\varepsilon^s(E), \quad \text{for all } 0 < \varepsilon < 1, \tag{3.7}$$

and

$$\mathcal{H}^s(E) \leq \mathcal{M}^s(E) \leq c(n)\mathcal{H}^s(E). \tag{3.8}$$

Proof. Since in the definition of $\mathcal{M}^s$ the infimum is taken over a smaller class of sets, the left-hand inequality in Eqs. (3.7) and (3.8) is clear.

To show the right-hand inequality, suppose that C is any set whose diameter is less than or equal to ε. It is possible to find an integer k such that $2^{-(k+1)} \leq |C| < 2^{-k}$. Let N be an interval in $\mathcal{N}$ of side 2^{-k} that intersects C. Then C is contained in a collection of 3^n of such intervals of side 2^{-k} and diameter $2^{-k}\sqrt{n}$ consisting of N and its nearest neighbors. Subdividing these intervals into 2^{n^2} smaller ones shows that C is contained in $3^n 2^{n^2}$ intervals of diameter $2^{-k}\sqrt{n}2^{-n} \leq 2^{1-n}\sqrt{n}|C| \leq |C| \leq \varepsilon$.

Given a countable ε-cover $\{C_{i,\varepsilon}\}$ of E, then for each $C_{i,\varepsilon}$ one has $C_{i,\varepsilon} \subseteq \bigcup_{j=1}^{3^n 2^{n^2}} N_{ij}$, where $\{N_{ij}\}_{j=1}^{3^n 2^{n^2}}$ is a collection of $3^n 2^{n^2}$ intervals of diameter at most ε. Hence,

$$E \subseteq \bigcup_{i \in \mathbb{N}} \bigcup_{j=1}^{3^n 2^{n^2}} N_{ij}$$

and

$$\sum_{i \in \mathbb{N}} \sum_{j=1}^{3^n 2^{n^2}} |N_{ij}|^s \leq (3^n 2^{n^2}) \sum_{i \in \mathbb{N}} |C_{i,\varepsilon}|^s.$$

Setting $c(n) := 3^n 2^{n^2}$, Inequality (3.7) follows.

Inequality (3.8) is obtained by letting $\varepsilon \to 0+$.   ∎

This result is sometimes used to calculate the Hausdorff dimension of fractals, since by Eq. (3.8), $\mathcal{H}^s$ and $\mathcal{M}^s$ exhibit the same 0-∞-behavior.

The calculation of Hausdorff dimension of a fractal set is in general a difficult problem. One of the difficulties arises from the fact that one is required to consider coverings by balls of diameter *less than or equal* to ε. For most practical purposes it is therefore natural to weaken this requirement. By doing so, one arrives at slight modifications of Hausdorff dimension. These modified Hausdorff dimensions will be called *box dimensions*, although several other terms are in use. An incomplete list of such terms is *Kolmogorov entropy or dimension, capacity, fractal dimension,* or *information dimension.*

In what follows, E will always denote a non-empty bounded subset of a complete metric space (X, d).

Definition 3.4 For $\varepsilon > 0$, let $C_\varepsilon(E)$ be a cover of E by sets of diameter equal to ε. Denote the class of all such covers of E by $\mathcal{C}_\varepsilon(E)$. Let $N_\varepsilon(E) := \min\{\|C_\varepsilon(E)\|_c \mid C_\varepsilon(E) \in \mathcal{C}_\varepsilon(E)\}$, where $\|C_\varepsilon(E)\|_c$ denotes the cardinality of $C_\varepsilon(E)$. The lower and upper box dimensions of E are defined by

$$\underline{\dim}_\beta E := \liminf_{\varepsilon \to 0+} \frac{\log N_\varepsilon(E)}{-\log \varepsilon}, \tag{3.9}$$

and

$$\overline{\dim}_\beta E := \limsup_{\varepsilon \to 0+} \frac{\log N_\varepsilon(E)}{-\log \varepsilon}, \tag{3.10}$$

respectively. If both limits are equal, their common value is called the box dimension of E:

$$\dim_\beta E = \lim_{\varepsilon \to 0+} \frac{\log N_\varepsilon(E))}{-\log \varepsilon}. \tag{3.11}$$

There are several equivalent definitions for the lower and upper box dimension of a set $E \subseteq X$. To establish these equivalent definitions, certain classes of covers of E have to be introduced.

Let $U_{B;\varepsilon}(E)$ denote a cover of E whose covering elements are closed balls of radius ε, and let $\mathcal{U}_{B;\varepsilon}(E)$ denote the class of all such covers. Let $U_{C;\varepsilon}(E)$ be a cover of E by cubes of side ε. Denote by $\mathcal{U}_{C;\varepsilon}(E)$ the collection of all these covers. Now consider a cover $V_\varepsilon(E)$ of E by sets of diameter at most ε and the corresponding class $\mathcal{V}_\varepsilon(E)$ of all such covers. Finally, let $W_\varepsilon(E)$ represent a cover consisting of disjoint balls of radius ε whose centers are in E, and $\mathcal{W}_\varepsilon(E)$ its class. Set

$$N_\varepsilon(E) := \begin{cases} \min\{\|U_{B;\varepsilon}(E)\|_c \,|\, U_{B;\varepsilon}(E) \in \mathcal{U}_{B;\varepsilon}(E)\} \\[2mm] \min\{\|U_{C;\varepsilon}(E)\|_c \,|\, U_{C;\varepsilon}(E) \in \mathcal{U}_{C;\varepsilon}(E)\} \\[2mm] \min\{\|V_\varepsilon(E)\|_c \,|\, V_\varepsilon(E) \in \mathcal{V}_\varepsilon(E)\} \\[2mm] \max\{\|W_\varepsilon(E)\|_c \,|\, W_\varepsilon(E) \in \mathcal{W}_\varepsilon(E)\} \end{cases} \tag{3.12}$$

Then the lower, upper, and box dimensions, respectively, of E are given by

$$\underline{\dim}_\beta E = \liminf_{\varepsilon \to 0+} \frac{\log N_\varepsilon(E)}{-\log \varepsilon}, \tag{3.13}$$

$$\overline{\dim}_\beta E = \limsup_{\varepsilon \to 0+} \frac{\log N_\varepsilon(E)}{-\log \varepsilon}, \tag{3.14}$$

$$\dim_\beta E = \lim_{\varepsilon \to 0+} \frac{\log N_\varepsilon(E)}{-\log \varepsilon}. \tag{3.15}$$

The proofs of these equivalences are easy exercises left for the reader.

It is worth noting that in (3.12) it suffices to take the limit through a discrete sequence $\{\varepsilon_\nu\}_{\nu \in \mathbb{N}}$ with the property that $\varepsilon_{\nu+1} \le c\varepsilon_\nu$, for some constant

$0 < c < 1$. For, if $\varepsilon_{\nu+1} \leq \varepsilon < \varepsilon_\nu$, then

$$\log N_\varepsilon(E)/-\log\varepsilon \quad \leq \quad \log N_{\varepsilon_{\nu+1}}(E)/-\log\varepsilon_\nu$$

$$\leq \quad \log N_{\varepsilon_{\nu+1}}(E)/[-\log\varepsilon_{\nu+1} + \log\varepsilon_{\nu+1}/\varepsilon_\nu]$$

$$\leq \quad \log N_{\varepsilon_{\nu+1}}(E)/[-\log\varepsilon_{\nu+1} + \log c].$$

Hence,

$$LIM_{\varepsilon\to0+}\frac{\log N_\varepsilon(E)}{-\log\varepsilon} \leq LIM_{\varepsilon\to0+}\frac{\log N_{\varepsilon_\nu}(E)}{-\log\varepsilon_\nu},$$

where LIM represents any one of the three limits in (3.13), (3.14), and (3.15). That the reverse inequality holds is clear, since the limit is taken over a subset of sequences $\varepsilon \to 0+$.

Definition 3.5 Suppose that $E \subseteq \mathbb{R}^n$. The lower and upper Minkowski dimensions of E are defined by

$$\underline{\dim}_M E := n - \limsup_{\eta\to0+}\frac{\log \operatorname{vol}^n(E_\eta)}{\log\varepsilon}, \tag{3.16}$$

and

$$\overline{\dim}_M E := n - \liminf_{\eta\to0+}\frac{\log \operatorname{vol}^n(E_\eta)}{\log\varepsilon}, \tag{3.17}$$

respectively.

In Definition 3.5 the *η-parallel body* E_η of a non-empty bounded subset of $\mathbb{R}^n$ was introduced. It is defined by

$$E_\eta := \{x \in \mathbb{R}^n \mid d_E(x,y) \leq \eta, \text{ for some } y \in E\}, \tag{3.18}$$

where $\eta > 0$.

The next proposition relates the Minkowski dimensions to the box dimensions.

Proposition 3.5 *Let $E \subseteq \mathbb{R}^n$. Then*

$$\underline{\dim}_\beta E = \underline{\dim}_M E,$$

and

$$\overline{\dim}_\beta E = \overline{\dim}_M E.$$

Proof. Only the second equality will be proven. Let v_n denote the volume of the unit ball in $\mathbb{R}^n$, and let $N_\varepsilon(E)$ be the largest number of balls of radius ε whose centers are in E. Then $\mathrm{vol}^n(E) \geq (v_n(2\varepsilon)^n)N_\varepsilon(E)$. Thus, $\overline{\dim}_\beta E \leq \overline{\dim}_M E$.

Now suppose that E has been covered by $N_\varepsilon(E)$ balls of radius ε. Then the ε-parallel body of E can be covered by the concentric balls of radius 2ε. Therefore, $\mathrm{vol}^n(E_\varepsilon) \leq N_\varepsilon(E)(v_n(2\varepsilon)^n)$. Hence, after taking logarithms,

$$\liminf_{\varepsilon \to 0+} \frac{\log \mathrm{vol}^n(E_\varepsilon)}{\log \varepsilon} \geq n + \overline{\dim}_\beta E.$$

∎

Now Hausdorff dimension is compared to box dimension. For argument's sake it is assumed that $\underline{\dim}_\beta E = \overline{\dim}_\beta E$. It follows then immediately from the definition of Hausdorff dimension (Section 2.1.1) that if E is covered by $N_\varepsilon(E)$ sets of diameter ε,

$$\mathcal{H}_\varepsilon^s(E) \leq N_\varepsilon(E)\varepsilon^s,$$

and thus

$$\dim_H E \leq \dim_\beta E, \tag{3.19}$$

since the infimum is taken over a smaller class of covers of E. The inequality in (3.19) can be strict! A simple example is provided by the set $E := \{x \in \mathbb{R} | \forall n \in \mathbb{N} : x = 1/n\}$. Obviously, $\dim_H E = 0$ but $\dim_\beta E = 1$.

Analogous to Hausdorff measure a *box or fractal content* may be defined. This may be done via the following procedure.

Definition 3.6 Suppose that Ω is a non-empty set in $\mathbb{R}^n$ and $\mathcal{R}$ a collection of subsets of Ω satisfying

    (a) $\forall A, B \in \mathcal{R}$: $A \cup B \in \mathcal{R}$.

    (b) $\forall A, B \in \mathcal{R}$: $A \setminus B \in \mathcal{R}$.

Such a collection of subsets of Ω is called a ring (of subsets of Ω).

Definition 3.7 A mapping $\iota : \mathcal{R} \to \mathbb{R}_0^+$ is called a content (on $\mathcal{R}$) iff it satisfies the following two conditions:

Congruence invariance: $A, B \in \mathcal{R} \wedge A \equiv B \Longrightarrow \iota A = \iota B$.

Additivity: $A, B \in \mathcal{R} \wedge A \cap B \neq \emptyset \Longrightarrow \iota(A \cup B) = \iota A + \iota B$.

It is not difficult to verify that if $\mathcal{R}$ is a ring and ι a content, then the following additional properties hold:

Finite additivity: Let $\{A_\nu\}_{\nu=1}^n$ be a *finite* collection of elements of $\mathcal{R}$. Then $\iota(\bigcup_{\nu=1}^n A_\nu) = \bigcup_{\nu=1}^n \iota A_\nu$.

Monotonicity: For all $A, B \in \mathcal{R}$ with $A \subseteq B$, $\iota A \leq \iota B$.

Let s be a non-negative number and let $\{D_{i,\varepsilon}(E)\}$ be a *finite* ε-cover of E. Denote by $\mathcal{D}_\varepsilon(E)$ the class of all such covers of E. Define

$$\iota_\varepsilon^s(E) := \inf\left\{\sum_i \varepsilon^s \,|\, D_{i,\varepsilon}(E) \in \mathcal{D}_\varepsilon(E)\right\} = N_\varepsilon(E)\varepsilon^s. \qquad (3.20)$$

Now let the s-dimensional content $\iota^s(E)$ of E be defined by

$$\iota^s(E) := \lim_{\varepsilon \to 0+} \iota_\varepsilon^s. \qquad (3.21)$$

The box dimension is then that value of s for which

$$\iota^s(E) = \begin{cases} \infty & \text{if } s < \dim_\beta E \\ 0 & \text{if } s > \dim_\beta E. \end{cases}$$

Since for the calculation of the box dimension of a set E covers $W_\varepsilon(E)$ consisting of disjoint balls of radius $\varepsilon > 0$ whose centers are in E can be used, one might try to define an analogue to Hausdorff measure and Hausdorff dimension for these *dense packings* of E by disjoint balls.

Therefore, let $\varepsilon > 0$ be given, let $s \in \mathbb{R}_0^+$, and let E be a set in $\mathbb{R}^n$. Define

$$\mathcal{P}_\varepsilon^s(E) := \sup\left\{\sum_i |W_i|^s \,|\, W_i \in \mathcal{W}_\varepsilon(E)\right\} \qquad (3.22)$$

((3.12)). As $\mathcal{P}_\varepsilon^s(E)$ decreases as ε decreases, the limit

$$\tilde{\mathcal{P}}^s(E) := \lim_{\varepsilon \to 0+} \mathcal{P}_\varepsilon^s(E) \qquad (3.23)$$

exists. However, $\tilde{\mathcal{P}}^s(E)$ is *not* a measure on $\mathbb{R}^n$, but one can be obtained by setting

$$\mathcal{P}^s(E) := \inf\left\{\sum_i \tilde{\mathcal{P}}^s(E_i) \,\middle|\, E \subseteq \bigcup_{i=1}^\infty E_i\right\}. \qquad (3.24)$$

This measure is called the *s-dimensional packing measure on* $\mathbb{R}^n$. Now one proceeds in the usual way to define the associated dimension: Let

$$
\begin{aligned}
\dim_P E &:= \sup\{s \in \mathbb{R}_0^+ \,|\, \mathcal{P}^s(E) = \infty\} \\
&= \inf\{s \in \mathbb{R}_0^+ \,|\, \mathcal{P}^s(E) = 0\}.
\end{aligned}
\tag{3.25}
$$

This dimension is called the *packing dimension of E*. In the next section it will be related to the box dimension.

3.3 Probabilistic Dimensions

The previous sections dealt primarily with dimension from a geometric point of view. However, as already seen, a fractal has a much richer structure; namely, it is the support of an invariant measure. It is therefore natural to look at the "size" of the support of this measure when defining a dimension for the fractal.

Suppose then that A is the attractor for an IFS (with probabilities). As before, the **p**-balanced measure of the IFS is denoted by μ. Recall that $\mu A = 1$.

Definition 3.8 The Hausdorff dimension of the measure μ is defined by

$$
\dim_H \mu := \inf\{\dim_H E \,|\, E \subseteq A \land \mu E = 1\}.
\tag{3.26}
$$

Remark. Since $A = \operatorname{supp}(\mu)$ it is clear that $\dim_H \mu \leq \dim_H A$. In general one may have strict inequality. However, equality of both dimensions is assured if there exist numbers $s > 0$ and $c > 0$ such that for all sets C

$$
\mu(E \cap C) \leq c|C|^s.
$$

To see this let $\{C_\nu\}$ be any cover of A. Then

$$
0 < \mu A = \mu\left(\bigcup_\nu E \cap C_\nu\right) \leq \sum_\nu \mu C_\nu \leq c\sum_\nu |C_\nu|^s,
$$

and thus, choosing $\varepsilon > 0$ small enough and taking the infimum, $\mathcal{H}_\varepsilon^s(A) \geq \mu A/c$. Hence, $\mathcal{H}^s(A) \geq \mu A/c$, and so $\dim_H A \geq s$ implying $\dim_H \mu \geq s$.

The Hausdorff dimension of a measure will be considered again in the next chapter when *dynamical systems* are introduced.

Another type of a probabilistic dimension was introduced by P. Billings-
ley in [26, 27]. He starts with a probability space $(\Omega, \mathcal{B}, \mu)$ on which there
is defined a stochastic process $\{X_n\}_{n \in \mathbb{N}_0}$ with countable state space S and
defines an outer measure analogous to Hausdorff outer measure but allows
only cylinders defined by $\{X_n\}_{n \in \mathbb{N}_0}$ as covering sets. It will be shown that
for $\Omega = [0, 1]$ and under a certain condition the *Billingsley dimension* agrees
with Hausdorff dimension.

The set $Z := \{\omega \in \Omega \mid X_n(\omega) = i_n \wedge i_n \in S \wedge n = 1, \ldots, N\}$ is called an
N-cylinder of Ω. The unique cylinder Z_n that contains $\omega \in \Omega$ is denoted by
$Z_n(\omega)$. Let s be any non-negative number. Define an s-dimensional outer
measure $\mathcal{H}_B^s$ of a subset $\Omega_0 \subseteq \Omega$ by

$$\mathcal{H}_B^s(\Omega_0) := \liminf_{\varepsilon \to 0+} \left\{ \sum_{n=1}^{\infty} \mu Z_i^s \,\Big|\, \Omega_0 \subseteq \bigcup Z_i \wedge \mu(Z_i) < \varepsilon \right\}. \tag{3.27}$$

Note that if Ω_0 does not possess a covering of the required form, then
$\inf \emptyset := \infty$.

Definition 3.9 The Billingsley dimension of a subset Ω_0 of Ω is defined by

$$\dim_B \Omega_0 := \sup\{s \mid \mathcal{H}_B^s(\Omega_0) = \infty\}. \tag{3.28}$$

Furthermore, if it is assumed that the following condition

$$\text{(A):} \qquad \lim_{n \to \infty} \mu Z_n(\omega) = 0$$

holds for all $\omega \in \Omega$, then

$$0 \le \dim_B \Omega_0 \le 1, \text{ for all } \Omega_0 \subseteq \Omega$$

and

$$\dim_B \Omega_0 = 1, \text{ if } \mu^*(\Omega_0) > 0,$$

where μ^* denotes the outer measure corresponding to μ.

Next it is shown that — by defining an appropriate semi-metric on Ω — the
Hausdorff dimension induced by this semi-metric agrees with the Billingsley
dimension.

Let $\mathcal{Z} := \{Z_n(\omega) \mid \forall\, n \in \mathbb{N} \,\forall\, \omega \in \Omega : \mu Z_n(\omega) > 0\}$ and denote by $\mathcal{T}$ the
topology on Ω generated by the cylinders in $\mathcal{Z}$. Furthermore, for all $\omega_1, \omega_2 \in \Omega$
define a mapping $d : \Omega \times \Omega \to \mathbb{R}_0^+$ by

$$d(\omega_1, \omega_2) := \inf\{\mu Z \mid Z \text{ cylinder} \wedge \omega_1, \omega_2 \in Z\}. \tag{3.29}$$

Then the following result holds ([178]).

Theorem 3.9 *Suppose $(\Omega, \mathcal{B}, \mu)$ is a probability space and $\{X_n\}_{n\in\mathbb{N}_0}$ a stochastic process satisfying condition (A). Then the function d defined by Eq. (3.29) is a semi-metric on Ω and the topology $\mathcal{T}$ is semi-metrizable with respect to d. Furthermore, for all $\Omega_0 \subseteq \Omega$,*

$$\dim_B \Omega_0 = \dim_H \Omega_0,$$

where $\dim_H$ denotes the Hausdorff dimension induced by the semi-metric d.

Proof. Conditions 1 and 2 for a semi-metric (Definition 1.4) are readily verified. To show the triangle inequality assume that $d(\omega_1, \omega_2) > 0$. Then there is a cylinder $Z_n(\omega_1)$ such that $\omega_2 \in Z_n(\omega_1)$ but $\omega_2 \notin Z_{n+1}(\omega_1)$. Hence, $d(\omega_1, \omega_2) = \mu Z_n(\omega_1)$. If $\omega_3 \in Z_{n+1}(\omega_1)$, then $d(\omega_1, \omega_3) = \mu Z_{n+1}(\omega_1)$, and if $\omega_3 \notin Z_{n+1}(\omega_1)$, then $d(\omega_1, \omega_3) \geq \mu Z_n(\omega_1)$. Therefore, $d(\omega_1, \omega_2) \leq \max\{d(\omega_1, \omega_3), d(\omega_2, \omega_3)\}$.

To show that $\mathcal{T}$ is semi-metrizable by d, it suffices to note that if $\mu Z_n(\omega) > 0$, then

$$\{\omega_2 \in \Omega \mid d(\omega_1, \omega_2) < \mu Z_n(\omega_1)\} \subseteq Z_n(\omega_1)$$

$$\subseteq \{\omega_2 \in \Omega \mid d(\omega_1, \omega_2) \leq \mu Z_n(\omega_1)\},$$

and if $\mu(Z_n(\omega_1)) = 0$, then

$$Z_n(\omega_1) = \{\omega_2 \in \Omega \mid d(\omega_1, \omega_2) < \mu Z_{n-1}(\omega_1)\}.$$

To prove the last statement observe that any cover of Ω_0 by cylinders of diameter $\mu Z < \varepsilon$ is clearly a cover by sets of diameter less than ε. Hence, $\dim_H \Omega_0 \leq \dim_B \Omega_0$. The reverse inequality follows from the fact that every covering set of diameter less than ε in any cover of Ω_0 is contained in a cylinder of equal diameter. Thus, $\dim_H \Omega_0 \geq \dim_B \Omega_0$. ∎

Example 3.1 Let $\Omega := [0, 1]$, let $\mathcal{B}$ be the σ-algebra of Borel sets, and let μ be Lebesgue measure. Define a stochastic process $\{X_n\}$ by

$$\omega = \sum_{n=1}^{\infty} X_n(\omega) 2^{-n},$$

where $X_n(\omega) = 0$ or 1 and $X_n(\omega) = 1$ infinitely often. Define $\mathcal{T}$ to be the topology that is generated by all half-open intervals with dyadic endpoints. Then the hypotheses of Theorem 3.9 are clearly satisfied, and thus the Billingsley dimension agrees with the Hausdorff dimension.

Now suppose that Ω is a set, $\mathcal{B}$ a σ-algebra of subsets of Ω, and μ and ν are two probability measures on $\mathcal{B}$. Also, let $\{X_n\}_{n\in\mathbb{N}_0}$ and $\{Y_n\}_{n\in\mathbb{N}_0}$ be two stochastic processes and d_1 and d_2 the induced semi-metrics. Then, if

$$K_i(\omega, r) := \{\omega' \in \Omega \mid d_i(\omega, \omega') < r\},$$

$K_1(\omega, r)$ is a cylinder that contains ω. If in addition $X_n = Y_n$ for all $n \in \mathbb{N}_0$, then $d_1(K_1(\omega, r))$ is the μ-measure and $d_2(K_1(\omega, r))$ the ν-measure of this cylinder. Proposition 3.3 gives the next result (also [27]).

Theorem 3.10 *Let* $\Omega_0 \subseteq \Omega$. *Assume that*

$$\Omega_0 = \left\{ \omega \,\middle|\, \liminf_{n\to\infty} \frac{\log \nu Z_n(\omega)}{\log \mu(Z_n(\omega))} \geq \delta \right\}.$$

Then

$$\dim_H^{(1)} \Omega_0 \geq \delta \, \dim_H^{(2)} \Omega_0.$$

∎

Remark. In the preceding theorem the following conventions about ratios of logarithms are used. Let $0 < \xi, \eta < 1$, then

(a) $\log \xi / \log 0 = \log 1 / \log \eta = \log 1 / \log 0 := 0.$

(b) $\log 0 / \log \eta = \log \xi / \log 1 = \log 0 / \log 1 := \infty.$

(c) $\log 0 / \log 0 = \log 1 / \log 1 := 1.$

3.4   Dimension Results for Self-Affine Fractals

This section is concerned with the calculation of the Hausdorff and box dimension of *self-affine* fractal sets generated by ordinary and recurrent IFSs. A fractal set F in a complete (semi-)metric space (X, d) is called *self -affine* iff it is generated by a finite collection of maps w_i, $i = 1, \ldots, N$, of the form

$$w_i(x) := A_i x + v_i, \tag{3.30}$$

for some bounded, linear, and contractive operators $A_i : X \to X$ and some $v_i \in X$. The fractal F is called *self-similar* iff it is constructed using contractive similitudes.

Remark. In the case $X = \mathbb{R}^n$ it can be shown that ([98]) $S : \mathbb{R}^n \to \mathbb{R}^n$ is a similitude iff $S = H_s \circ \tau_v \circ O$, for some *homothety* $H_s : \mathbb{R}^n \to \mathbb{R}^n$,

$H_s(x) := sx$, $s \in \mathbb{R}$, some *translation operator* $\tau_v : \mathbb{R}^n \to \mathbb{R}^n$, $\tau_v x := x + v$, and some *orthonormal operator* $O : \mathbb{R}^n \to \mathbb{R}^n$.

The space of all similitudes $S : \mathbb{R}^n \to \mathbb{R}^n$ will be denoted by $\mathcal{S}(\mathbb{R}^n, \mathbb{R}^n)$, and the subspace consisting of contractive similitudes by $\mathcal{S}^*(\mathbb{R}^n, \mathbb{R}^n)$.

3.4.1   Dimension of self-similar fractals

The first result concerning the Hausdorff dimension of self-similar fractal sets is essentially due to P. A. P. Moran ([140]). In [98] and [65] Moran's theorem and proof are presented in the language of IFSs. K. Falconer has perhaps the most general theorem for the Hausdorff and box dimension of self-affine fractal sets ([66]) generated by IFSs. It basically states that the Hausdorff dimension and box dimension of a self-affine fractal are "equal almost surely," a notion that will be made more precise later. A number of other dimension results, such as the box dimension of an attractor of a recurrent IFS, are also given.

The results will be presented in $\mathbb{R}^n$, although most of them hold in more general (semi-)metric spaces. In order to obtain the formula for the Hausdorff and box dimension of self-similar fractal sets, one has to impose what Hutchinson called the *open set condition*. This condition ensures that the components $S_i(\mathsf{F})$ of a self-similar fractal F do not overlap "too much."

***Definition* 3.10 (Open Set Condition)** Assume that F is the attractor of a hyperbolic IFS $(\mathbb{R}^n, \mathbf{S}, \mathbf{p})$ (with probabilities), where $\mathbf{S} := \{S_i \mid i = 1, \ldots, N\}$ is a collection of similitudes with contractivities $s_i \in [0, 1)$. The family $\mathbf{S}$ is said to satisfy the open set condition iff there exists a non-empty bounded open set $G \subseteq \mathbb{R}^n$ such that

$$G \supseteq \sum_{i=1}^{N} S_i(G), \tag{3.31}$$

where $\sum$ denotes the disjoint union of sets.

Remarks.

1. The rather lengthy expression : "$\mathbf{S}$ satisfies the open set condition" will be replaced by "$\mathbf{S}$ satisfies OSC."

2. A simple volume argument shows that if $\mathbf{S}$ satisfies OSC, then

$$\sum_{i=1}^{N} s_i^n < 1.$$

The following lemma is needed:

Lemma 3.2 *Suppose that $\{G_i\} \subseteq \mathbb{R}^n$ is a countable collection of disjoint open sets with the property that each G_i contains a ball of radius $\rho_1 r$ and is contained in a ball of radius $\rho_2 r$. Then any ball B of radius r intersects at most $(1 + 2\rho_2)^n \rho_1^{-n}$ of the closures $\overline{G_i}$.*

Proof. Suppose that $\overline{G_i} \cap B \neq \emptyset$. Then $\overline{G_i}$ is contained in a concentric ball of radius $(1 + 2\rho_2)r$. Denote by m the number of sets $\overline{G_i}$ that intersect B. The sum over the volumes of the interior balls of radii $\rho_1 r$ yields

$$m(\rho_1 r)^n \leq (1 + 2\rho_2)^n r^n.$$

This gives the required result.   ∎

Theorem 3.11 *Suppose that F is the attractor of a hyperbolic i.f.s $(\mathbb{R}^n, \mathbf{S}, \mathbf{p})$ and that $\mathbf{S}$ satisfies OSC. Then*

$$\dim_H \mathsf{F} = \dim_\beta \mathsf{F} = d, \tag{3.32}$$

where d is the unique positive solution of

$$\sum_{i=1}^{N} s_i^d = 1. \tag{3.33}$$

Moreover, F is a d-set.

Proof. Let $n \in \mathbb{N}$. The fixed-point property of F under the set-valued map $\mathfrak{S} : \mathfrak{H}(\mathbb{R}^n) \to \mathfrak{H}(\mathbb{R}^n)$, $\mathfrak{S}(E) = \bigcup_{i=1}^{N} S_i(E)$, implies that

$$\mathsf{F} = \bigcup_{\mathbf{i}(n) \in \Sigma_n} \mathsf{F}_{\mathbf{i}(n)},$$

where $\Sigma_n = \{1, \ldots, N\}^{\mathbb{N}_n}$ (Section 1.2 and (2.29)). Since the composition of the similitudes $S_{\mathbf{i}(n)}$ is a similitude with contractivity $s_{\mathbf{i}(n)}$, Eq. (3.33) implies that

$$\begin{aligned}
\sum_{\mathbf{i}(n) \in \Sigma_n} |\mathsf{F}_{\mathbf{i}(n)}|^d &= \sum_{\mathbf{i}(n) \in \Sigma_n} (s_{\mathbf{i}(n)})^d |\mathsf{F}|^d = \left(\sum_{i_1} s_{i_1}^d \right) \cdots \left(\sum_{i_n} s_{i_n}^d \right) |\mathsf{F}|^d \\
&= |\mathsf{F}|^d,
\end{aligned}$$

where $\mathbf{i}(n) := (i_1\, i_2\, \ldots i_n)$. Now given any $\varepsilon > 0$, one can always find an $n \in \mathbb{N}$ large enough so that $|F_{\mathbf{i}(n)}|^d \leq (\max_i s_i)^d \leq \varepsilon$. Thus, $\mathcal{H}^d_\varepsilon(F) \leq |F|^d$, and consequently $\mathcal{H}^d(F) \leq |F|^d$.

To obtain a lower bound a measure ν on n-cylinders $Z_{\mathbf{i}(n)} := \{\mathbf{i} \in \Sigma \,|\, \mathbf{i} = \mathbf{i}(n) \wedge \mathbf{j}\}$ is introduced, where $\wedge : \Sigma \to \Sigma$ denotes the *concatenation* of codes: $(i_1\, i_2\, \ldots i_n) \wedge (j_1\, j_2\, \ldots j_m\, \ldots) := (i_1\, i_2\, \ldots i_n\, j_1\, \ldots j_m\, \ldots)$. Define

$$\nu Z_{\mathbf{i}(n)} := \left(s_{\mathbf{i}(n)}\right)^d.$$

It follows directly from Eq. (3.33) that $\nu Z_{\mathbf{i}(n)} = \sum_i \nu Z_{\mathbf{i}(n) \wedge i}$ and therefore $\nu \Sigma = 1$. Using Eq. (2.29), ν is extended to a measure $\tilde{\nu}$ on F by setting

$$\tilde{\nu} := \gamma^{\#} \circ \nu.$$

Clearly, $\tilde{\nu} F = 1$.

Let G be the non-empty bounded open set whose existence is guaranteed by the open set condition. The fact that every compact set converges (in the Hausdorff metric) to the attractor F implies

$$\overline{G} \supseteq \mathfrak{S}(\overline{G}) \supseteq \cdots \supseteq \mathfrak{S}^n(\overline{G}) \to F.$$

Clearly, $\overline{G} \supseteq F$, and therefore $\overline{G}_{\mathbf{i}(n)} \supseteq F_{\mathbf{i}(n)}$, for all $\mathbf{i}(n)$, $n \in \mathbb{N}$.

Now let B be a ball of radius $r < 1$ intersecting F. Let $\mathbf{i} \in \Sigma$ and let n be the first integer for which

$$(\min_i s_i)r \leq s_{\mathbf{i}(n)} \leq r.$$

Denote by Σ^* the set of all such codes. Observe that for any code $\mathbf{i} \in \Sigma$ there is exactly one integer n such that $\mathbf{i}(n) \in \Sigma^*$. Since $\{G_1, \ldots, G_N\}$ is disjoint, so is $\{G_{\mathbf{i}(n) \wedge 1}, \ldots, G_{\mathbf{i}(n) \wedge N}\}$, for all $\mathbf{i}(n) \in \Sigma_n$. Hence, the collection $\{G_{\mathbf{i}(n)} \,|\, \mathbf{i}(n) \in \Sigma^*\}$ is disjoint, and therefore

$$F \subseteq \bigcup_{\mathbf{i}(n) \in \Sigma^*} F_{\mathbf{i}(n)} \subseteq \bigcup_{\mathbf{i}(n) \in \Sigma^*} \overline{G}_{\mathbf{i}(n)}.$$

Now choose two real numbers ρ_1 and ρ_2 such that G contains a ball of radius ρ_1 and is contained in a ball of radius ρ_2. If $\mathbf{i}(n) \in \Sigma^*$, the set $G_{\mathbf{i}(n)}$ contains a ball of radius $s_{\mathbf{i}(n)}\rho_1$ and thus one of radius $(\min_i s_i)\rho_1 r$, and is contained in a ball of radius $s_{\mathbf{i}(n)}\rho_2$ and hence in one of radius $\rho_2 r$. Now denote by Σ^{**} the set of all codes in Σ^* for which $G_{\mathbf{i}(n)} \cap B \neq \emptyset$. By the preceding lemma, there are at most $m = (1 + 2\rho_2)^n \rho_1^{-n} (\min_i s_i)^{-n}$ codes in Σ^{**}. Then

$$\tilde{\nu} B = \tilde{\nu}(B \cap \mathsf{F}) \leq \nu\{\mathbf{i}(n) \in \Sigma | \gamma(\mathbf{i}) \in B \cap \mathsf{F}\} \leq \nu\left(\bigcup_{\mathbf{i}(n) \in \Sigma^{**}} Z_{\mathbf{i}(n)}\right).$$

Thus,

$$\tilde{\nu} B \leq \sum_{\mathbf{i}(n) \in \Sigma^{**}} \nu Z_{\mathbf{i}(n)} = \sum_{\mathbf{i}(n) \in \Sigma^{**}} s_{\mathbf{i}(n)}^d \leq \sum_{\mathbf{i}(n) \in \Sigma^{**}} r^d \leq mr^d.$$

As every set U is contained in a ball of radius $|U|$, $\tilde{\nu}(U) \leq m|U|^d$. The remark following Definition 3.8 implies that $\mathcal{H}^d(\mathsf{F}) \geq m^{-1} > 0$, and thus $\dim_H \mathsf{F} = d$.

It remains to show that $\dim_\beta \mathsf{F} = d$. To this end, notice that the cardinality $\|\Sigma^*\|_c$ of Σ^* is at most $(\min_i s_i)^{-d} r^{-d}$. (This follows immediately from $\sum_{\mathbf{i}(n) \in \Sigma^*} s_{\mathbf{i}(n)}^d = 1$ and the definition of Σ^*.) Hence, if $\mathbf{i}(n) \in \Sigma^*$, then $|\overline{G}_{\mathbf{i}(n)}| = s_{\mathbf{i}(n)} |\overline{G}| \leq r|\overline{G}|$. Therefore, F can be covered by $\|\Sigma^*\|_c$ sets of diameter $r|\overline{G}|$, for any $r < 1$. Eq. (3.12) together with Eq. (3.14) then implies that $\overline{\dim}_\beta \mathsf{F} \leq d$. Hence, $\dim_\beta \mathsf{F} = d$. ∎

Now the packing dimension can be related to the box dimension. In order to achieve this, the *strong open set condition* has to be introduced ([110]).

Definition 3.11 A collection of similitudes **S** satisfies the strong open set condition (SOSC) iff there exists an open set $G \subseteq \mathbb{R}^n$ such that Eq. (3.31) is satisfied and G can be chosen such that $G \cap \mathsf{F} \neq \emptyset$.

The following theorem is proven in [110].

Theorem 3.12 *Suppose that* F *is the attractor of a hyperbolic IFS whose collection of similitudes satsifies SOSC. Then*

$$\dim_P \mathsf{F} = \dim_\beta \mathsf{F}. \tag{3.34}$$

∎

3.4.2 Dimension of self-affine fractals

The calculation of dimension for self-affine fractal sets in $\mathbb{R}^n$ is far more difficult. One of the most general results concerning the equality of Hausdorff and box dimension for such fractals is due to K. Falconer ([66]). His theorem is stated without presenting the full proof since it requires potential-theoretic arguments and methods that will not be developed in this monograph.

Recall that an affine mapping $w : \mathbb{R}^n \to \mathbb{R}^n$ is of the form

$$w(x) = Ax + v,$$

where $A \in \mathcal{L}(\mathbb{R}^n, \mathbb{R}^n)$ and $v \in \mathbb{R}^n$. Let $\mathcal{L}^*(\mathbb{R}^n, \mathbb{R}^n)$ be that subspace of $\mathcal{L}(\mathbb{R}^n, \mathbb{R}^n)$ consisting of all contractive and non-singular linear mappings A. Denote the eigenvalues of A^*A by $\lambda_1, \ldots, \lambda_n$. The *singular values of A* are defined as

$$\alpha_i := \sqrt{\lambda_i}, \quad i = 1, \ldots, n,$$

and — after possibly reindexing — order them according to

$$1 > \alpha_1 \geq \alpha_2 \ldots \geq \alpha_n > 0.$$

For a real number $0 \leq s \leq n$, define the *singular value function φ^s of A* by

$$\varphi^s(A) := \alpha_1 \, \alpha_2 \cdots \alpha_{r-1} \, \alpha_r^{s-r+1}, \tag{3.35}$$

where r is that integer for which $r - 1 < s \leq r$. Clearly, $\varphi^s(A)$ is a continuous function and strictly decreasing function of s.

Proposition 3.6 *For a fixed $s \in [0, n]$ the function $\varphi^s : \mathcal{L}^*(\mathbb{R}^n, \mathbb{R}^n) \to \mathbb{R}^+$ is sub-multiplicative, i.e., for $A, B \in \mathcal{L}^*(\mathbb{R}^n, \mathbb{R}^n)$,*

$$\varphi^s(AB) \leq \varphi^s(A)\, \varphi^s(B).$$

Proof. Easy exercise. ∎

Let $k \in \mathbb{N}$, let $s \in [0, n]$ be fixed, and let w_i, $i = 1, \ldots, N$ be a collection of affine mappings whose linear parts $A_i \in \mathcal{L}^*(\mathbb{R}^n, \mathbb{R}^n)$. For a finite code $\mathbf{i}(k) \in \Sigma_k$, define

$$S_k^s := \sum_{\mathbf{i}(k) \in \Sigma_k} \varphi^s(A_{\mathbf{i}(k)}). \tag{3.36}$$

Then, for any $m \in \mathbb{N}$,

$$
\begin{aligned}
S_{k+m}^s &= \sum_{\mathbf{i}(k+m) \in \Sigma_{k+m}} \varphi^s(A_{\mathbf{i}(k+m)}) \\
&\leq \sum_{\mathbf{i}(k+m) \in \Sigma_{k+m}} \varphi^s(A_{\mathbf{i}(k)}) \varphi^s(A_{\mathbf{i}(k+m)}) \\
&= \sum_{\mathbf{i}(k) \in \Sigma_k} \varphi^s(A_{\mathbf{i}(k)}) \sum_{\mathbf{i}(m) \in \Sigma_m} \varphi^s(A_{\mathbf{i}(m)}) = S_k^s \, S_m^s.
\end{aligned}
$$

Hence the sequence $\{S_k^s\}_{k\in\mathbb{N}}$ is sub-multiplicative, and therefore the sequence $\{(S_k^s)^{1/k}\}_{k\in\mathbb{N}}$ converges to a limit S_∞^s. Note that since the function φ^s is strictly decreasing in s, so is S_∞^s. This then implies that if $S_\infty^n \le 1$, there exists a unique $s^* \in [0,n]$ so that

$$1 = S_\infty^s = \lim_{k\to\infty} \left(\sum_{i(k)\in\Sigma_k} \varphi^s(A_{i(k)}) \right)^{1/k}. \qquad (3.37)$$

Using the sub-multiplicativity of S_k^s, this last equation can also be expressed in the form

$$s^* = \inf \left\{ s \in [0,n] \,\middle|\, \sum_{k=1}^{\infty} \sum_{i(k)\in\Sigma_k} \varphi^s(A_{i(k)}) < \infty \right\}. \qquad (3.38)$$

Theorem 3.13 *Let* F *be the attractor of a hyperbolic IFS* $(\mathbb{R}^n, \mathbf{w}, \mathbf{p})$, *where* $\mathbf{w}$ *is a collection of affine maps* $w_i(x) = A_i x + v_i$, *with* $A_i \in \mathcal{L}^*(\mathbb{R}^n, \mathbb{R}^n)$ *and* $v_i \in \mathbb{R}^n$, $i = 1, \ldots, N$. *Then for almost all* $(v_1, \ldots, v_N) \in \mathbb{R}^{nN}$

$$\dim_H \mathsf{F} = \dim_\beta \mathsf{F} = s^*. \qquad (3.39)$$

Remark. "For almost all . . .." refers to nN-dimensional Lebesgue measure.

Proof. An argument quite similar to the one given in the proof of Theorem 3.11 gives $\dim_H \mathsf{F} \le s^*$. The lower bound requires potential-theoretic methods and is therefore omitted. ∎

3.4.3   Recurrent IFSs and dimension

The next objective is the calculation of box dimension of a fractal set F generated by a recurrent IFS. But first it has to be clarified what is meant by the box dimension of a set in $\mathsf{X}\mathfrak{H}(X)$, where X is a compact (semi-) metrizable topological space (X, d).

Definition 3.12 Suppose that $\mathsf{X}E := (E_1, E_2, \ldots, E_N) \subseteq \mathsf{X}\mathfrak{H}(X)$. The box dimension of $\mathsf{X}E$ is defined as

$$\dim_\beta \mathsf{X}E := \max_{k\in\{1,\ldots,N\}} \left\{ \dim_\beta^{(k)} E_k \right\}. \qquad (3.40)$$

Now suppose a hyperbolic recurrent IFS $(X, \mathbf{w}, P)$, whose collection of maps $\mathbf{w}$ consists of similitudes with contractivities s_i, $i = 1, \ldots, N$, is given. Recall that the connection matrix $C = (c_{ij})$ is defined by $c_{ij} := 1$ if $p_{ji} > 0$ and $c_{ij} := 0$ if $p_{ji} = 0$.

Definition 3.13 The attractor $A = \bigcup_{i=1}^{N} A_i$ of an IFS $(X, \mathbf{w}, P)$ is called non-overlapping iff $A_j \cap A_k = \emptyset$, for all $j, k \in \mathbf{I}(i)$ such that $j \neq k$, and $i = 1, \ldots, N$.

Theorem 3.14 *Let $(X, \mathbf{w}, P)$ be a recurrent IFS and F its attractor. Furthermore, assume that the maps w_i, $i = 1, \ldots, N$, in the collection $\mathbf{w}$ belong to $\mathcal{S}^*(\mathbb{R}^n, \mathbb{R}^n)$. Let C be an irreducible $N \times N$ connection matrix. For all $t \in \mathbb{R}^+$ define a diagonal matrix $D(t)$ by*

$$D(t) := \begin{pmatrix} s_1^t & 0 & \cdots & 0 \\ 0 & s_2^t & \cdots & 0 \\ 0 & 0 & \ddots & 0 \\ 0 & 0 & \cdots & s_N^t \end{pmatrix}, \tag{3.41}$$

where s_i is the contractivity constant of the map w_i, $i = 1, \ldots, N$. Assume also that F is non-overlapping. Let t^* be the unique positive number such that 1 is an eigenvalue of $D(t^*)C$ of greatest absolute value. Then

$$\dim_\beta \mathsf{F} = t^*. \tag{3.42}$$

Proof. Let $\underline{s} := \min_{i=1,\ldots,N}\{s_i\}$ and $\overline{s} := \max_{i=1,\ldots,N}\{s_i\}$. The compactness of each $\mathsf{F}_j \subseteq \mathsf{F}$ implies the existence of an $\varepsilon_0 > 0$ such that

$$\forall i = 1, \ldots, N \; \forall (j, k \in \mathbf{I}(i) \wedge j \neq k) : \; d(w_i(\mathsf{F}_j), w_i(\mathsf{F}_k)) > \varepsilon_0. \tag{3.43}$$

For $i = 1, \ldots, N$, denote by $\mathcal{N}_{i,\varepsilon}(\mathsf{F}_i)$ the cardinality of a minimal cover of F_i by balls of radius $\varepsilon > 0$. Since $\mathsf{F}_i = \bigcup_{j \in \mathbf{I}(i)} w_i(\mathsf{F}_j)$ and $w_i \in \mathcal{S}^*(\mathbb{R}^n, \mathbb{R}^n)$, together with Eq. (3.43) they give a system of functional equations of the form

$$\mathcal{N}_{i,\varepsilon}(\mathsf{F}_i) = \bigcup_{j \in \mathbf{I}(i)} \mathcal{N}_{j, \varepsilon/s_i}(\mathsf{F}_j), \tag{3.44}$$

for all $0 < \varepsilon < \underline{s}\,\varepsilon_0$.

By the Perron-Frobenius Theorem there exists a strictly positive eigenvector of $D(t^*)C$ corresponding to the eigenvalue 1. This eigenvector is denoted by $e = (e_1, \ldots, e_N)$. There exist positive constants c_1 and c_2 such that

$$\forall (\underline{s}\,\varepsilon_0 < \varepsilon \leq \varepsilon_0) \; \forall i = 1, \ldots, N : \; c_1 \, e_i \, \varepsilon^{-t^*} \leq \mathcal{N}_{i,\varepsilon}(\mathsf{F}_i) \leq c_2 \, e_i \, \varepsilon^{-t^*}. \tag{3.45}$$

For $n \in \mathbb{N}$ suppose that Eq. (3.45) holds for all $\overline{s}^{\,n}\underline{s}\varepsilon_0 \leq \varepsilon \leq \varepsilon_0$. Now assume that ε is chosen in the interval $[\overline{s}^{\,n}\underline{s}\varepsilon_0, \underline{s}\varepsilon_0]$. Then $\overline{s}^{\,n}\underline{s}\varepsilon_0 \leq \varepsilon/s_i \leq \varepsilon_0$, and thus

$$\mathcal{N}_{i,\varepsilon}(\mathsf{F}_i) = \sum_{j \in \mathbf{I}(i)} \mathcal{N}_{j, \varepsilon/s_i}(\mathsf{F}_i) \leq c_2 \, \varepsilon^{-t^*} s_i^{t^*} \sum_{j \in \mathbf{I}(i)} e_j = c_2 \, e_2 \, \varepsilon^{-t^*}, \tag{3.46}$$

and analogously,

$$c_1 \, e_i \, \varepsilon^{t^*} \leq \mathcal{N}_{i,\,\varepsilon}(\mathsf{F}_i). \tag{3.47}$$

By induction on n the following inequalities which prove the theorem are obtained:

$$\forall (0 < \varepsilon \leq \varepsilon_0) \; \forall i = 1, \dots, N : \; c_1 \, e_i \, \varepsilon^{-t^*} \leq \mathcal{N}_{i,\,\varepsilon}(\mathsf{F}_i) \leq c_2 \, e_i \, \varepsilon^{-t^*}.$$

∎

3.4.4   Recurrent sets and Mauldin-Williams fractals

Next a dimension result for a class of recurrent sets is stated. This result was first conjectered by M. Dekking ([47]) and then fully proved by T. Bedford ([15]). The full proof, which involves pressure-theoretic arguments, is omitted. First a few definitions are needed.

Definition 3.14   1. Let $w \in S[X]$. Define $| \cdot |_x : S[X] \to \mathbb{N}_0$ as the number of occurences of the letter x in the word w, and $|\cdot|_E : S[X] \to \mathbb{N}_0$ as the number of essential letters in w.

2. A recurrent set $K_\vartheta(w)$ is called resolvable iff there exists an $\eta > 0$ so that

$$\liminf_{\nu \to \infty} \frac{\lambda K_\eta(\vartheta^\nu w)}{|\vartheta^\nu w|_E} > 0, \tag{3.48}$$

where λ denotes n-dimensional Lebesgue measure and $K_\eta(\vartheta^\nu w)$ the η-parallel body of $K(\vartheta^\nu(w))$.

Let ϑ be a semigroup endomorphism and let $x \in X$. Define a matrix $M_E := (m_{xy})_{x,y \in E}$ on the essential letters by $m_{xy} := |\vartheta(x)|_y$. Let λ_E be the largest non-negative eigenvalue of M_E.

Definition 3.15 A semigroup endomorphism ϑ is called essentially mixing iff the matrix $M_E := (m_{xy})_{x,y \in E}$ is mixing, in other words, iff there exists a $\nu \in \mathbb{N}$ such that for all $x, y \in E$, x occurs in $\vartheta^\nu(y)$.

Theorem 3.15 *Suppose that $\vartheta : S[X] \to S[X]$ is essentially mixing and that L_ϑ is an expansive similitude with eigenvalue λ. Then*

$$\dim_H K_\vartheta(w) = \frac{\log \lambda_E}{\log |\lambda|} \tag{3.49}$$

iff $K_\vartheta(w)$ is resolvable.

∎

For the sake of completeness, a formula for the Hausdorff dimension of Mauldin-Williams fractals is given. Recall that by Theorem 2.12 there exists a unique element $(X_v)_{v \in V} \in \prod_{v \in V} \mathfrak{H}(X_v)$ such that

$$X_u = \bigcup_{\substack{v \in V \\ e \in E_{uv}}} S_e X_v.$$

Let $X := \bigcup_{u \in V} X_u$. Given a positive number d, define a matrix $M(d) := (M_{uv}(d))_{1 \leq u, v \leq N}$, where

$$M_{uv}(d) := \sum_{e \in E_{uv}} s(e)^d,$$

and $N := \|V\|_c$. The spectral radius of $M(d)$ is denoted by $r(d)$. By the Perron-Frobenius Theorem, $r(d)$ is the largest non-negative eigenvalue of $M(d)$. The following theorem is proven in [135]:

Theorem 3.16 *Let* (G, s) *be a strongly connected and contracting Mauldin-Williams graph. The Hausdorff dimension of* X *is that number* d^* *for which* $r(d) = 1$. *Furthermore,* X *is a* d^**-set.* ∎

3.5 The Box Dimension of Projections

In this section the box dimension of a self-affine set in $\mathbb{R}^2$ is related to the box dimension of the orthogonal projection onto a coordinate axis. For this purpose let (x, y) be a cartesian coordinate system of $\mathbb{R}^2$; let $\mathsf{X}A = (A_1, \ldots, A_N)$ be the attractor of the recurrent IFS $(X, \mathbf{w}, P)$, where X is — without loss of generality — the unit square in $\mathbb{R}^2$, and the maps $w_i : X \to X$ are affine and of the form

$$w_i(x, y) := \begin{pmatrix} a_i & 0 \\ 0 & b_i \end{pmatrix} \begin{pmatrix} x \\ y \end{pmatrix} + \begin{pmatrix} c_i \\ d_i \end{pmatrix}, \qquad (3.50)$$

with $0 < |a_i|, |b_i| < 1$, $i = 1, \ldots, N$. Furthermore, it is assumed that the connection matrix $C = (c_{ij})$ associated with the recurrent IFS $(X, \mathbf{w}, P)$ is irreducible, i.e., $c_{ij} = 0$ or 1.

For a set $E \subset \mathbb{R}^2$, denote the orthogonal projection of E onto the y-axis by E_y. Note that $(\mathsf{X}A)_y$ is the attractor of the recurrent IFS (X_y, w_y, P) with $(w_i)_y : X_y \to X_y$ given by

$$(w_i)_y(y) = b_i y + d_i, \qquad i = 1, \ldots, N.$$

It follows from a result in [68] that $\dim_\beta (\mathsf{X}A)_y$ exists and equals $\dim_H (\mathsf{X}A)_y$. The next theorem relates the box dimension of $(\mathsf{X}A)_y$ to that of A.

Theorem 3.17 *Let* $(X, \mathbf{w}, P)$ *be a recurrent IFS as defined earlier. Assume* $\mathbf{w}$ *satisfies the OSC*

$$\forall\, i, j \in \{1, \dots, N\},\ i \neq j : w_i \overset{\circ}{X} \cap w_j \overset{\circ}{X} = \emptyset.$$

Let d^* *be the unique positive number such that* $r(\mathrm{diag}(|a_i|^{d^*})C) = 1$ *and let* d *be determined by the formula*

$$r(\mathrm{diag}(|b_i|^{d_y}\,|a_i|^{d - d_y})\, C) = 1,$$

where $d_y := \dim_\beta (\mathsf{X}A)_y$. *If*

$$r(\mathrm{diag}(|b_i|^{d_y}\,|a_i|^{d^* - d_y})\, C) > 1, \tag{3.51}$$

then $\dim_\beta \mathsf{X}A = d$.

Proof. Since $\mathsf{X}A = \bigcup_{i=1}^N A_i$ and $A_i = \bigcup_{j \in \mathbf{I}(i)} w_i A_j$, it follows that

$$\dim_\beta \mathsf{X}A = \max_{i=1,\dots,N} \dim_\beta A_i = \dim_\beta A_j.$$

Furthermore, the irreducibility of the connection matrix C implies that each component A_i contains a non-singular affine image of A_j, and thus $\dim_\beta A_i = \dim_\beta A_j = \dim_\beta \mathsf{X}A$, $i = 1, \dots, N$.

Now let $0 < \varepsilon < 1$ be given. Denote by Σ_ε the set of all finite codes $\mathbf{i}(n) = \mathbf{i}(n)(\varepsilon)$ such that

$$|a_{i_{-1}} \cdots a_{i_{-n}}| \leq \varepsilon, \tag{3.52}$$

$$|a_{i_{-1}} \cdots a_{i_{-n+1}}| > \varepsilon, \tag{3.53}$$

and $\mathbf{i} = i_{-1} \in \Sigma_\varepsilon$ if $|a_{i_{-1}}| \leq \varepsilon$. It is easy to see that Σ_ε generates a partition of the code space Σ into cylinder sets $\mathbf{i}(n)$. To simplify notation, let

$$a_{\mathbf{i}(n)} := |a_{i_{-1}} \cdots a_{i_{-n}}| \quad \text{and} \quad b_{\mathbf{i}(n)} := |b_{i_{-1}} \cdots b_{i_{-n}}|,$$

and

$$w_{\mathbf{i}(n)} := w_{i_{-1}} \circ \dots \circ w_{i_{-n}}, \qquad A_{\mathbf{i}(n)} := w_{i_{-1}} \circ \dots \circ w_{i_{-n+1}} A_{i_{-n}}.$$

Note that the codes are "time-reversed." For $i = 1, \dots, N$, let $U_i := \bigcup_{c_{ij}>0} A_j$ and note that $A_{\mathbf{i}(n)} = w_{\mathbf{i}(n)} U_{i_{-n}}$.

Let E be a bounded set in $\mathbb{R}^2$, and let $\mathcal{C}_\varepsilon$ be a class of covers such that each $C_\varepsilon \in \mathcal{C}_\varepsilon$ consists of $\varepsilon \times \varepsilon$-squares with sides parallel to the coordinate axis. Denote by $\mathcal{N}_\varepsilon(E)$ the cardinality of a minimal cover from $\mathcal{C}_\varepsilon$, and by $\mathcal{N}_\varepsilon(E_y)$ the minimum number of compact intervals of length ε needed to cover E_y. The proof is based upon the following geometrical observations: Let $\varepsilon > 0$ and let $\mathbf{i}(n) \in \Sigma_\varepsilon$. Then

1. $A_{\mathbf{i}(n)} \subset w_{\mathbf{i}(n)} X$ and $w_{\mathbf{i}(n)} X$ is a rectangle of width $a_{\mathbf{i}(n)}$ and height $b_{\mathbf{i}(n)}$.

2. Since $a_{\mathbf{i}(n)} \leq \varepsilon$, any cover of $A_{\mathbf{i}(n)}$ that is in $\mathcal{C}_\varepsilon$ may be arranged in such a way that each $\varepsilon \times \varepsilon$-square meets both vertical sides of $w_{\mathbf{i}(n)} X$. (This is possible since each $\varepsilon \times \varepsilon$-square is wider than $w_{\mathbf{i}(n)} X \supset A_{\mathbf{i}(n)}$.) If Q is such a square then $w_{\mathbf{i}(n)}^{-1} Q$ is a rectangle of height $\varepsilon / b_{\mathbf{i}(n)}$ that meets the lines $x = 0$ and $x = 1$. Hence, $Q \cap A_{\mathbf{i}(n)} \neq \emptyset$ iff $w_{\mathbf{i}(n)}^{-1} Q \cap (U_{i_{-n}})_y \neq \emptyset$. This now defines a one-to-one correspondance between covers of $A_{\mathbf{i}(n)}$ in $\mathcal{C}_\varepsilon$ and covers of $(U_{i_{-n}})_y$ consisting of compact intervals of length $\varepsilon / b_{\mathbf{i}(n)}$. Consequently,

$$\mathcal{N}_\varepsilon(A_{\mathbf{i}(n)}) = \mathcal{N}_{\varepsilon/b_{\mathbf{i}(n)}}((U_{i_{-n}})_y).$$

In order to establish the dimension result, the cardinality of Σ_ε is needed. For this purpose the following probability measure on Σ_ε is introduced. Let $v = (v_1, \dots, v_N)^t$ be a right positive eigenvector of $\operatorname{diag}(|a_i|^{d^*} C)$ and define a row-stochastic irreducible matrix $M = (m_{ij})$ by

$$m_{ij} := |a_i|^{d^*} c_{ij} \frac{v_j}{v_i}.$$

Let $\delta = (\delta_1, \dots, \delta_N)^t$ be the unique stationary distribution associated with the matrix M, in other words, $M\delta = \delta$ and $\sum_{i=1}^{N} \delta_i = 1$. Let μ_M denote the probability measure on Σ generated by M with initial distribution δ. Then, for any $\mathbf{i}(n) \in \Sigma_\varepsilon$, one has

$$\frac{\min v_i}{\max v_i} |a_{\mathbf{i}(n)}|^{d^*} \leq \mu_M \mathbf{i}(n) \leq \frac{\max v_i}{\min v_i} |a_{\mathbf{i}(n)}|^{d^*}.$$

Equations (3.52) and (3.53) now imply that

$$c_1 \varepsilon^{d^*} \leq \mu_M \mathbf{i}(n) \leq c_2 \varepsilon^{d^*},$$

for some positive constants c_1 and c_2. Consequently,

$$\mu_M \Sigma_\varepsilon = 1 \geq c_1 \varepsilon^{d^*} \|\Sigma_\varepsilon\|_c.$$

The remainder of the proof involves estimates on certain sums. To obtain these estimates a second row-stochastic matrix is now introduced. For $0 \leq \beta \leq 1$, let α be the unique positive number so that

$$r(\operatorname{diag}(b_i^\beta a_i^{\alpha - \beta} C)) = 1, \tag{3.54}$$

and let u be a positive left eigenvector of $\operatorname{diag}(b_i^\beta a_i^{\alpha-\beta} C)$ with eigenvalue one. Let $\tilde{M} := (\tilde{m}_{ij})$ be defined by

$$\tilde{m}_{ij} := b_i^\beta a_i^{\alpha-\beta} c_{ji} \frac{u_j}{u_i}.$$

Given the initial distribution $\tilde{\delta} = (1/N, \ldots, 1/N)^t$, the induced probability measure $\tilde{\mu}_{\tilde{M}}$ satisfies

$$\tilde{\mu}_{\tilde{M}} = \frac{1}{N} \tilde{m}_{i_{-1}i_{-2}} \cdot \ldots \cdot \tilde{m}_{i_{-n+1}i_{-n}} = \frac{1}{N} \frac{u_{i_{-n}}}{u_{i_{-1}}} b_{\mathbf{i}(n)}^\beta a_{\mathbf{i}(n)}^{\alpha-\beta}.$$

Hence,

$$\frac{1}{N} \frac{\min u_i}{\max u_i} \tilde{\mu}_{\tilde{M}} \mathbf{i}(n) \le b_{\mathbf{i}(n)}^\beta a_{\mathbf{i}(n)}^{\alpha-\beta} \le \frac{1}{N} \frac{\max u_i}{\min u_i} \tilde{\mu}_{\tilde{M}} \mathbf{i}(n). \qquad (3.55)$$

Now all the tools are available to prove the dimension result. First it is shown that d is an upper bound for $\dim_\beta \mathsf{X}A$. Note that, since Σ_ε generates a partition of Σ,

$$\mathsf{X}A = \bigcup_{\mathbf{i}(n)\in\Sigma_\varepsilon} A_{\mathbf{i}(n)}.$$

Thus,

$$\mathcal{N}_\varepsilon(\mathsf{X}A) \le \sum_{\mathbf{i}(n)\in\Sigma_\varepsilon} \mathcal{N}_\varepsilon(A_{\mathbf{i}(n)}) = \sum_{\mathbf{i}(n)\in\Sigma_\varepsilon} \mathcal{N}_{\varepsilon/b_{\mathbf{i}(n)}}((U_{i_{-n}})_y).$$

Let $\beta > d_y$ and α be as in Eq. (3.54). Then it follows from the Perron-Frobenius Theorem that $\lim_{\beta\to d_y}\alpha = d$. As $\beta > d_y$, there exists a constant $c > 0$ with the property that

$$\mathcal{N}_\varepsilon((U_j)_y) \le c\varepsilon^{-\beta},$$

for all $j = 1, \ldots, N$. Let $\Sigma_\varepsilon^* := \{\mathbf{i}(n)(\varepsilon) \in \Sigma_\varepsilon \mid \varepsilon/b_1 > 1\}$. Then

$$\mathcal{N}_{\varepsilon/b_{\mathbf{i}(n)}}((U_j)_y) = 1,$$

for all $j = 1, \ldots, N$, whereas if $\mathbf{i}(n) \in \Sigma_\varepsilon^{**} := \Sigma_\varepsilon \setminus \Sigma^*$, then $\varepsilon/b_{\mathbf{i}(n)} \le 1$ and thus, for some $c > 0$,

$$\mathcal{N}_{\varepsilon/b_{\mathbf{i}(n)}}((U_j)_y)) \le c\varepsilon^{-\beta} b_{\mathbf{i}(n)}^\beta, \qquad j = 1, \ldots, N.$$

Therefore,

$$\mathcal{N}_\varepsilon(\mathsf{X}A) \le \sum_{\mathbf{i}(n)\in\Sigma_\varepsilon} \mathcal{N}_\varepsilon(A_{\mathbf{i}(n)}) = \sum_{\mathbf{i}(n)\in\Sigma_\varepsilon} \mathcal{N}_{\varepsilon/b_{\mathbf{i}(n)}}((U_{i_{-n}})_y)$$

$$\le \; c \sum_{\mathbf{i}(n) \in \Sigma_\varepsilon^*} \varepsilon^{-\beta} b_{\mathbf{i}(n)}^\beta + \|\Sigma_\varepsilon^{**}\|_c \le c \sum_{\mathbf{i}(n) \in \Sigma_\varepsilon} \varepsilon^{-\beta} b_{\mathbf{i}(n)}^\beta + \|\Sigma_\varepsilon^{**}\|_c$$

$$\le \; c \left(\sum_{\mathbf{i}(n) \in \Sigma_\varepsilon} a_{\mathbf{i}(n)}^{\alpha-\beta} b_{\mathbf{i}(n)}^\beta \right) \varepsilon^{-\alpha} + \|\Sigma_\varepsilon^{**}\|_c.$$

The last inequality follows from the fact that $a_{\mathbf{i}(n)} \le \varepsilon$ for $\mathbf{i}(n) \in \Sigma_\varepsilon$. Since $\|\Sigma_\varepsilon^{**}\|_c \le \|\Sigma_\varepsilon\|_c$ and since by Eq. (3.55)

$$\frac{1}{N} \frac{\max u_i}{\min u_i} = \frac{1}{N} \frac{\max u_i}{\min u_i} \tilde{\mu}_{\tilde{M}} \Sigma_\varepsilon > \sum_{\mathbf{i}(n) \in \Sigma_\varepsilon} a_{\mathbf{i}(n)}^{\alpha-\beta} b_{\mathbf{i}(n)}^\beta,$$

the preceding inequality becomes

$$\mathcal{N}_\varepsilon(\mathsf{X}A) \le \varepsilon^\alpha \left(\frac{c}{N} \frac{\max u_i}{\min u_i} + \frac{\varepsilon^{\alpha - d^*}}{c_1} \right).$$

Letting $\beta \to d_y$, yields $\dim_\beta \mathsf{X}A \le d$.

Finally it is established that d is also a lower bound for $\beta \to d_y$. Note that the OSC implies that $w_i \overset{\circ}{X} \cap w_j \overset{\circ}{X} = \emptyset$, for all $\mathbf{i}(n), \mathbf{j}(n) \in \Sigma_\varepsilon$ with $\mathbf{i}(n) \ne \mathbf{j}(n)$. Thus, any $\varepsilon \times \varepsilon$-square in a cover belonging to the class $\mathcal{C}_\varepsilon$ meets at most $L = 2a^{-1} + 4$ of the rectangles $\{w_{\mathbf{i}(n)} X \,|\, \mathbf{i}(n) \in \Sigma_\varepsilon^*\}$. Hence,

$$\mathcal{N}_\varepsilon(\mathsf{X}A) \; \ge \; L^{-1} \sum_{\mathbf{i}(n) \in \Sigma_\varepsilon^*} \mathcal{N}_\varepsilon(A_{\mathbf{i}(n)})$$

$$= \; L^{-1} \left(\sum_{\mathbf{i}(n) \in \Sigma_\varepsilon} \mathcal{N}_\varepsilon(A_{\mathbf{i}(n)}) - \|\Sigma_\varepsilon^{**}\|_c \right).$$

The last equality holds since $\mathcal{N}_\varepsilon(A_{\mathbf{i}(n)}) = 1$ for $\mathbf{i}(n) \in \Sigma_\varepsilon^{**}$. Choosing $\beta < d_y$ and proceding as above yields the result. ∎

In the case $(X, \mathbf{w}, P)$ is an IFS, i.e., $C = (1)$, Eq.(3.51) reduces to

$$\sum_{i=1}^N |a_i|^{d^* - d_y} |b_i|^{d_y} > 1,$$

with $\sum_{i=1}^N |a_i|^{d^*} = 1$. Hence, the following corollary holds.

Corollary 3.2 *Let $(X, \mathbf{w})$ be an IFS with attractor A and maps $w_i X \to X$ of the form (3.50). Suppose that the OSC*

$$\forall\, i, j \in \{1, \ldots, N\},\, i \neq j : w_i \overset{\circ}{X} \cap w_j \overset{\circ}{X} = \emptyset$$

and $\sum_{i=1}^{N} |a_i|^{d^* - d_y} |b_i|^{d_y} > 1$ hold. If d is given by $\sum_{i=1}^{N} |a_i|^{d - d_y} |b_i|^{d_y} = 1$, then $\dim_\beta A = d$.

Chapter 4

Dynamical Systems and Dimension

This chapter introduces the concept of a *dynamical system*. It will be seen how the geometric theory of dynamical systems can be used to describe attractors of IFSs. In particular, the *Lyapunov dimension* of an attractor of a dynamical system is defined, and it is shown how it relates to the Hausdorff and box dimension. Again, the limited scope of this monograph allows only the presentation of the most basic aspects of the theory, thus giving the reader a general overview of this fascinating subject. Also, results will be presented in $\mathbb{R}^n$ rather than on finite-dimensional Riemannian manifolds. Some of the references in the bibliography give a more precise and general introduction to dynamical systems ([51, 52, 147, 158, 177]).

Let X be a subset of $\mathbb{R}^n$ and let $\mu : 2^X \to \mathbb{R}$ be a measure on X. The σ-algebra of all μ-measurable subsets of X is denoted by $\mathcal{B}(X)$.

***Definition* 4.1** Let $f : X \to X$ be a μ-measurable map. Assume that f is μ-invariant, i.e., $f^{\#}\mu B = \mu f^{-1}B = \mu B$, for all $B \in \mathcal{B}(X)$. Then the quadruple $(X, \mathcal{B}(X), f, \mu)$ is called an (n-dimensional) dynamical system.

Remark. There are two types of dynamical systems: discrete and continuous. Here only discrete dynamical systems are considered, i.e., only the behavior of the *iterates* f^m of f as $m \to \infty$ is studied. Examples of continuous dynamical systems are provided by the solutions of initial value problems for autonomous ordinary and partial differential equations.

Definition 4.2 Suppose that $(X, \mathcal{B}(X), f, \mu)$ is a dynamical system and that $\mu X = 1$, i.e., μ is a probability measure on $\mathcal{B}$. Such a dynamical system is called ergodic iff

$$\forall E \subseteq X : f^{-1}E = E \implies \mu E = 0 \text{ or } 1. \tag{4.1}$$

Remark. Definition 4.2 expresses the simple fact that an ergodic dynamical system cannot be decomposed into non-trivial subsystems which do not interact with each other.

The sequence of iterates $\{f^m(x)\}_{m \in \mathbb{N}}$ of f for a fixed $x \in X$ is called the *orbit of f*.

Definition 4.3 (Attractor of a dynamical system)   Suppose that $(X, \mathcal{B}(X), f, \mu)$ is a dynamical system. A closed subset A of X is said to be the attractor of the dynamical system $(X, \mathcal{B}(X), f, \mu)$ iff

1. A is invariant under f, i.e., $fA = A$, and no proper subset of A satisfies this requirement.

2. There exists an open set G, called the basin of attraction, such that $\forall x \in G$: $d_E(f^m(x), A) \to 0$ as $k \to \infty$.

Examples of attractors of dynamical systems are fixed points, limit cycles, and orbits of *p-periodic points* $\{y, f(y), \ldots, f^{p-1}(y)\}$, where p is the least positive integer satisfying $f^p(y) = y$ and $d_E(f^m(x), f^i(y)) \to 0$ as $m \to \infty$. Besides these classical attractors, A may be a fractal set. There are numerous examples of such fractal attractors in the literature: the Lorentz attractor, the Hénon attractor, the Rössler attractor, etc. If a dynamical system $(X, \mathcal{B}(X), f, \mu)$ exhibits such a fractal attractor A, the map f is very often *chaotic* in the following sense:

(a) There exists an $x \in X$ such that $\overline{\{f^m(x)\}_{m \in \mathbb{N}}} = A$.

(b) $\overline{\mathrm{Per}(f)} = A$, where $\mathrm{Per}(f)$ denotes the set of periodic points of f (points for which there exists a positive integer p such that $f^p(x) = x$).

(c) f has *sensitive dependence on initial conditions*, that is,
$\forall \varepsilon > 0 \, \exists \delta > 0 \, \forall x \in A \, \exists y \in A \, \exists m \in \mathbb{N}$:

$$d_E(x, y) < \varepsilon \implies d_E(f^m(x), f^m(y)) \geq \delta.$$

An example of a chaotic map is the *logistic map*, which has its origin in population dynamics. It is the archetype of a one-parameter family of maps associated with a one-dimensional dynamical system.

***Example* 4.1 (The logistic map)** Let $X := \mathbb{R}$, let μ be Lebesgue measure and let $c \in \mathbb{R}^+$. Define a map $f_c : \mathbb{R} \to \mathbb{R}$ by

$$f_c(x) := cx(1 - x). \tag{4.2}$$

As the parameter c increases, the p-periodic orbits of f_c exhibit different behavior, and as c reaches the value $c_\infty \approx 3.855$ a Cantor-like attractor appears. For a more detailed description of the behavior of f_c the reader is, for instance, referred to [51, 52].

Next the geometric structure of dynamical systems is briefly investigated. For this purpose, the *tangent map* of f and also Oseledec's Multiplicative Ergodic Theorem have to be introduced.

***Definition* 4.4** Let $f : X \subseteq \mathbb{R}^n \to \mathbb{R}^k$ be a function and suppose that $p \in X$ is a point of accumulation of X. Then f is called differentiable at p iff there exists a map $T_p f \in \mathcal{L}(\mathbb{R}^n, \mathbb{R}^k)$ such that

$$\lim_{x \to p} \frac{f(x) - f(p) - T_p f(x - p)}{d_E(x, y)} = 0. \tag{4.3}$$

The function f is called differentiable (on X) if it is differentiable at each point $p \in X$.

Note that the linear map $T_p f$ is a $k \times n$ matrix whose entries are the univariate $\mathbb{R}$-valued functions $\partial f_i / \partial x_j(p)$, $i = 1, \ldots, k$ and $j = 1, \ldots, n$.

The partial derivatives $\partial_j(p) := \partial / \partial x_j(p)$, $j = 1, \ldots, n$, form an $\mathbb{R}$-vector space basis for the *tangent space at p*; every *tangent vector $v(p)$ at p* is of the form

$$v(p) = \sum_{j=1}^{n} c_j \partial_j(p),$$

for scalars $c_j \in \mathbb{R}$, $j = 1, \ldots, n$. The tangent space at $p \in X$ is denoted by $T_p X$. The *tangent space of X* is then defined as

$$TX := \bigcup_{p \in X} T_p X.$$

Now suppose that $f : X \subseteq \mathbb{R}^n \to Y \subseteq \mathbb{R}^m$ is a differentiable function. One can "lift" f to a mapping $Tf : TX \to TY$ between tangent spaces via $T_p f : T_p X \to T_{f(p)} Y$, where $T_p f$ maps tangent vectors at $p \in X$ to tangent vectors at $f(p) \in Y$. More precisely,

$$\frac{\partial}{\partial x_j}(p) \overset{T_p f}{\longmapsto} \sum_{i=1}^{n} \frac{\partial f_i}{\partial x_j}(p) \frac{\partial}{\partial y_i}(f(p)),$$

where $\{\partial/\partial y_i(f(p)) \mid i = 1, \ldots, n\}$ is the tangent space basis induced by $\{\partial/\partial x_j(p) \mid j = 1, \ldots, m\}$.

Remark. The mapping Tf is an element of $\mathcal{L}(\mathbb{R}^n, \mathcal{L}(\mathbb{R}^n, \mathbb{R}^m))$ which is isomorphic to $\mathcal{L}(\mathbb{R}^n, \mathbb{R}^{m+n}))$.

Definition 4.5 Let $f : X \subseteq \mathbb{R}^n \to X$ be a bijective differentiable function. Then f is said to be a diffeomorphism (on X) iff its inverse f^{-1} is also differentiable (on X). A C^k-diffeomorphism is a bijective k-times continuously differentiable function whose inverse is also k-times continuously differentiable.

The next result is one of the most important theorems in ergodic theory.

Theorem 4.1 (Birkhoff's Ergodic Theorem)   *Suppose that $(X, \mathcal{B}(X), f, \mu)$ is a dynamical system and $\phi : X \to \mathbb{R}$ an integrable function, i.e., $\int_X \phi \, d\mu < \infty$. Then the time-average*

$$\frac{1}{k} \sum_{m=0}^{k-1} \phi(f^m(x)) \overset{k \to \infty}{\longrightarrow} \overline{\phi} \tag{4.4}$$

converges for a.e. $x \in X$. Furthermore, if $(X, \mathcal{B}(X), f, \mu)$ is ergodic then $\overline{\phi}$ equals the space average:

$$\overline{\phi} = \int_X \phi(x) \, d\mu(x). \tag{4.5}$$

■

The matrix version of Birkhoff's Ergodic Theorem is due to Y. I. Oseledec (cf. [146]).

Theorem 4.2 (Oseledec's Multiplicative Ergodic Theorem) *Let f be a diffeomorphism of $X \subseteq \mathbb{R}^n$ whose derivative is uniformly bounded and let*

μ be an invariant Borel probability measure on X. Then at a.e. $x \in X$, the tangent space $T_x X$ can be decomposed into a direct sum of subspaces

$$T_x X = \bigoplus_{j=1}^{r(x)} W_j(x),$$

such that

(a) $(T_x f)W_j(x) = W_j(f(x))$, $\forall j = 1, \ldots, r(x)$.

(b) $\forall x \in X \; \forall j = 1, \ldots, r(x) \; \exists \lambda_j(x) \in \mathbb{R} \; \forall w \in W_j(x)$:

$$\lim_{k \to \infty} \frac{1}{k} \log \|T_x f^{\pm 1} w\| = \pm \lambda_j(x). \tag{4.6}$$

(c) If $\theta_x(i,j)$ denotes the angle between $W_i(x)$ and $W_j(x)$ then

$$\lim_{k \to \infty} \frac{1}{k} \log |\sin(\theta_{f^{\pm k}(x)}(i,j))| = 0. \tag{4.7}$$

The numbers λ_j counted with multiplicity $\dim W_j(x)$ are called the *Lyapunov exponents* at $x \in X$.
If the dynamical system $(X, \mathcal{B}(X), f, \mu)$ is ergodic, then the Lyapunov exponents are independent of x. $\blacksquare$

Remarks.

1. The Lyapunov exponents measure the exponential growth of an infinitesimal volume element dV in the tangent space $T_x X$.

2. If f is an affine map, $T_x f$ is constant, and therefore the Lyapunov exponents are independent of x. The eigenvalues $\ell_i(k)$ of $(T_x f)^k = T_x^k f$ are then related to the Lyapunov exponents via

$$\lambda_i = \lim_{k \to \infty} \frac{1}{k} \log |\ell_i(k)|.$$

In order to predict the long-term behavior of a dynamical system, information has to be gathered through a series of experiments. A measure of the gain or loss of information is provided by the *entropy*. This invariant was first introduced in information theory by Shannon and in its present form by

Kolmogorov. A simple example from probability theory will motivate Kolmogorov's definition.

Suppose an experiment $\mathcal{E}$ that has k possible outcomes $E_1, \ldots, E_k$ occurring with probabilities $p_1, \ldots, p_k$ is performed. In order to predict *in advance* the outcome of $\mathcal{E}$, one has to associate a function $H(\mathcal{E})$ with $\mathcal{E}$ that measures the amount of uncertainty in the prediction. It has been proven that the only such function that agrees with intuition is — up to a multiplicative constant — given by

$$H(p_1, \ldots, p_k) = -\sum_{i=1}^{k} p_i \log p_i. \tag{4.8}$$

In order to define an entropy for a dynamical system, the concept of a *partition* of a set in $\mathcal{B}$ needs to be introduced.

Definition 4.6 Suppose that $(X, \mathcal{B}(X), f, \mu)$ is a dynamical system. A finite collection $\{E_1, \ldots, E_k\} \subseteq \mathcal{B}$ is called a partition of X iff

$$\sum_{i=1}^{k} E_i = X, \tag{4.9}$$

where Σ denotes the disjoint union of sets.

For two partitions $\mathcal{P}$ and $\mathcal{Q}$ of X, the *join* $\mathcal{P} \vee \mathcal{Q}$ is defined by

$$\mathcal{P} \vee \mathcal{Q} := \{P \cap Q \,|\, P \in \mathcal{P} \wedge Q \in \mathcal{Q}\}, \tag{4.10}$$

and the pre-image of $\mathcal{P}$ under f by

$$f^{-1}\mathcal{P} := \{f^{-1}P \,|\, P \in \mathcal{P}\}. \tag{4.11}$$

It is easy to see that $\mathcal{P} \vee \mathcal{Q}$ and $f^{-1}\mathcal{P}$ are again partitions of X. The k-fold join of partitions $\mathcal{P}_1 \vee \ldots \vee \mathcal{P}_k$ is written as $\bigvee_{i=1}^{k} \mathcal{P}_i$.

The preceding example can be related to the present setup by asking for a measure of uncertainty in the prediction of the location of a point $x \in X$. Giving partition $\mathcal{P}$ of X is equivalent to predicting the location of x in a specific element of $\mathcal{P}$. This gives rise to the definition of *entropy of the partition* $\mathcal{P}$ *of* X:

$$H(\mathcal{P}) := -\sum_{P \in \mathcal{P}} \mu P \log \mu P. \tag{4.12}$$

Now suppose $\mathcal{Q}$ is another partition of X. The *conditional entropy of* $\mathcal{Q}$ *given* $\mathcal{P}$ is defined by

$$H(\mathcal{Q}|\mathcal{P}) := H(\mathcal{P} \vee \mathcal{Q}) - H(\mathcal{P}). \tag{4.13}$$

Next the entropy for a dynamical system is defined. The basic idea is as follows: For a given $\varepsilon > 0$, choose a partition $\mathcal{P}$ of X whose elements have diameter less than or equal to ε. To predict the location of the iterate $f^k(x)$ of a point $x \in X$, one has to know to which element of $f^{-k}(\mathcal{P})$ x initially belonged. More generally, to predict the approximate location of the orbit $\{x, f(x), \ldots, f^k(x)\}$, one has to know the element of $\bigvee_{i=0}^k f^{-i}(x)$ initially containing x. Therefore, it is natural to define the *entropy of f with respect to the partition $\mathcal{P}$* by

$$h_\mu(f, \mathcal{P}) := \lim_{k \to \infty} H\left(f^{-(k+1)}\mathcal{P} \,\middle|\, \bigvee_{i=0}^k f^{-i}\mathcal{P} \right). \tag{4.14}$$

A rather lengthy calculation shows that

$$h_\mu(f, \mathcal{P}) = \lim_{k \to \infty} \frac{1}{k} H\left(\bigvee_{i=0}^k f^{-i}\mathcal{P} \right). \tag{4.15}$$

The advantage of Eq. (4.15) lies in its interpretation: The entropy $h_\mu(f, \mathcal{P})$ can be thought of as the average uncertainty in predicting the first k iterates of $x \in X$.

Finally, the entropy of a map is defined.

Definition 4.7 Let $(X, \mathcal{B}(X), f, \mu)$ be a dynamical system and let $\mathfrak{P}(X)$ denote the collection of all partitions $\mathcal{P}$ of X. Then the entropy $h_\mu(f)$ is defined by

$$h_\mu(f) := \sup_{\mathcal{P} \in \mathfrak{P}(X)} h_\mu(f, \mathcal{P}). \tag{4.16}$$

Suppose that μ and ν are two measures defined on the same σ-algebra $\mathcal{B}(X)$ of a space X. Recall that μ is said to be *absolutely continuous with respect to ν*, in symbols $\mu \ll \nu$, iff there exists a positive ν-measurable function $g : X \to X$ so that

$$\forall B \in \mathcal{B}(X) : \mu B = \int_B g(x) \, d\nu(x).$$

The next result is kown as *Pesin's Formula* ([152]).

Theorem 4.3 *Let $(X, \mathcal{B}(X), f, \mu)$ be a dynamical system, where X is compact and μ an invariant Borel probability measure. Assume that f is a C^2-diffeomorphism on X and that μ is absolutely continuous with respect to Lebesgue measure. Then*

$$h_\mu(f) = \int_X \left(\sum_{\lambda_i(x)>0} \lambda_i(x) \right) d\mu(x). \tag{4.17}$$

$\blacksquare$

Remark. Note that if $(X, \mathcal{B}(X), f, \mu)$ is ergodic, then Pesin's formula simplifies to

$$h_\mu(f) = \sum_{\lambda_i>0} \lambda_i. \tag{4.18}$$

Since partitions are a special class of covers, one might expect that there exists a relationship between the Hausdorff dimension of the invariant measure μ and the Lyapunov exponents and the entropy of f. The existence of such a relationship was established by L.-S. Young ([185]).

Theorem 4.4 *Let $(X, \mathcal{B}(X), f, \mu)$ be an ergodic two-dimensional dynamical system and let f be a C^2-diffeomorphism such that $f^\# \mu = \mu$, where μ is a Borel probability measure. Also, assume that the Lyapunov exponents of f satisfy $\lambda_1 \geq \lambda_2$. Then*

$$dim_H(\mu) = h_\mu(f) \left(\frac{1}{\lambda_1} - \frac{1}{\lambda_2} \right), \tag{4.19}$$

whenever the right side of Eq. (4.19) is not equal to 0/0. $\blacksquare$

In the proof of this theorem one uses a result that is worth mentioning.

Proposition 4.1 *Let $X \subseteq \mathbb{R}^n$ be μ-measurable with $\mu X > 0$. Let $B(x, r)$ denote the closed ball of radius $r > 0$ centered at $x \in X$. Assume that for every $x \in X$*

$$\underline{d} \leq \liminf_{r \to 0+} \frac{\log \mu B(x, r)}{\log r} \leq \limsup_{r \to 0+} \frac{\log \mu B(x, r)}{\log r} \leq \bar{d}. \tag{4.20}$$

Then

$$\underline{d} \leq \dim_H X \leq \bar{d}.$$

$\blacksquare$

Remarks.

1. There is no loss of generality if in the preceding limits the continuous variable r is replaced by a sequence $\{r_\nu\}_{\nu \in \mathbb{N}}$ with $\lim r_\nu = 0$ and $\lim \log r_{\nu+1} / \log r_\nu = 1$.

2. It can be shown ([185]) that if μ is a Borel probability measure, X compact, and

$$\lim_{r \to 0+} \frac{\log B(x,r)}{\log r} = d, \qquad (4.21)$$

for μ-a.e. $x \in X$, then

$$\dim_H \mu = \dim_\beta \mu = d, \qquad (4.22)$$

where $\dim_\beta \mu$ denotes the box dimension of the invariant measure μ. This dimension is defined as follows:

$$\dim_\beta \mu := \sup_{\delta \to 0+} \inf \{\dim_\beta Y \,|\, Y \subseteq X \wedge \mu Y \geq 1 - \delta\}. \qquad (4.23)$$

Theorem 4.4 gives a relationship between the Hausdorff dimension of the invariant measure μ and the Lyapunov exponents and entropy of f. However, there exists another notion of dimension associated with the invariant measure μ that relates it to the box dimension of the attractor A of a dynamical system. This dimension, which was introduced by J. Kaplan and J. A. Yorke in [103], is called the *Lyapunov dimension* of the invariant measure μ on A. The definition of Lyapunov dimension can be motivated by looking at a simple example.

Suppose that $(X, \mathcal{B}(X), f, \mu)$ is an ergodic n-dimensional dynamical system and that f is an affine map leaving the Borel probability measure μ invariant. Assume that the Lyapunov exponents of f are such that

$$\lambda_1 \geq \lambda_2 \geq \ldots \geq \lambda_n,$$

and that $\lambda_1 > 0$. Let A be the attractor of the dynamical system. To estimate the box dimension of A, one has to find a minimal cover of A consisting of n-cubes of side $\varepsilon > 0$. Denote by $N_\varepsilon(A)$ the cardinality of this cover. Applying f k-times to a cube in this cover yields a parallelepiped with sides $(\exp \lambda_1 k)\varepsilon, \ldots, (\exp \lambda_n k)\varepsilon$. Hence, the volume of this ε-cube has changed by a factor of $(\exp(\lambda_1 + \ldots + \lambda_n)k)$. Now let k_0 be the largest integer such that $\lambda_1 + \ldots + \lambda_{k_0} > 0$.

Now consider a cover of A by cubes of side $\exp(\lambda_{k_0+1})\varepsilon$. Then every $(\exp \lambda_1 k)\varepsilon \times \ldots \times (\exp \lambda_n k)\varepsilon$-parallelepiped can be covered by approximately

$$\frac{\exp(\lambda_1 + \ldots + \lambda_n)k}{\exp \lambda_{k_0+1} k} = \exp(\lambda_1 + \ldots + \lambda_n - n\lambda_{k_0+1})k$$

such cubes. Since $\lambda_{k_0+1} \geq \lambda_{k_0+2} \geq \ldots \geq \lambda_n$, it can also be covered by approximately

$$\exp(\lambda_1 + \ldots + \lambda_{k_0} - k_0\lambda_{k_0})k$$

cubes of side $(\exp \lambda_{k_0+1} k)\varepsilon$. Denoting the cardinality of the associated minimal cover of A by $N_{(\exp \lambda_{k_0+1} k)\varepsilon}(A)$, one thus obtains

$$N_{(\exp \lambda_{k_0+1} k)\varepsilon}(A) \approx (\exp(\lambda_1 + \ldots + \lambda_{k_0} - k_0\lambda_{k_0})k)N_\varepsilon(A).$$

If it is *assumed* that $N_\varepsilon(A) \propto \varepsilon^{-d_\beta}$, where $d_\beta := \dim_\beta A$, then

$$((\exp \lambda_{k_0+1} k)\varepsilon)^{-d_\beta} \approx (\exp(\lambda_1 + \ldots + \lambda_{k_0} - k_0\lambda_{k_0})k)\varepsilon^{-d_\beta}.$$

Taking logarithms and solving for d_β yields

$$d_\beta \approx k_0 - \frac{\lambda_1 + \ldots + \lambda_{k_0}}{\lambda_{k_0+1}} = k_0 + \frac{\lambda_1 + \ldots + \lambda_{k_0}}{|\lambda_{k_0+1}|}, \qquad (4.24)$$

since $\lambda_{k_0+1} < 0$. The right side of Eq. (4.24) is called the Lyapunov dimension of μ.

Definition 4.8 Let $(X, \mathcal{B}(X), f, \mu)$ be an ergodic n-dimensional dynamical system. Assume that μ is a Borel probability measure μ and that $f^\# \mu = \mu$. Let $\lambda_1 \geq \lambda_2 \geq \ldots \geq \lambda_n$ be the Lyapunov exponents of f. Let $k_0 := \max_{1 \leq k \leq n}\{\lambda_1 + \ldots + \lambda_k > 0\}$. If $1 \leq k_0 < n$, the Lyapunov dimension of μ is defined by

$$\dim_\Lambda \mu := k_0 + \frac{\lambda_1 + \ldots + \lambda_{k_0}}{|\lambda_{k_0+1}|}. \qquad (4.25)$$

If no such k_0 exists, $\dim_\Lambda \mu := 0$, and if $k_0 = n$, then $\dim_\Lambda \mu := n$.

Now Theorem 4.4 can be brought into the picture. Suppose that a two-dimensional ergodic dynamical system defined on a compact $X \subseteq \mathbb{R}^2$ is given and that $\lambda_1 > 0$. Then, under the hypotheses of Theorem 4.4,

$$\dim_\Lambda \mu = 1 + \dim_H \mu = 1 + \dim_\beta \mu. \qquad (4.26)$$

The next goal is to relate the attractor of an IFS to the attractor of a dynamical system. For $X \in \mathfrak{H}(\mathbb{R}^n)$, let $(X, \mathbf{w}, \mathbf{p})$ be a hyperbolic IFS with probabilities. Its attractor is denoted by F and the $\mathbf{p}$-balanced measure by μ. There is a natural way of representing $(X, \mathbf{w}, \mathbf{p})$ as a dynamical system. Namely, let

$$X^* := X \times [0, 1],$$

$$\mu^* := \mu \times m,$$

where m denotes uniform Lebesgue measure on $[0, 1]$. Furthermore, let $\mathcal{B}^*(X^*)$ be the smallest σ-algebra containing $\mathcal{B}(X)$ and the Borel sets on $[0, 1]$. Finally, define a map $f^* : X^* \to X^*$ by

$$f^*(x, y) := \begin{cases} \left(w_i(x), p_i^{-1}(y - \mathsf{S}_{j=1}^{i-1})\right), & (x, y) \in X \times [\mathsf{S}_{i-1}, \mathsf{S}_i), \\ & i = 1, \dots, N-1 \\ \left(w_N(x), p_N^{-1}(y - \mathsf{S}_{j=1}^{N-1})\right), & (x, y) \in X \times [\mathsf{S}_{N-1}, \mathsf{S}_N], \end{cases} \tag{4.27}$$

where $\mathsf{S}_i := \sum_{j=1}^i p_j$ and $\mathsf{S}_0 := 0$. For similar constructions see [69, 151]. The result below is almost immediate.

Proposition 4.2 *The quadruple* $(X^*, \mathcal{B}^*(X^*), \mu^*, f^*)$ *is a dynamical system.* ∎

Note that the attractor F^* of $(X^*, \mathcal{B}^*(X^*), \mu^*, f^*)$ is $\mathsf{F} \times [0, 1]$.

For the remainder of this section it is assumed that $\mathbf{w} := \{\mathsf{S}_i := s_i A_i : i = 1, \dots, N\} \subseteq \mathcal{S}^*(X, X)$ and that $\mathbf{w}$ satisfies OSC.

Under the preceding assumption, the tangent map $T_{x^*} f^*$ exists for μ^*-a.e. $x^* \in X^*$ and equals a constant:

$$T_{x^*} f^* = (s_i A_i) \oplus \left(\frac{1}{p_i}\right),$$

for $x^* \in X \times [\mathsf{S}_{i-1}, \mathsf{S}_i)$, $i \in \{1, \dots, N-1\}$, respectively $x^* \in X \times [\mathsf{S}_{N-1}, \mathsf{S}_N]$. Here $\oplus : \mathsf{M}_{a,b} \times \mathsf{M}_{c,d} \to \mathsf{M}_{a+c,b+d}$ is defined by

$$A \oplus B := \begin{pmatrix} A & 0 \\ 0 & B \end{pmatrix},$$

where $\mathsf{M}_{r,s}$ denotes the algebra of all $r \times s$ matrices $(a, b, c, d, r, s \in \mathbb{N})$.

Now let $k \in \mathbb{N}$. The kth iterate of the tangent map, $T_{x^*}^k f^*$, is given by

$$T_{x^*}^k f^* = \left(\prod_{j=1}^k s_j^{k_j} \prod_{j=1}^k A_j^{k_j}\right) \oplus \left(\frac{1}{\prod_{j=1}^k p_j^{k_j}}\right),$$

where k_j denotes the number of times map w_i is chosen. The moduli of the eigenvalues of $T_{x*}^k f^*$ are

$$|\ell_1| = \prod_{i=1}^{N} \left(\frac{1}{p_i}\right)^{k_i}$$

and

$$|\ell_2| = \ldots = |\ell_{n+1}| = \prod_{i=1}^{N} s_i^{k_i}.$$

Using the Law of Large Numbers (Theorems 1.11 and 1.12) to find the Lyapunov exponents yields

$$\lambda_1 = -\sum_i^N p_i \log p_i > 0,$$

and

$$\lambda_2 = \ldots = \lambda_{n+1} = \sum_i^N p_i \log s_i > 0.$$

The Lyapunov dimension of the invariant measure μ^* of the attractor F^* is now given by

$$\dim_{\Lambda(\mathbf{p})} \mu^* = k + \frac{\sum_{i=1}^{N} p_i \log s_i^{k-1}/p_i}{\sum_{i=1}^{N} p_i \log 1/s_i} \tag{4.28}$$

Here k is the largest integer such that $\sum_{i=1}^{N} s_i^k \leq 1$, which implies that $\lambda_1 + \ldots + \lambda_{k+1} \leq 0$ and $\lambda_1 + \ldots + \lambda_k > 0$.

In order to proceed the following lemma is needed.

Lemma 4.1 *Let* $g : [0,1]^N \to \mathbb{R}$, $N \in \mathbb{N}$, *be defined by*

$$g(x) := \frac{\sum_{i=1}^{N} x_i \log a_i/x_i}{\sum_{i=1}^{N} x_i \log 1/b_i},$$

where $0 < a_i, b_i \leq 1$, $i = 1, \ldots, N$ *and* $\sum_{i=1}^{N} a_i > 1$. *Assume that* $\sum_{i=1}^{N} x_i = 1$. *Then there exists an* $\overset{\circ}{x}$ *in the interior of* $[0,1]^N$ *at which* g *attains its maximum value* $\overset{\circ}{g}$. *Furthermore, this maximum value satisfies* $\sum_{i=1}^{N} a_i b_i^{\overset{\circ}{g}} = 1$.

Proof by calculus.   ∎

The next result is an immediate consequence of Lemma 4.1 and the preceding arguments.

Theorem 4.5 *Let* $(X^*, \mathcal{B}^*(X^*), \mu^*, f^*)$ *be the dynamical system associated with the IFS* $(X, \mathbf{w}, \mathbf{p})$. *Then there exists a set* $\overset{\circ}{\mathbf{P}}$ *of probabilities that maximizes the Lyapunov dimension* $\dim_{\Lambda(\mathbf{p})} \mu^*$. *This maximized Lyapunov dimension* $\overset{\circ}{\Lambda}$ *satisfies*

$$\overset{\circ}{\Lambda} = 1 + \dim_H \mathsf{F} = 1 + \dim_\beta \mathsf{F}. \tag{4.29}$$

Proof. Use Lemma 4.1 with $a_i := s_i^{k-1}$ and $b_i := s_i$. ∎

Remarks.

1. Clearly, since $\mathsf{F}^* = \mathsf{F} \times [0, 1]$,

$$\dim_H \mathsf{F}^* = \dim_\beta \mathsf{F}^* = \overset{\circ}{\Lambda}.$$

2. The set $\overset{\circ}{\mathbf{P}}$ of probabilities that maximizes the Lyapunov dimension is given by $\overset{\circ}{p}_i = s_i^d$, where d stands for both $\dim_H \mu^*$ and $\dim_\beta \mu^*$.

Theorem 4.6 *Let* $(X, \mathbf{w}, \mathbf{p})$ *be a hyperbolic IFS with probabilities and let* $(X^*, \mathcal{B}^*(X^*), \mu^*, f^*)$ *be its associated dynamical system. Let* $\overset{\circ}{\mathbf{P}}$ *be the unique set of probabilities that maximizes the Lyapunov dimension of* $\mu^* = \mu^*(\mathbf{p})$. *Then the box and Hausdorff dimension of the* $\overset{\circ}{\mathbf{P}}$-*balanced measure supported on the attractor* F *of* $(X, \mathbf{w}, \mathbf{p})$ *equals* $\Lambda(\mu^*(\overset{\circ}{\mathbf{P}})) - 1$.

Proof. By Remark 2 following Proposition 4.1, it suffices to show that

$$\lim_{r \to 0+} \frac{\log B(x, r)}{\log r} = \Lambda(\mu^*(\overset{\circ}{\mathbf{P}})) - 1$$

for μ-a.e. $x \in X$.

Let $m \in \mathbb{N}$ and let $\mathbf{i}(m) \in \Sigma_m$. Remark 2 above implies $\overset{\circ}{p}_i = s_i^d$, and thus

$$\mu w_{\mathbf{i}(m)} \mathsf{F} \;\geq\; \prod_{k=1}^{m} \overset{\circ}{p}_{i_k} \mu(w_{\mathbf{i}(m)}^{-1} \, w_{\mathbf{i}(m)} \mathsf{F})$$

$$= \; \prod_{k=1}^{m} \overset{\circ}{p}_{i_k} \mu \mathsf{F} = \prod_{k=1}^{m} \overset{\circ}{p}_{i_k}.$$

Now suppose $x \in \mathsf{F}$. Then by Theorem 2.4 there exists a code $\mathbf{i} \in \Sigma$ such that $x = \gamma(\mathbf{i})$. Let $0 < r < 1$ be given and let $q = q(r)$ be the least integer such that $w_{\mathbf{i}(q)}(\mathsf{F}) \subseteq B(x, r)$. Order the s_i in the following way:

$$s_1 \leq s_2 \leq \ldots \leq s_N.$$

Then

$$\left(\prod_{k=1}^{q} s_{i_k}\right) |\mathsf{F}| \geq r s_1,$$

for if the reverse inequality held then

$$\left(\prod_{k=1}^{q} s_{i_k}\right) |\mathsf{F}| < r s_1 \leq r s_{i_q},$$

contradicting the choice of q. Hence,

$$\mu B(x, r) < \mu w_{\mathbf{i}(q)} \mathsf{F} \leq \prod_{k=1}^{q} \overset{\circ}{p}_{i_k}.$$

Thus,

$$\frac{\log \mu B(x, r)}{\log r} \geq \frac{\log \prod_{k=1}^{q} \overset{\circ}{p}_{i_k}}{\log r} = \frac{\log \prod_{k=1}^{q} \overset{\circ}{p}_{i_k}}{\log \prod_{k=1}^{q} s_{i_k}} \frac{\log \prod_{k=1}^{q} s_{i_k}}{\log r}.$$

But, since

$$\prod_{k=1}^{q} s_{i_k} \geq (s_1 r)/|\mathsf{F}|,$$

one obtains

$$\frac{\log \prod_{k=1}^{q} s_{i_k}}{\log r} \leq 1 + \frac{\log(s_1 r)/|\mathsf{F}|}{\log r}.$$

Therefore,

$$\lim_{r \to 0+} \frac{\log \mu B(x, r)}{\log r} \leq \lim_{r \to 0+} \left(\frac{\log \prod_{k=1}^{q} \overset{\circ}{p}_{i_k}}{\log \prod_{k=1}^{q} s_{i_k}}\right) \left(1 + \frac{\log(s_1 r)/|\mathsf{F}|}{\log r}\right).$$

As $q \to \infty$ as $r \to 0+$, Theorems 1.11 and 1.12 again imply

$$\lim_{r \to 0+} \frac{\log \prod_{k=1}^{q} \overset{\circ}{p}_{i_k}}{\log \prod_{k=1}^{q} s_{i_k}} = \lim_{q \to \infty} \frac{\sum_{i=1}^{N} (q_i/q) \log \overset{\circ}{p}_i}{\sum_{i=1}^{N} (q_i/q) \log s_i} = \frac{\sum_{i=1}^{N} \overset{\circ}{p}_i \log \overset{\circ}{p}_i}{\sum_{i=1}^{N} \overset{\circ}{p}_i \log s_i}$$

$$= \Lambda(\mu^*(\overset{\circ}{\mathbf{P}})) - 1.$$

Here q_i denotes the number of times maps w_i have been applied. Finally,

$$\lim_{r \to 0+} \frac{\log \mu B(x, r)}{\log(r)} \leq \Lambda(\mu^*(\overset{\circ}{\mathbf{P}})) - 1.$$

To show the opposite inequality, one has to use Lemma 3.2. Denote by G the open set whose existence is guaranteed by OSC. Suppose that G contains a ball of radius ρ_1 and is contained in a ball of radius ρ_2. Let $r > 0$ be given. For each code $\mathbf{j} = (j_1 \, j_2 \, \ldots) \in \Sigma$ choose the least integer q such that $s_1 r \le s_{j_1} r \le \ldots \le s_{j_q} \le r$. The set $\{\mathbf{j}(q) = (j_1 \, j_2 \ldots j_q) \in \Sigma \mid s_1 r \le s_{j_1} r \le \ldots \le s_{j_q} \le r\}$ of all such codes is denoted by $\mathcal{J}$. Note that if $\mathbf{i} = (i_1 \, i_2 \, \ldots) \in \Sigma$, there exists exactly one $\mathbf{j}(q) \in \mathcal{J}$ such that $i_1 = j_1$, $i_2 = j_2$, ..., $i_q = j_q$. This then implies that $\{w_{\mathbf{j}(q)}(G) \mid \mathbf{j} \in \mathcal{J}\}$ is a finite disjoint collection of open sets. Furthermore, each $w_{\mathbf{j}(q)}(G)$ contains a ball of radius $s_{\mathbf{j}(q)} \rho_1$ and hence of radius $s_1 r \rho_1$, and is contained in a ball of radius $s_{\mathbf{j}(q)} \rho_2$ and thus in one of radius $r \rho_2$. It follows from Lemma 3.2 that at most $(1 + 2\rho_2)^n (s_1 \rho_1)^{-n}$ of the closures of the $w_{\mathbf{j}(q)}(G)$ can meet $B(x, r)$, and therefore at most $(1 + 2\rho_2)^n (s_1 \rho_1)^{-n}$ of the $w_{\mathbf{j}(q)}(\mathsf{F})$ can possibly intersect $B(x, r)$.

Now, for all $\mathbf{j}(q) \in \mathcal{J}$,

$$\mu w_{\mathbf{j}(q)}(\mathsf{F}) = \prod_{k=1}^{q} \overset{\circ}{p}_{j_k} = \prod_{k=1}^{q} s_{j_k}^d \le r^{d_\Lambda},$$

where $d_\Lambda := \Lambda(\mu^*(\overset{\circ}{\mathbf{p}})) - 1$. Thus,

$$\lim_{r \to 0+} \frac{\log B(x, r)}{\log r} \ge \frac{\log(1 + 2\rho_2)^n (s_1 \rho_1)^{-n} r^{d_\Lambda}}{\log r} = d_\Lambda.$$

Part II

Fractal Functions and Fractal Surfaces

Chapter 5

Fractal Function Construction

In this chapter a special class of continuous functions is considered. These functions are referred to as *fractal functions*, since their graphs usually have non-integral dimension. It is shown that these fractal functions may be used for interpolation and approximation purposes, and are in this way analogous to splines. In Section 2.1 examples of classical fractal functions have already been encountered. Here a general construction based on a *Read-Bajraktarević operator* acting on $L^\infty(X, Y)$ is presented. The Read-Bajraktarević operator provides a natural framework for the description of fractal functions in terms of hyperbolic IFSs. Several other related constructions are studied and set into perspective. Properties of these functions are presented and discussed. Our focus is primarily on univariate $\mathbb{R}$-valued fractal functions defined on subsets of $\mathbb{R}$, although one section is devoted to *Peano curves* and their relation to so-called *hidden variable fractal functions*.

5.1 The Read-Bajraktarević Operator

The fractal functions considered in this section are fixed points of a contractive operator. This operator has its origin in the theory of functional equations and was first investigated by Read ([154]) and then later by Bajraktarević ([4]). Lately, these operators have been used by Dubuc ([55, 56]) and Bedford ([19]) to construct fractal functions.

Definition 5.1 Let X be a set and Y a complete metric space. Suppose that $b : X \to X$ is an arbitrary function on X. Furthermore, assume that mappings $v(x, \, \cdot \,) : Y \to Y$, $x \in X$, are given. The Read-Bajraktarević operator associated with b and $\{v(x, \, \cdot \,)|\, x \in X\}$ is defined by

$$\Phi : L^{\infty}(X, Y) \to Y^{X}, \quad f \mapsto v(\, \cdot \,, f(b(\, \cdot \,))). \tag{5.1}$$

In order to construct fractal functions, some additional conditions have to be put on the mappings $v(x, \, \cdot \,)$. It is required that $v(x, \, \cdot \,)$ satisfy:

(A) For all $x \in X$ the mapping $v(x, \, \cdot \,) \in \mathrm{Lip}^{(<1)}(Y)$.

(B) There exists a family $\mathfrak{F}(X, Y)$ of continuous functions from X into Y such that for all $\mathfrak{f} \in \mathfrak{F}(X, Y)$ the function $v(\, \cdot \,, \mathfrak{f} \circ b(\, \cdot \,))$ is a member of $\mathfrak{F}(X, Y)$.

Then the following result holds:

Theorem 5.1 *If condition (A) holds, then Φ is a contractive operator on $L^{\infty}(X, Y)$, and hence its fixed point is a unique bounded function $f_{\Phi} : X \to Y$. If condition (B) also holds, then $f_{\Phi} \in C(X, Y)$.*

Proof. The operator Φ is well-defined and by condition (A) contractive on $L^{\infty}(X, Y)$. Hence its unique fixed point is an element of $L^{\infty}(X, Y)$.
If also condition (B) holds, let $\mathfrak{f}_{0} \in \mathfrak{F}(X, Y)$ and define $\mathfrak{f}_{n} := \Phi \mathfrak{f}_{n-1}$, $n \in \mathbb{N}$. Thus, by the Banach Fixed-Point Theorem, $\lim_{n \to \infty} \mathfrak{f}_{n} = f_{\Phi} \in C(X, Y)$. ∎

Now suppose X is a complete metric space and $u_{i} : X \to X$ is a given collection of N bijections such that $\{X_{i} := u_{i}X\,|\, i = 1, \ldots, N\}$ is a partition of X, i.e., $\bigcup_{i=1}^{N} X_{i} = X$ and $\overset{\circ}{X}_{i} \cap \overset{\circ}{X}_{j} = \emptyset$, for $i \neq j$. Furthermore, assume that N uniform contractions $v_{i}(x, \, \cdot \,) : Y \to Y$, $x \in X$, are given. A function $b : X \to X$ is defined by

$$b(x) := \sum_{i=1}^{N} u_{i}^{-1}(x) \chi_{X_{i}}(x), \tag{5.2}$$

and a mapping $v(x, \, \cdot \,) : Y \to Y$, $i = 1, \ldots, N$, by

$$v(x, \, \cdot \,) := \sum_{i=1}^{N} v_{i}(x, \, \cdot \,) \chi_{X_{i}}(x). \tag{5.3}$$

Clearly, $v(x, \, \cdot \,)$ satisfies condition (A) above. Instead of requiring that (B) hold, the following stronger condition is needed:

(B⁰) There exists a family $\mathfrak{F}(X,Y)$ of continuous functions from X into Y such that for all $\mathfrak{f} \in \mathfrak{F}(X,Y)$ the functions $v_i(\,\cdot\,,\mathfrak{f} \circ b(\,\cdot\,))$ are members of $\mathfrak{F}(X,Y)$ and

$$v_i(u_i^{-1}(x), \mathfrak{f} \circ u_i^{-1}(x)) = v_{i+1}(u_{i+1}^{-1}(x), \mathfrak{f} \circ u_{i+1}^{-1}(x)), \qquad (5.4)$$

for $x \in X_i \cap X_{i+1}$.

Let s_i denote the contractivity constant of $v_i(x, \cdot)$ and let $s := \max\{s_i \mid i = 1, \ldots, N\}$. Let $B(X,Y)$ be the complete metric space of all bounded functions $f : X \to Y$ endowed with the metric $d : B(X,Y) \times B(X,Y) \to \mathbb{R}$, given by

$$d(f,g) := \sup_{x \in X} d_Y(f(x), g(x)). \qquad (5.5)$$

Theorem 5.2 *Let b and $v(x, \cdot)$ be defined as in (5.2) and (5.3), respectively. Suppose that $v(x, \cdot)$ satisfies condition (B⁰). Let Φ be the Read-Bajraktarević operator associated with b and $v(x, \cdot)$. Then Φ is contractive on $B(X,Y)$ and its unique fixed point is an element of $C(X,Y)$.*

Proof. Let $f \in B(X,Y)$. Condition (B⁰) implies that Φf is well-defined on X. Furthermore, if $f, g \in B(X,Y)$, then

$$
\begin{aligned}
d_Y(\Phi f(x), \Phi g(x)) \;&\leq\; d_Y(v(x, f(b(x))), v(x, g(b(x)))) \\[2mm]
&\leq\; s\, d_Y(f(b(x)), g(b(x))) \\[2mm]
&\leq\; s\, d(f,g).
\end{aligned}
$$

Thus, $d(\Phi f, \Phi g) \leq s\, d(f,g)$. The continuity of f_Φ follows from Theorem 5.1 and (B⁰). ∎

The unique fixed point $f_\Phi \in C(X,Y)$ of the Read-Bajraktarević operator Φ is called a *fractal function*. This terminology has its origin in [8] and refers to the fact that the graph of f_Φ is in general a fractal set. The metric space (endowed with the sup-norm) of all fractal functions is denoted by $\mathcal{F}(X,Y)$.

Example 5.1 Let $X := [0,1]$ and let $Y \in \mathfrak{H}(\mathbb{R})$. Define functions $u_i : [0,1] \to [0,1]$ by

$$u_i(x) := (1/2)(-1)^{i-1}x + (i-1), \qquad i = 1, 2.$$

Let $s \in (-1, 1)$ be fixed. Let $\lambda_1, \lambda_2 \in C([0, 1])$ satisfy $\lambda_1(1) = \lambda_2(1)$. Define mappings $v_i(x, \cdot) : [0, 1] \times Y \to Y$ by

$$v_i(x, y) := \lambda_i(x) + sy, \quad i = 1, 2.$$

It is easy to verify that $\{u_i[0, 1] | \, i = 1, \ldots, N\}$ is a partition of $[0, 1]$ and that $v(x, \cdot)$ satisfies condition (B^0) for all $\mathfrak{f} \in C([0, 1], Y)$. Hence the associated Read-Bajraktarević operator defines a continuous fractal function on $[0, 1]$. Fig. 5.1 shows the graph of such a fractal function for $\lambda_1(x) := x$, $\lambda_2(x) := x^2$, and $s := 3/4$.

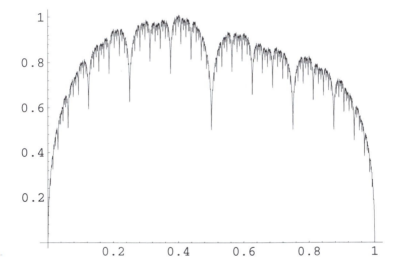

Figure 5.1: The graph of the fractal function defined in Example 5.1.

The function in Example 5.1 represents an important class of fractal functions. These fractal functions will be encountered later in connection with *wavelet expansions*. It is therefore worthwhile now to introduce this class of fractal functions.

***Example* 5.2** Let $N \in \mathbb{N} \setminus \{1\}$, let $X := [0, 1]$, and let $Y := \mathbb{R}$. For a given collection of knots $\{0 = x_0 < x_1 < \ldots < x_N = 1\}$, contractive homeomorphisms $u_i : X \to X$, $i = 1, \ldots, N$, are defined by

$$u_i^{-1}(x) = x_N, \quad u_{i+1}^{-1}(x) = x_0,$$

for $x \in X_i \cap X_{i+1}$ and $i \in \{1, \ldots, N\}$. Let $\lambda_i \in \mathrm{Lip}(X)$ and let $s \in (-1, 1)$. Define mappings $v_i(x, \cdot) : Y \to Y$ by

$$v_i(x, y) := \lambda_i(x) + sy, \tag{5.6}$$

for $i \in \{1, \ldots, N\}$. Furthermore, require that

$$\lambda_i \circ u_i^{-1} = \lambda_{i+1} \circ u_{i+1}^{-1}, \tag{5.7}$$

for $x \in X_i \cap X_{i+1}$. Clearly, conditions (A) and (B^0) are satisfied for all $\mathfrak{f} \in \mathfrak{F}(X, Y) := \{f \in C(X, Y) \mid f(x_0) = f(x_N)\}$. Hence, the Read-Bajraktarević operator Φ, now given explicitly by

$$\Phi f = \lambda_i \circ u_i^{-1} + s \, f \circ u_i^{-1}, \quad x \in X_i, \tag{5.8}$$

has as its unique fixed point a continuous fractal function f_Φ with the property $f_\Phi(x_0) = f_\Phi(x_N)$.

Remark. *Any continuous function g with $g(x_0) = g(x_N)$ is the fixed point of a Φ: Set $s = 0$ and $\lambda_i := g \circ u_i$. In other words, every such continuous function is a* fractal *function.*

The free parameter s in the preceding example can be chosen so that the resulting fractal function is *smooth*. This is shown for a particular choice of maps λ_i.

Example 5.3 Let $X := [0, 1]$ and let $N := 2$. Define mappings $u_i : [0, 1] \to [0, 1]$ by $u_i(x) := 1/2(x + i - 1)$, and maps $v_i(x, y) : [0, 1] \times \mathbb{R} \to \mathbb{R}$ by $v_i(x, y) = 1/2(-1)^{i-1}(x - i + 1) + sy$, $i = 1, 2$. Now let us choose $s := 1/4$. The Read-Bajraktarević operator Φ — if defined on $\{f \in C(X, Y) \mid f(0) = f(1) = 0\}$ — is then given by

$$\Phi f (x) = \begin{cases} x + 1/4 f(2x) & x \in [0, 1/2] \\ 1 - x + 1/4 f(2x - 1) & x \in [1/2, 1]. \end{cases}$$

It is easy to verify that the unique fixed point of this Φ operator is the polynomial $f_\Phi(x) = 2x(1 - x)$, a smooth function on $[0, 1]$.

The fact that s is a free parameter will be used again in Chapter 7. There it will be shown that an appropriate choice of s will yield so-called continuous, compactly supported, and orthogonal (fractal) wavelets on $\mathbb{R}$.

Next an IFS whose attractor is the graph G of a fractal function $f_\Phi \in C(X, Y)$ is constructed. For the remainder of this section it is assumed that the bijections $u_i : X \to X$ are uniformly contractive, and that the mappings $v_i(x, \cdot)$ given in (5.3) not only satisfy (B^0) but also

(**C**) The mappings $v_i(\cdot, y)$ are uniformly Lipschitz.

Define maps $w_i : X \times Y \to X \times Y$ by

$$w_i(x, y) := (u_i(x), v_i(x, y)), \tag{5.9}$$

for $i = 1, \dots, N$.

Let a_i denote the contractivity constant of u_i and let $a := \max\{a_i | \; i = 1, \dots, N\}$. Denote the uniform Lipschitz constant of the $v_i(\, \cdot \,, y)$ by L. Now let $\theta := (1 - a)/2L$, and define a metric d_θ on $X \times Y$ by

$$d_\theta := d_X + \theta \, d_Y, \tag{5.10}$$

where d_X and d_Y denotes the metric on X and Y, respectively. Then the following result holds:

Theorem 5.3 *The pair $(X \times Y, \mathbf{w})$ is a hyperbolic IFS in the metric d_θ and its attractor is the graph of a continuous fractal function $f_\Phi : X \to Y$.*

Proof. Clearly, $(X \times Y, \mathbf{w})$ is an IFS. To show hyperbolicity note that for $(x, y), (x', y') \in X \times Y$ and $i = 1, \dots, N$, one has

$$d_\theta(w_i(x, y), w_i(x', y')) \;=\; d_X(u_i(x), u_i(x')) + \theta \, d_Y(v_i(x, y), v_i(x', y'))$$

$$\begin{aligned} \leq \;\; & a \, d_X(x, x') + \theta \, d_Y(v_i(x, y), v_i(x', y)) \\ & + \theta \, d_Y(v_i(x', y), v_i(x, y')) \end{aligned}$$

$$\leq \;\; (a + \theta \, L) \, d_X(x, x') + \theta \, s \, d_Y(y, y')$$

$$\leq \;\; q \, d_\theta((x, y), (x', y')),$$

where $q := \max\{a + \theta \, L, s\} < 1$.

Denote the unique attractor of $(X \times Y, \mathbf{w})$ by A. Note that $(X, \mathbf{u})$ is also a hyperbolic IFS whose unique attractor is X. The graph G of f_Φ is also an attractor of $(X \times Y, \mathbf{w})$, for

$$\begin{aligned} \mathfrak{w}(G) \;\; &= \;\; \bigcup_{i=1}^{N} w_i(G) \\[2mm] &= \;\; \bigcup_{i=1}^{N} w_i(\{(x, f_\Phi(x)) | \, x \in X\}) \\[2mm] &= \;\; \bigcup_{i=1}^{N} \{(u_i(x), v_i(x, f_\Phi(x))) | \, x \in X\} = G. \end{aligned}$$

Here the fact that f_Φ is the fixed point of the Read-Bajraktarević operator Φ, or equivalently, that $v_i(x, f_\Phi(x)) = f_\Phi(u_i(x))$, for $x \in X$, was used. Hence, $A = G$. ∎

Remark. If $\text{proj}_X : X \times Y \to X$ denotes the orthogonal projection operator $\text{proj}_X(x, y) := x$, then $\text{proj}_X A = X$. The imposed conditions on u_i and v_i guarantee that A is the graph of a function $f_\Phi : X \to Y$. Note that X itself may be a fractal set. This setting is studied in more detail in Section 5.5.

Let Σ be the code space associated with the hyperbolic IFS $(X \times Y, \mathbf{w})$ on $(X \times Y, d_\theta)$, and let $E \in \mathfrak{H}(X \times Y)$ be such that $\mathfrak{w}(E) \subseteq E$ (such a set exists by Proposition 2.4). Suppose that $f \in \mathcal{F}(X, Y)$ and its graph G is entirely contained within E. If $\mathbf{i}(k) \in \Sigma$, $k \in \mathbb{N}$, is the initial segment of $\mathbf{i} \in \Sigma$, then by Theorem 2.4,

$$G \ni (x, f(x)) = \lim_{k \to \infty} \mathfrak{w}_{\mathbf{i}(k)} E = \gamma(\mathbf{i}).$$

Here points on G were identified with codes $\mathbf{i}$ in Σ. Recall that, for a fixed $k \in \mathbb{N}$, Σ_k denotes the set of all codes $\mathbf{i} \in \Sigma$ of length k. It follows then immediately from the preceding arguments that $\bigcup_{k=1}^{\infty} \{\mathfrak{w}_{\mathbf{i}(k)} \,|\, \mathbf{i}(k) \in \Sigma_k\} = G$. Thus, for any $k \in \mathbb{N}$, $\{\mathfrak{w}_{\mathbf{i}(k)} \,|\, \mathbf{i}(k) \in \Sigma_k\}$ is a partition of G. This partition is depicted in Fig. 5.3 for $k = 1, \ldots, 6$.

Next a subset of $\mathcal{F}(X, Y)$ consisting of fractal functions that satisfy a certain *interpolation property* is defined. M. F. Barnsley introduced this class of functions in [8] and called them *fractal interpolation functions*.

Let $\{x_0, x_1, \ldots, x_N\}$ be a given set of $N + 1$ distinct points in X and let $\{y_0, y_1, \ldots, y_N\}$ be a set of points in Y. The collection $\Delta := \{(x_0, y_0),$ $(x_1, y_1), \ldots, (x_N, y_N)\}$ is called a set of *interpolation points in $X \times Y$* or an *interpolation set in $X \times Y$*. An element f_Φ on $\mathcal{F}(X, Y)$ is said to have the *interpolation property with respect to Δ* iff

$$\forall\, j = 0, 1, \ldots, N : \; f_\Phi(x_j) = y_j. \tag{5.11}$$

The pair (f_Φ, Δ) is called an *interpolation pair*. The set of all interpolation pairs is denoted by $\mathcal{F}\Delta(X, Y)$. If the interpolation set Δ is fixed, the set of all $f_\Phi \in \mathcal{F}(X, Y)$ that interpolate Δ will be denoted by $\mathcal{F}|\Delta(X, Y)$.

Suppose an interpolation set $\Delta := \{(x_0, y_0), \ldots, (x_N, y_N)\} \subset X \times Y$ is given. Now define contractive homeomorphisms $u_i : X \to X$ such that

$$u_i(x_0) := x_{i-1} \quad \text{and} \quad u_{i-1}(x_N) := x_i, \tag{5.12}$$

for all $i = 1, \ldots, N$. The mappings $v_i : X \times Y \to Y$ are to satisfy condition (A) as well as

$$v_i(x_0, y_0) := y_{i-1} \quad \text{and} \quad v_{i-1}(x_N, y_N) := y_i, \qquad (5.13)$$

for all $i = 1, \ldots, N$. Setting $C^*(X, Y) := \{f \in C(X, Y) | \forall\, j = 0, 1, \ldots, N :$ $f(x_j) = y_j\}$, one readily verifies that v_i also satisfies condition (B^0) with $\mathfrak{F}(X, Y) := C^*(X, Y)$, and that the Read-Bajraktarević operator Φ is contractive on $C^*(X, Y)$. Moreover, its unique fixed point f_Φ satisfies

$$f_\Phi(x_j) = y_j, \quad \forall j = 0, 1, \ldots, N,$$

i.e., f_Φ has the interpolation property with respect to Δ. Hence, for a given Δ there exists a unique $f_\Phi \in \mathcal{F}(X, Y)$ so that (f_Φ, Δ) is an interpolation pair. On the other hand, a given $f_\Phi \in \mathcal{F}(X, Y)$ does in general determine more than one interpolation pair: If f_Φ interpolates $\Delta := \{(x_0, y_0), \ldots, (x_N, y_N)\}$, then it also interpolates
$\Delta_{i(n)} := \{(u_{i(n)}(x_0), f_\Phi(u_{i(n)}(x_0))), \ldots, (u_{i(n)}(x_N), f_\Phi(u_{i(n)}(x_N)))\}$, for any $i(n) \in \Sigma$.

Example **5.4** Let $X := [0, 1]$ and let $Y \in \mathfrak{H}(X)$ be sufficiently large. Let N be an integer larger than one and let $\Delta := \{(x_j, y_j) \in [0, 1] \times Y | 0 = x_0 < x_1 < \ldots < x_N = 1 \wedge j = 0, 1, \ldots, N\}$. Define

$$u_i(x) := a_i x + d_i, \qquad (5.14)$$

and

$$v_i(x, y) := c_i x + s_i y + e_i, \qquad (5.15)$$

where $|s_i| < 1$ is given, and the a_i, c_i, d_i, and e_i are determined by the conditions

$$u_i(x_0) := x_{i-1} \quad \text{and} \quad u_{i-1}(x_N) := x_i,$$

$$v_i(x_0, y_0) := y_{i-1} \quad \text{and} \quad v_{i-1}(x_N, y_N) := y_i.$$

Thus,

$$a_i = \frac{x_i - x_{i-1}}{x_N - x_0}, \qquad (5.16)$$

$$c_i = \frac{y_i - y_{i-1}}{x_N - x_0} - \frac{s_i(y_N - y_0)}{x_N - x_0}, \qquad (5.17)$$

$$d_i = \frac{x_N x_{i-1} - x_0 x_i}{x_N - x_0}, \qquad (5.18)$$

and

$$e_i = \frac{x_N y_{i-1} - x_0 y_i}{x_N - x_0} - \frac{s_i(x_N y_0 - x_0 y_N)}{x_N - x_0}. \tag{5.19}$$

Fig. 5.2 shows the graph of the fractal function interpolating the set $\Delta :=$ $\{(0,0), (1/2, -1/3), (3/4, 1/2), (1, 1/3)\}$ with $s_1 := 2/3$, $s_2 := -1/2$, and $s_3 :=$ $1/3$.

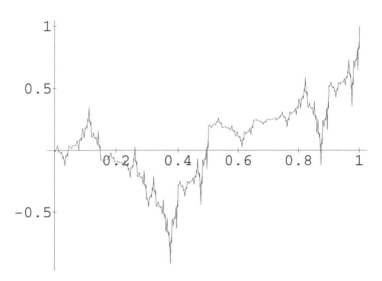

Figure 5.2: The graph of a fractal function with interpolation property.

The preceding example can be extended to interpolation in $\mathbb{R}^n$:

***Example* 5.5** Let $X := [0,1]$ and let $Y = \mathbb{R}^n$, $n > 1$. If the hypotheses of Theorem 5.1 are satisfied in this special case, then the unique fixed point $f_\Phi : [0,1] \to \mathbb{R}^n$ of the Read-Bajraktarević operator Φ is called a *vector-valued fractal function* (also [129]). A special class of such vector-valued fractal functions is considered. Let $\Delta := \{(x_j, \boldsymbol{y}_j) | 0 = x_0 < x_1 < \cdots < x_N = 1 \wedge j = 0, 1, \ldots, N\} \subseteq [0,1] \times \mathbb{R}^n$ and let $\mathfrak{F}([0,1], \mathbb{R}^n) := \{f \in C([0,1], \mathbb{R}^n) \,|\, \forall\, j = 0, 1, \ldots, N : f(x_j) = \boldsymbol{y}_j\}$. Define mappings $u_i : [0,1] \to [0,1]$ by

$$u_i(x) = (x_i - x_{i-1})\, x + x_{i-1},$$

and $\boldsymbol{v}_i : \mathbb{R}^n \to \mathbb{R}^n$ by

$$\boldsymbol{v}_i(\boldsymbol{y}) := \begin{pmatrix} \delta y_{i,1} & s_{i,1} & \cdots & 0 \\ \vdots & & \ddots & \\ \delta y_{i,n} & 0 & \cdots & s_{i,n} \end{pmatrix} \boldsymbol{y} + \boldsymbol{t}_i,$$

where

$$\delta y_{i,k} := y_{i,k} - y_{i-1,k} - s_{i,k}(y_{N,k} - y_{0,k}),$$

and

$$t_{i,k} := y_{i-1,k} - s_{i,k}y_{0,k},$$

$i = 1, \ldots, N$ and $k = 1, \ldots, n$. Here, $\boldsymbol{y}_j := (y_{j,1}, \ldots, y_{j,n}) \in \mathbb{R}^n$, $j = 0, 1, \ldots, N$, and $\boldsymbol{t}_i := (t_{i,1}, \ldots, t_{i,N}) \in \mathbb{R}^n$, $i = 1, \ldots, N$. The $s_{i,k} \in (-1, 1)$ are free parameters. The Read-Bajraktarević operator Φ on $\mathfrak{F}([0, 1], \mathbb{R}^n)$ is then given by

$$\Phi\mathfrak{f}(x) := \boldsymbol{v}_i(\mathfrak{f} \circ u_i^{-1}(x)), \qquad x \in u_i[0, 1]. \tag{5.20}$$

The unique fixed point of Φ is a continuous fractal function $\boldsymbol{f}_\Phi : [0, 1] \to \mathbb{R}^n$ interpolating Δ.

This special class of vector-valued fractal functions will be considered again in Chapter 6, where dimension formulae for the graphs of fractal functions are presented.

Theorem 5.3 can be used to give a geometric interpretation of the construction of a fractal function possessing the interpolation property. The hyperbolic IFS associated with such a fractal function is given by

$$w_i(x, y) = \begin{pmatrix} a_i & 0 \\ c_i & s_i \end{pmatrix} \begin{pmatrix} x \\ y \end{pmatrix} + \begin{pmatrix} d_i \\ e_i \end{pmatrix}, \tag{5.21}$$

where the a_i, c_i, d_i, and e_i are defined as in Eqs. (5.16) — (5.19), and the $s_i \in (-1, 1)$ are free parameters $(i = 1, \ldots, N)$. Recall that by Proposition 2.4 one can find a $\hat{K} \in \mathfrak{H}(X)$ such that $\mathfrak{w}(\hat{K}) \subseteq \hat{K}$. This is used in Fig. 5.3 to depict the convergence of $\hat{K} \subseteq \mathbb{R}^2$ to the graph of a fractal function with interpolation property.

In Section 2.1.2 Weierstraß-like fractal functions were introduced, in particular, the functions f_ϕ ((2.9)) defined by Besicovitch and Ursell. Now it will be seen that they can also be defined via IFSs.

Recall that the function $\phi|_{[0,1]}$ interpolates the points $(0, 0)$, $(\frac{1}{2}, 1)$, and $(1, 0)$. In the notation of Section 2.1.2, let $b_j := 2^j$ and $\alpha := -\log|s|/\log 2$, for some $0 < |s| < 1$. Defining $u_1(x) := 1/2\,x$, $u_2(x) := (1/2)(x + 1)$, $\lambda_1(x) := x$, and $\lambda_2(x) := 1 - x$ it is easily seen that, with v_1 and v_2 defined as in (5.6), the operator $\Phi := \lambda_i \circ u_i^{-1} + s(\,\cdot\,) \circ u_i^{-1}$ is contractive on the Banach space $\{f \in C(X, Y)|\, f(0) = f(1) = 0\}$ endowed with the sup-norm. Its unique fixed point is a continuous fractal function f_Φ satisfying $f_\Phi(0) = f_\Phi(1) = 0$.

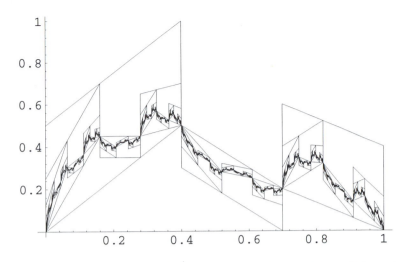

Figure 5.3: The convergence of $\hat{K}$ to the graph of a fractal function with interpolation property.

Claim. $f_\Phi \equiv f_\phi$ on $[0,1]$.

Proof. It suffices to show that f_ϕ satisfies the fixed-point equation

$$f_\Phi = \lambda_i \circ u_i^{-1} + s\, f_\Phi \circ u_i^{-1},$$

for all $x \in u_i([0,1])$. This, however, follows easily from the fact that

$$\phi|_{u_i[0,1]} = \lambda_i \circ u_i^{-1},$$

for all $x \in u_i[0,1]$. ∎

The preceding arguments indicate that there might exist an infinite sum representation of a continuous fractal function. Such a representation will be derived for fractal functions generated by operators of the form (5.8).

For the unique fixed point f_Φ of Φ, one has

$$f_\Phi = \lambda_i \circ u_{i_1}^{-1} + s\, f_\Phi \circ u_{i_1}^{-1},$$

for all $x \in u_{i_1}[0,1]$. Hence, for $x \in u_{i_1} \circ u_{i_2}[0,1]$,

$$f_\Phi \circ u_{i_1}^{-1}(x) = \lambda_{i_2} \circ u_{i_2}^{-1}(x) + s f_\Phi \circ (u_{i_1} \circ u_{i_2})^{-1}(x).$$

Since $(X, \mathbf{u})$ is a hyperbolic IFS with attractor X, for each $x \in X$ there is a code $\mathbf{i}_x \in \Sigma$ such that $x = \gamma(\mathbf{i}_x)$. Now, let $x \in X$ be arbitrary and let

$\mathbf{i}_x := (i_1\, i_2\, \ldots i_n\, \ldots)$ be its code. Clearly, by the preceding argument,

$$f_\Phi(x) = \sum_{\nu=1}^{n} s^{\nu-1} \lambda_{i_\nu} \circ u_{i_\nu}^{-1}(x) + s^n f_\Phi \circ u_{\mathbf{i}_x(n)}^{-1}(x),$$

for all $x \in u_{\mathbf{i}_x(n)}[0,1]$. Whence,

$$f_\Phi(x) = \sum_{\nu=1}^{\infty} s^{\nu-1} \lambda_{i_\nu} \circ u_{i_\nu}^{-1}(x), \qquad (5.22)$$

for $x = \gamma(\mathbf{i}_x)$.

5.2   Recurrent Sets as Fractal Functions

The recurrent set formalism of M. Dekking can also be used to construct fractal functions. A specific example will illustrate how this may be done in general ((4.3) in [45]).

Let $X := \{a, b\}$ and let $S[X]$ be the free semigroup generated by X. Define a free semigroup endomorphism $\vartheta : S[X] \to S[X]$ by

$$\vartheta(a) := a\,b\,b\,b, \qquad \vartheta(b) := b\,a\,a\,a.$$

Let $f : S[X] \to \mathbb{R}^2$ be defined by

$$f(a) := (1,1), \qquad f(b) := (1,-1).$$

Since $\mathbb{R}^X \cong \mathbb{R}^2$, L_ϑ is the full representation of ϑ and is given by

$$L_\vartheta = \begin{pmatrix} 4 & 0 \\ 0 & -2 \end{pmatrix}.$$

Finally, define $K : S[X] \to \mathbb{R}^2$ by

$$K(x) := \{tf(x)\,|\,t \in [0,1]\}.$$

It is not hard to convince oneself that the recurrent set $K_\vartheta(a)$ is the graph of a continuous (fractal) function passing through the points $(0,0, (1/4, -1/2)$, $(1/2, 0)$, $(3/4, 1/2)$, and $(1,1)$. This function is known as *Kiesswetter's fractal function* or *Kiesswetter's curve* ([105]). Fig. 5.4 displays its graph.

Now it is shown how univariate $\mathbb{R}^n$-valued fractal functions can be generated using the recurrent set formalism. But first some definitions.

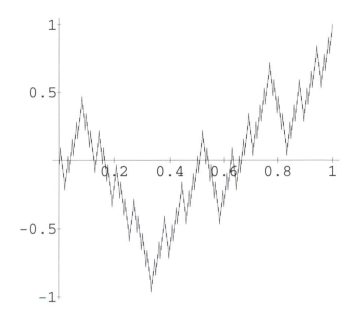

Figure 5.4: Kiesswetter's fractal function.

Definition 5.2 Let $k : [0, 1] \to \mathbb{R}^n$ be a continuous mapping. The image of k in $\mathbb{R}^n$ is called a Jordan curve.

Definition 5.3 A free group endomorphism $\vartheta : F[X] \to F[X]$ is called null-free iff for all $x \in X$, $\vartheta(x) \neq e$, where e denotes the empty word.

The next result is proven in [45].

Theorem 5.4 *Let ϑ be a null-free endomorphism of $S[X]$, and let L_ϑ be an expansive representation of ϑ. Let $f : S[X] \to \mathbb{R}^n$, $K : S[X] \to \mathfrak{H}(\mathbb{R}^n)$, and $K_\nu(\vartheta) : S[X] \to \mathfrak{H}(\mathbb{R}^n)$ be defined as in (2.49), (2.51), (2.52), and (2.53), respectively. Then, for any non-empty word $w \in S[X]$, there exists a set $K_\vartheta(w) \in \mathfrak{H}(\mathbb{R}^n)$ such that*

$$L_\vartheta^{-\nu} K_\nu(\vartheta)(w) \xrightarrow{h} K_\vartheta(w),$$

as $\nu \to \infty$. Moreover, the set $K_\vartheta(\cdot)$ does not depend upon the choice of the mapping K, and it is a curve.

Proof. The existence of the set $K_\vartheta(w)$ follows from Theorem 2.8. It remains to be shown that $K_\vartheta(w)$ is a curve.

Let $w \in S[X]$, and for all $\nu \in \mathbb{N}$ let $\vartheta^\nu := w_{\nu 1} \ldots, w_{\nu \mu(\nu)}$. For $\kappa \in \{1, \ldots, \mu(\nu)\}$, define line segments

$$I_{\nu,\kappa} := [(\kappa - 1)/\mu(\nu), \kappa/\mu(\nu)].$$

Since $K_\nu(w)$ is a curve, there exist functions $k_\nu : [0, 1] \to \mathbb{R}^n$ such that

$$k_\nu(I_{\nu,\kappa}) = L_\vartheta^{-\nu} K(w_{\nu \kappa}) + f(w_{\nu 1} \ldots w_{\nu \kappa - 1}).$$

For each $t \in [0, 1]$, choose a $t' = t'(t)$ in $I_{\nu,\kappa} \ni t$ such that the Euclidean distance $d_E(k_{\nu+1}(t), k_\nu(t'))$ is minimized. Then

$$
\begin{aligned}
d_E(k_{\nu+1}(t), k_\nu(t)) \;\leq\;& d_E(k_{\nu+1}(t), k_\nu(t')) + d_E(k_\nu(t'), k_\nu(t')) \\[2mm]
\leq\;& \mathfrak{h}(L_\vartheta^{-\nu} K(w_{\nu \kappa}), L_\vartheta^{-(\nu+1)} K(w_{\nu \kappa})) \\
& + d_E(k_\nu(\kappa/\mu(\nu)), k_\nu((\kappa - 1)/\mu(\nu))) \\[2mm]
\leq\;& \max_{x \in X} \mathfrak{h}(L_\vartheta^{-\nu} K(x), L_\vartheta^{-(\nu+1)} K(x)) \\
& + \max_{x \in X} \| L_\vartheta^{-\nu}(f(x)) \| \\[2mm]
\leq\;& c\,(\delta + \max_{x \in X} \| f(x) \|) \lambda^{-\nu}.
\end{aligned}
$$

Here Lemma 2.1 was used, as well as the notation introduced in the proof of Theorem 2.8. Hence, the functions k_ν converge uniformly to a continuous function $k : [0, 1] \to \mathbb{R}^n$. Since $[0, 1] = \bigcup_{\kappa,\nu} I_{\nu,\kappa}$, it is easy to see that $k[0, 1]$ is indeed $K_\vartheta(w)$. Whence, $K_\vartheta(w)$ is a (Jordan) curve in $\mathbb{R}^n$.

To show the independence of $K_\vartheta(w)$ on the mapping K, assume that K and K' are two mappings defined by Eq. (2.51). Let c and λ be as in Lemma 2.1. Then, by the properties of the Hausdorff metric h,

$$
\begin{aligned}
\mathfrak{h}(L_\vartheta^{-\nu} K(\vartheta^\nu(w)), L_\vartheta^{-\nu} K'(\vartheta^\nu(w))) \;\leq\;& \max_{x \in X} \mathfrak{h}(L_\vartheta^{-\nu} K(x), L_\vartheta^{-\nu} K'(x)) \\[2mm]
\leq\;& \max_{x \in X} c\lambda^{-\nu} \mathfrak{h}(K(x), K'(x))
\end{aligned}
$$

Thus, as $\nu \to \infty$, $L_\vartheta^{-\nu} K'(\vartheta^\nu(w)) \to K_\vartheta(w)$.   ■

Kiesswetter's fractal function f_K can also be generated using IFSs: Define $X := [0, 1]$ and $Y := [-1, 1]$. The maps $w_i : X \times Y \to X \times Y$ are given by

$$w_1(x, y) = \begin{pmatrix} 1/4 & 0 \\ 0 & -1/2 \end{pmatrix} \begin{pmatrix} x \\ y \end{pmatrix} \tag{5.23}$$

and

$$w_i(x, y) = \begin{pmatrix} 1/4 & 0 \\ 0 & 1/2 \end{pmatrix} \begin{pmatrix} x \\ y \end{pmatrix} + \begin{pmatrix} (i-1)/4 \\ (i-3)/2 \end{pmatrix}, \tag{5.24}$$

for $i = 2, 3, 4$, or equivalently,

$$u_i(x) = (1/4)x + (i-1)/4, \quad i = 1, 2, 3, 4,$$

$$v_1(x, y) = (-1/2)y, \quad v_i(x, y) = (1/2)y + (i-3)/2,$$

for $i = 2, 3, 4$, and

$$\Phi_K f := (\varepsilon_i / 2) f \circ u_i^{-1},$$

for $x \in u_i X$ and $f \in C^*(X, Y) := \{ f \in C(X, Y) \mid f(0) = 0 \wedge f(1) = 1 \}$, where

$$\varepsilon_i := \begin{cases} -1 & \text{for } i = 1 \\ 1 & \text{for } i = 2, 3, 4. \end{cases}$$

The graph G_K of Kiesswetter's fractal function f_K is then the unique attractor of the preceding hyperbolic IFS, or equivalently, f_K is the unique fixed-point of the Read-Bajraktarević operator Φ_K.

5.3 Iterative Interpolation Functions

In a series of papers ([50, 55, 56]), S. Dubuc introduced an *iterative interpolation process* for interpolation data Δ defined on a closed discrete subgroup of $\mathbb{R}^n$. For $n = 1$, this interpolation process defines univariate fractal functions having the interpolation property with respect to Δ. This section is devoted to the study of this case. However, the definitions and results are stated in such a way that the generalization to $\mathbb{R}^n$ is immediate. But first some definitions are needed.

***Definition* 5.4** Let G be a group and $*$ its group operation. A topology on the set G is said to be compatible with the group structure iff the mappings $\alpha : G \times G \to G$, $(g_1, g_2) \overset{\alpha}{\mapsto} g_1 * g_2$, and $\iota : G \to G$, $g \overset{\iota}{\mapsto} g^{-1}$, are continuous. The pair $(G, \mathcal{T})$ is called a topological group iff G is a group and $\mathcal{T}$ a topology on G that is compatible with the group structure.

A rather simple example of a topological group is provided by the additive group $(\mathbb{R}, +)$ whose topology $\mathcal{T}_{|\cdot|}$ is defined by the metric $|\cdot|$. Compatibility with respect to the group structure means that $r_n \to r$ and $s_n \to s$ imply $r_n + s_n \to r + s$ and $-r_n \to -r$.

Definition 5.5 A subset Y of a topological space $(X, \mathcal{T})$ is called discrete iff every $y \in Y$ has a neighborhood N such that $N \cap Y = \{y\}$.

The integers $\mathbb{Z}$ form a discrete subset of $\mathbb{R}$. Indeed, $(\mathbb{Z}, \mathcal{T}_{|\cdot|})$ is a closed discrete topological subgroup of $(\mathbb{R}, \mathcal{T}_{|\cdot|})$.

For the remainder of this section it is assumed that X is a closed discrete subgroup of $\mathbb{R}$ whose metric topology is inherited from $\mathbb{R}$ and that the vector subspace $\mathrm{vec}(X)$ generated by X is dense in $\mathbb{R}$ (recall that a vector space is an abelian group over a field satisfying certain compatibility conditions). Furthermore, assume that a set Δ of complex-valued interpolation points over G is given, i.e., $\Delta = \{(x, y) \mid x \in X \wedge y \in Y\}$, where Y is a $\|X\|_c$-element subset of $\mathbb{C}$. The set Δ is considered as the graph of a univariate complex-valued function $f : X \to Y$.

The basic idea of Dubuc's interpolation process is to extend f to functions f_k whose domains are subgroups $X_k \supset X_{k-1} \supset \cdots \supset X$ such that their union is dense in $\mathbb{R}$. These extensions are defined via a linear transformation $T : \mathbb{R} \to \mathbb{R}$, and a complex-valued weight function p satisfying the following conditions:

1. $TX \supset X$;

2. $\mathrm{dom}(p) = TX$;

3. $p(0) = 1$ and $p(x) = 0$ for all $0 \neq x \in X$.

4. $\|\mathrm{supp}(p)\|_c < \infty$.

Here dom and supp denote the domain and support of p, respectively. A nested increasing sequence of subgroups is defined by $X_0 := X$ and $X_k := T^k X$, for all $k \in \mathbb{N}$. Let $X_\infty := \bigcup_{k \in \mathbb{N}_0} X_k$. Note that X_∞ is the direct limit of the directed system $\{X_k, \hookrightarrow\}$, where $\hookrightarrow$ denotes the injection map $X_{k-1} \hookrightarrow X_k$:

$$X_\infty = \varinjlim X_k.$$

Proposition 5.1 *There exists a unique function* $g : X_\infty \to \mathbb{C}$ *such that* $g \equiv f$ *on* X *and for any* $k \in \mathbb{N}$

$$g(T^{k+1} \cdot) = \sum_{x' \in X} p(T \cdot - x') g(T^k x'). \tag{5.25}$$

Remark. The function g is called the *iterative interpolation of f with respect to T and p* and the quadruple (f, X, T, p) an *iterative interpolation process*.

Proof. Since $\mathrm{vec}(X)$ is dense in $\mathbb{R}$ and $TX \supset X$, it follows that $T\mathbb{R} = \mathbb{R}$ and, hence, that T is injective.

A sequence of functions $f_k : X_k \to \mathbb{C}$ is defined by $f_0 := f$ and

$$f_k(T^k x) := \sum_{x' \in X} p(Tx - x') f_{k-1}(T^{k-1} x'), \tag{5.26}$$

for all $x \in X$. The injectivity of T implies that f_k is well-defined on X_k. Note that the finiteness of the support of p implies that the sum in (5.26) is finite. Clearly, f_k is an extension of f_{k-1}. Thus, there is a unique function $g : X_\infty \to \mathbb{C}$ satisfying $g \equiv f_k$ on X_k for all $k \in \mathbb{N}$. ∎

It is worthwhile to note some simple properties of the iterative interpolation process:

Linearity: Let f_1 and f_2 be functions defined on X and g_1 and g_2 their respective iterative interpolations. If $f := f_1 + f_2$ and if g is its iterative interpolation, then $g = g_1 + g_2$.

Scalar multiplication: If f has g as its iterative interpolation and if $c \in \mathbb{R}$, then cg is the iterative interpolation of cf.

Translation: Let $y \in X$ and let g be the iterative interpolation of a function f. Then $g(\,\cdot\, + y)$ is the iterative interpolation of $f(\,\cdot\, + y)$.

The following two examples illustrate the previously defined interpolation process.

Example **5.6 (Dyadic interpolation process)** Let $X := \mathbb{Z}$, and define a linear transformation T by $Tx := x/2$ and a weight function $p : \mathbb{Z}/2 \to \mathbb{R}$ by

$$p(x) := \begin{cases} 1 & x = 0 \\ 9/16 & x = \pm 1/2 \\ -1/16 & x = \pm 3/2 \\ 0 & \text{otherwise.} \end{cases}$$

This interpolation process is referred to as *dyadic interpolation* ([55]). It will be seen shortly that the dyadic interpolation process can be interpreted as an IFS on $[-3, 3]$.

The next example shows how the famous *Koch curve* can be obtained from iterative interpolation.

Example 5.7 (Koch curve) Again let $X := \mathbb{Z}$. Define a linear transformation $T : \mathbb{R} \to \mathbb{R}$ by $Tx := x/4$ and a weight function $p : \mathbb{Z}/4 \to \mathbb{C}$ by

$$p(x) := \begin{cases} 1 & x = 0 \\ 1/3 & x = \pm 3/4 \\ 1/2 + i\sqrt{3}/6 & x = -1/2 \\ 2/3 & x = \pm 1/4 \\ 1/2 - i\sqrt{3}/6 & x = 1/2 \\ 0 & \text{otherwise.} \end{cases}$$

For f one may use any function as long as $f(0) = 0$ and $f(1) = 1$. The continuous extension of g to the unit interval yields the Koch curve (Fig. 5.5).

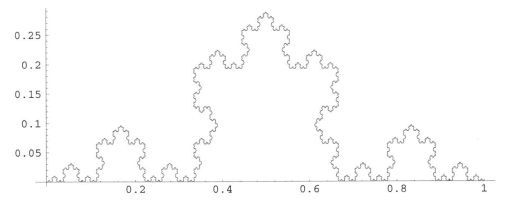

Figure 5.5: The Koch curve.

Next, conditions guaranteeing the density of X_∞ in $\mathbb{R}$ and the continuity of the iterative interpolation function g are derived. This is best done by introducing the *fundamental interpolating function F*.

Definition 5.6 Let (f, X, T, p) be an iterative interpolation process. Suppose that the function f is defined as follows:

$$f(x) := \begin{cases} 1 & x = 0 \\ 0 & \text{otherwise.} \end{cases}$$

The iterative interpolation of f is called the fundamental interpolating function F (with respect to T and p).

Remark. Setting $k = 0$ in Eq. (5.25) implies $F \equiv p$ on TX.

The next proposition justifies the terminology "fundamental":

Proposition 5.2 *Let (f, X, T, p) be an iterative interpolation process and g the extension of f. Then*

$$g(x) = \sum_{x' \in X} f(x')F(x - x').$$ (5.27)

For the proof of this proposition a lemma is needed.

Lemma 5.1 *Let $k \in \mathbb{N}_0$ and suppose $x \in T^k X \setminus \sum_{j=0}^{k} T^j \operatorname{supp}(p)$. Then $F(x) = 0$.*

Proof. The proof uses induction on k. The case $k = 1$ is obvious.

Suppose that $x \in X$ is such that $F(T^{k+1}x) \neq 0$. Then by Eq. (5.25) there exists an $x' \in X$ such that $p(Tx - x') \neq 0$ and $F(T^k x) \neq 0$. If $x'' := Tx - x'$, then $x'' \in \operatorname{supp}(p)$. Thus, by induction hypothesis, $T^k x' \in \sum_{j=0}^{k-1} T^j \operatorname{supp}(p)$. Since $T^{k+1}x = T^k x' - T^k x''$, $T^{k+1}x \in \sum_{j=0}^{k} T^j \operatorname{supp}(p)$. ■

Proof of Proposition 5.2. By the preceding lemma, for every $x \in X_k$ the sum $\sum_{x' \in X} f(x')F(x - x')$ is finite. The linearity and the translation property of the iterative interpolation process imply that the series $\sum_{x' \in X} f(x')F(x - x')$ is indeed an interpolation function g for f. Since this interpolation function is unique, $g \equiv \sum_{x' \in X} f(x')F(\cdot - x')$. ■

Taking $f(x) = F(Tx)$ in Proposition 5.2, using the linearity property of iterative interpolation and the fact that $F \equiv p$ on TX, one obtains the following useful functional equation for F:

$$F(Tx) = \sum_{x' \in X} p(Tx')F(x - x'), \quad x \in X_\infty.$$ (5.28)

The next result — which follows directly from Lemma 5.1 — gives an upper bound for the support of the fundamental interpolation function F.

Proposition 5.3 *Assume that (f, X, T, p) is a iterative interpolation process and that T has spectral radius less than one. Suppose that ρ is the length of the smallest interval centered at 0 containing the support of the weight function p. Then the support of F is contained in a closed interval of length $\rho \sum_{k \in \mathbb{N}_0} \|T^k\|$ centered at 0.* ■

More can be said about the support of F in a special case.

Proposition 5.4 *Assume that (f, X, T, p) is a iterative interpolation process for which $Tx = x/a$, for some $1 \neq a \in \mathbb{Z}$, and $X = \mathbb{Z}$. If the weights $p_n =*

$F(n/a)$ are real, then F vanishes outside the interval $(M_1/(a-1), M_2/(a-1))$, where M_1 denotes the smallest index for which $p_n \neq 0$ and M_2 the largest index for which $p_n \neq 0$.

Proof. Define a sequence of real numbers $\{t_n\}$ by

$$t_0 := 0, \quad t_n = t_{n-1} + M_1 b^{-n}, \quad n \in \mathbb{N}.$$

Clearly, $\{t_n\}$ is decreasing and converges to $M_1/(a-1)$. It now follows from Eq. (5.28) that $F|_{X_n} \equiv 0$ on $(-\infty, t_n)$, for all $n \in \mathbb{N}$. Hence, for every a-adique number $t \leq M_1/(a-1)$, $F(t) = 0$. Similarly, one shows that $F(t) = 0$ for $t \geq M_2/(a-1)$. ∎

At this point, the iterative interpolation scheme can be related to IFSs. This is best done by considering the following simple example, which nevertheless contains all the ingredients necessary to understand this relation.

Example 5.8 Let $X := \mathbb{Z}$, and let $T : \mathbb{Z} \to \mathbb{Z}$ be defined by $Tx := x/2$. Define the weight function $p : \mathbb{Z}/2 \to \mathbb{C}$ by $p(0) := 1$, $p(1/2) = p(-1/2) := 3/4$, and $p(x) := 0$ for $x \in \mathbb{Z} \setminus \{0, \pm 1/2\}$. Then by the above theorem, the iterative interpolation function g vanishes outside $(-1, 1)$. If f is chosen to be any function interpolating the set $\Delta := \{(n, F(n)) | n \in \{-1, 0, 1\}\}$, then graph$(g)$ is the attractor of the hyperbolic IFS $([-1, 1], \mathbf{w})$ with $w : [0, 1] \times [0, 1] \to [0, 1] \times [0, 1]$ given by

$$w_i(x, y) = \begin{pmatrix} 1/2 & 0 \\ 1/2(-1)^{i-1} & s_i \end{pmatrix} \begin{pmatrix} x \\ y \end{pmatrix} + \begin{pmatrix} 1/2(-1)^i \\ 1/2 \end{pmatrix},$$

for $i = 1, 2$. The contractivity constants are given by $s_1 = p(-1/2) = 3/4$ and $s_2 = p(1/2) = 3/4$. The graph of g is depicted in Fig. 5.6.
Now it should be clear how the dyadic interpolation process is interpreted as an IFS: The graph of the dyadic iterative interpolation function g is the unique attractor of a hyperbolic IFS defined on $[-3, 3] \times [-3, 3]$ consisting of four maps w_i which all scale horizontally by $1/2$ and vertically by $s_1 = p(-3/2) = -1/16$, $s_2 = p(-1/2) = 9/16$, $s_3 = p(1/2) = 9/16$, and $s_4 = p(3/2) = 1/16$. For f any function interpolating $\Delta = \{(n, F(n)) | n \in \{-3, -2, -1, 0, 1, 2, 3\}\}$ can be taken.

Next, necessary and sufficient conditions are given for the fundamental interpolation function to be continuous. Our presentation follows closely that in [56].

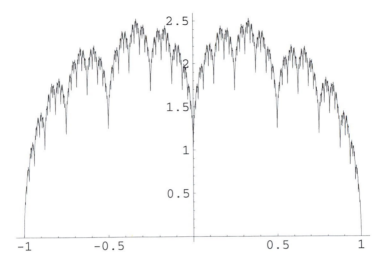

Figure 5.6: The iterative interpolation function g in Example 5.8.

Definition 5.7 The iterative interpolation process (f, X, T, p) is called continuous iff its fundamental interpolation function F is uniformly continuous.

Remark. The preceding definition is motivated by the following well-known fact from analysis: If $F : D \subseteq \mathbb{R}^n \to \mathbb{R}$ is uniformly continuous on its domain D, then it can be continuously extended to the closure of D. Clearly, if (f, X, T, p) is continuous then the iterative interpolation g of f is continuous.

At this point it is natural to ask under what conditions X_∞ is dense in $\mathbb{R}$. The proposition below gives the necessary conditions.

Proposition 5.5 *Suppose X is a subgroup of $\mathbb{R}^n$ and $\text{vec}(X) = \mathbb{R}^n$. If a linear transformation $T : \mathbb{R}^n \to \mathbb{R}^n$ is such that*

1. $X \subset TX$,

2. $\forall \lambda \in \sigma(T) : |\lambda| < 1$,

then X_∞ is dense in $\mathbb{R}^n$.

Proof. Let $\varepsilon > 0$ be given. Then there exists an integer n such that $\|T^n\| < \varepsilon$. By the surjectivity of T, one finds for an arbitrary $y \in \mathbb{R}$ a $z \in \mathbb{R}$ such that $T^n z = y$. Since $\text{vec}(X) = \mathbb{R}^n$, there is a $x \in X$ such that $d_E(x, z) < \delta$, for some $\delta > 0$. Hence, $d_E(T^n x, y) \le \delta \varepsilon$.   ∎

In the case $n = 1$ it is easy to see that T has to be a contractive linear transformation on $\mathbb{R}$ in order for X_∞ to be dense in $\mathbb{R}$. Thus, for such a T the continuous iterative interpolation process (f, X, T, p) defines a fundamental interpolation function F that can be continuously extended to all of $\mathbb{R}$. In what follows, no distinction is made between F and its continuous extension to $\mathbb{R}$.

Definition 5.8 A function $g : X_\infty \to \mathbb{C}$ is called an interpolation function iff there exists a function $f : X \to \mathbb{C}$ such that g is the iterative interpolation for an iterative interpolation process (f, X, T, p).

Two types of modulus of continuity have to be considered for the study of continuity of an interative interpolation process:

Definition 5.9 Let (f, X, T, p) be an iterative interpolation process and F its fundamental interpolation function. Let $g : X_\infty \to \mathbb{C}$ and $\eta \in \mathbb{R}^+$. Then, for $k \in \mathbb{N}$, define

$$C_k(\eta) := \max_{x, x' \in X} \left\{ \sum_{x'' \in X} |F(T^k x - x'') - F(T^k x' - x'')| \,\Big|\, |x - x'| \leq \eta \right\} \quad (5.29)$$

and

$$\omega_k(g, \eta) := \sup\{|g(T^k x) - g(T^k x')| \,|\, x, x' \in X \wedge |x - x'| \leq \eta\}. \quad (5.30)$$

Some more notation needs to be introduced: Let $S_k := \{x \in X_k | F(x) \neq 0\}$ and let ρ_k denote the radius of the smallest ball centered at 0 containing S_k.

The next three lemmas are needed to prove one of the main theorems.

Lemma 5.2 *Assume that T is contractive and $\eta \geq \rho_1/(1 - \|T\|)$. Furthermore, assume that $\xi \in \mathbb{R}$ and $x \in X$ are such that $|x - \xi| \leq \eta$. Then, if g is an interpolation function,*

$$g(Tx) = \sum_{\substack{x' \in X \\ |x' - T\xi| \leq \eta}} F(Tx - x')g(x').$$

Proof. Eq. (5.27) implies $g(Tx) = \sum_{x' \in X} F(Tx - x')g(x')$. Now suppose that $x \in X$ and $x' \in X$ is such that $F(Tx - x') \neq 0$. Then $x' - T\xi = T(x - \xi) - (Tx - x')$. But, since $Tx - x' \in S_1$, $|x' - T\xi| \leq \|T\| \eta + \rho_1 \leq \eta$. ∎

Lemma 5.3 *Assume that g is an interpolation function and that for any $x \in X_1$, $\sum_{x' \in X} F(x - x') = 1$. If $\|T\| < 1$ and $\eta \geq 2\rho_1/(1 - \|T\|)$, then*

$$\omega_k(g, \eta) \leq (C_1(\eta)/2)^k \, \omega_0(g, \eta).$$

Proof. First consider the case $k = 1$. For a given $\eta > 0$, let $x_1, x_2 \in X$ be such that $|x_1 - x_2| \leq \eta$. Setting $\xi = (x_1 + x_2)/2$ implies $|x_1 - \xi| = |x_2 - \xi| \leq \eta/2$. Hence, by Lemma 5.2,

$$g(Tx_i) = \sum_{\substack{x' \in X \\ |x' - T\xi| \leq \eta}} F(Tx_i - x')g(x'),$$

for $i = 1, 2$. Now let $m := \min\{g(x') \,|\, x' \in X \wedge |x' - T\xi| < \eta/2\}$, $M := \max\{g(x') \,|\, x' \in X \wedge |x' - T\xi| < \eta/2\}$, and $c := (m + M)/2$. Then, using the hypothesis that $\sum_{x' \in X} F(x - x') = 1$, for all $x \in X_1$, it follows that, for $i = 1, 2$,

$$g(Tx_i) - c = \sum_{\substack{x' \in X \\ |x' - T\xi| \leq \eta/2}} F(Tx_i - x')(g(x') - c).$$

Thus,

$$|g(Tx_1) - g(Tx_2)| \leq \sum_{\substack{x' \in X \\ |x' - T\xi| \leq \eta/2}} |F(Tx_1 - x') - F(Tx_2 - x')| \, |g(x') - c|,$$

and therefore,

$$|g(Tx_1) - g(Tx_2)| \leq \sum_{\substack{x' \in X \\ |x' - T\xi| \leq \eta/2}} |F(Tx_1 - x') - F(Tx_2 - x')|(M - m)/2.$$

This, however, implies that $\omega_1(g, \eta) \leq C_1(\eta)\omega_0(g, \eta)/2$. Induction on k now yields the general result. ∎

Lemma 5.4 *Assume that, for a given $\eta > 0$, the subgroup $Y[G]$ of X generated by $G := \{x \in X \,|\, |x| \leq \eta\}$ is all of X, and that $\eta' > 0$. Then there exists a positive integer K such that for every $k \in \mathbb{N}$ and every function $g : G_k \to \mathbb{C}$, $\omega_k(g, \eta') \leq K \, \omega(g, \eta)$ (K depends on η and η', but not on g and k).*

Proof. Let $G' := \{x \in X \,|\, |x| \leq \eta'\}$. Then every element of G' can be expressed as a sum of elements of G. Thus, for sufficiently large K, every element of G' can be written as a sum of K elements of G. If $x, x' \in X$ satisfy

$|x - x'| \le \eta'$, then $K + 1$ elements $\{x_0, x_1, \ldots, x_K\} \subset X$ can be found such that $x_0 = x$, $x_K = x'$ and $|x_\ell - x_{\ell-1}| \le \eta$, for $\ell = 1, \ldots, K$. Hence,

$$g(T^k x) - g(T^k x') = \sum_{\ell=1}^{K} g(T^k x_{\ell-1}) - g(T^k x_\ell),$$

and thus

$$|g(T^k x) - g(T^k x')| \le K \, \omega_k(g, \eta).$$

∎

Now a theorem that gives sufficient conditions for an iterative interpolation process to be continuous can be proven.

Theorem 5.5 *Let (f, X, T, p) be an interative interpolation process. Assume that T is contractive and that the fundamental interpolation function F satisfies $\sum_{x' \in X} F(x - x') = 1$, for all $x \in TX$. If there exists a positive number η such that $\eta \ge 2\rho_1/(1 - \|T\|)$ and $C_1(\eta) < 2$, and if $Y[G] = X$, then (f, X, T, p) is continuous.*

Proof. Note that by Lemmas 5.3 and 5.4, one can find a large enough integer n so that $2 \sum_{k>n} \omega_k(F, \|T\|^{-1}\rho_1) + \omega_n(F, 4\eta) < \varepsilon$, for any given $\varepsilon > 0$. Let $\delta := 2\eta/\|T^{-n}\|$. It suffices to be shown that, if $x, x' \in X_\infty$ satisfy $|x - x'| < \delta$, then $|F(x) - F(x')| < \varepsilon$.

To this end, then let, $x, x' \in X_\infty$ be so that $|x - x'| < \delta$ with $\delta := 2\eta/\|T^{-n}\|$. Since X_∞ is a direct limit, there exits an integer m which can be chosen larger than n so that $x, x' \in X_m$. A sequence $\{x_k\}_{n \le k \le m}$ is constructed recursively as follows: For $k = m$, let $x_m := x$. Since, for $k \le m$, $x_k \in X_k$, there exists a $y_k \in X$ with $T^k y_k = x_k$. Using the assumption that $\sum_{x' \in X} F(x - x') = 1$, for all $x \in TX$, there is a $z \in X$ such that $Ty_k - z \in S_1$. Let $y_{k-1} := z$ and $x_{k-1} := T^{k-1} y_{k-1}$. Note that $x_{k-1} \in X_{k-1}$ and — by definition of ρ_1 — $|Ty_k - y_{k-1}| \le \rho_1$. Therefore, Eq. (5.30) implies that $|F(x_k) - F(x_{k-1})| = |F(T^k y_k) - F(T^{k-1} y_{k-1})| \le \omega_k(F, \|T^{-1}\|\rho_1)$. Hence, $|F(x) - F(x_n)| \le \sum_{k>n} \omega_k(F, \|T\|^{-1}\rho_1)$.

In the same way one can define a sequence $\{x'_k\}_{n \le k \le m}$ for x', yielding $|F(x') - F(x'_n)| \le \sum_{k>n} \omega_k(F, \|T\|^{-1}\rho_1)$. Note that, if y'_k is defined for x' as y_k is for x, then

$$|y_n - y'_n| \le \sum_{k=0}^{m-n-1} |T^k y_{n+k} - T^{k+1} y_{n+k+1}| + |T^{m-n} y_m - T^{m-n} y'_m|$$

$$+ \sum_{k=0}^{m-n-1} |T^k y'_{n+k} - T^{k+1} y'_{n+k+1}|,$$

$$|y_n - y'_n| \leq \sum_{k=0}^{m-n-1} \|T^k\| |y_{n+k} - T y_{n+k+1}| + |T^{-n}x - T^{-n}x'|$$

$$+ \sum_{k=0}^{m-n-1} \|T^k\| |y'_{n+k} - T y'_{n+k+1}|,$$

$$|y_n - y'_n| \leq 2 \sum_{k=0}^{m-n-1} \|T^k\| \rho_1 + |T^{-n}x - T^{-n}x'|,$$

$$|y_n - y'_n| \leq 2\eta + \|T^{-n}\| |x - x'| \leq 4\eta.$$

So, $|F(x_n) - F(x'_n)| \leq \omega_n(F, 4\eta)$. Thus, combining the found results gives $|F(x) - F(x')| \leq 2 \sum_{k>n} \omega_k(F, \|T^{-1}\|\rho_1) + \omega_n(F, 4\eta) < \varepsilon$. ∎

Setting $\tilde{T} := T^n$ and $\tilde{p}(x) := F(x)$, for $x \in T^n X$, and applying the preceding theorem, yields

Corollary 5.1 *Suppose that the fundamental interpolation function F satisfies $\sum_{x' \in X} F(x - x') = 1$, for all $x \in TX$. If there exists an $n \in \mathbb{N}$ and an $\eta \in \mathbb{R}^+$ such that $\|T^n\| < 1$, $\eta \geq 2\rho_n/(1 - \|T^n\|)$, and $C_n(\eta) < 2$, and if $Y[G] = X$, then (f, X, T, p) is continuous.* ∎

The next goal is to obtain necessary conditions for the continuity of an iterative interpolation process (f, X, T, p). But first a lemma.

Lemma 5.5 *If the iterative interpolation process (f, X, T, p) is continuous and $r(T) < 1$, then, for all $x \in TX$, $\sum_{x' \in X} F(x - x') = 1$.*

Proof. In the functional equation (5.28), x is taken in TX and $k \to \infty$. This yields

$$1 = \lim_{k \to \infty} F(T^k x) = \lim_{k \to \infty} \sum_{x' \in X} F(T^k x') F(x - y) = \sum_{x' \in X} F(x - y).$$

(Here the fact was used that $F(T^k x') \to F(0) = 1$ and that, because of finite support of F, the sum is independent of k.) ∎

Finally, necessary conditions for the continuity of an iterative interpolation process are given.

Theorem 5.6 *Suppose the iterative interpolation process (f, X, T, p) is continuous and $r(T) < 1$. Then there exists a $k \in \mathbb{N}$ and an $\eta \in \mathbb{R}^+$ such that $\eta \geq 2\rho_k/(1 - \|T^k\|)$, $C_k(\eta) < 2$, and the subgroup $Y[G]$ of X generated by G is equal to X.*

Proof. Since the spectral radius of T is less than 1, the support of F is finite and the sequence ρ_k bounded. It is clear that there exists a real number η so that $Y[G] = X$ and that for all $k \in \mathbb{N}$, $\eta \geq 2\rho_k/(1 - \|T^k\|)$. It remains to be shown that $C_k(\eta) < 2$, for all $k \in \mathbb{N}$. Consider the sets $M_\xi := \{x \in X \,|\, F(\xi - x) \neq 0\}$, where $\xi \in \mathbb{R}$. Let $M := \max\{\|M_\xi\|_c \,|\, \xi \in \mathbb{R}\}$. Then by the uniform continuity of F, there exists a $\delta > 0$ such that whenever $x, x' \in X_\infty$ and $|x - x'| < \delta$, $|F(x) - F(x')| < 1/M$. Since $r(T) < 1$, there is a $k \in \mathbb{N}$ so that $\|T^k\|\eta < \delta$. But then $C_k(\eta) < 2$.   ∎

Note that together with Lemma 5.5 and Theorem 5.6, Corollary 5.1 gives necessary and sufficient conditions for the continuity of an iterative interpolation process.

Examples of continuous iterative interpolation processes have already been encountered: dyadic interpolation and the Koch curve. Using Corollary 5.1 one sets in the former case $k := 3$. Then $\rho_3 = 21/8$ and $\|T^3\| = 1/8$. The number η is chosen so that $2\rho_3/(1 - \|T^3\|) = 6$. Then, after some straightforward calculations, $C_3(\eta) = 7/4$. In the latter case, k is taken to be 1. Then $\rho_1 = 3/4$, $\|T\| = 1/4$, $\eta = 2\rho_1/(1 - \|T\|) = 2$, and $C_1(\eta) = 2/\sqrt{3}$ ([56]).

5.4   Recurrent Fractal Functions

It is natural to extend the definition of fractal functions from IFSs to recurrent IFSs. This will be the objective of the present section. The terminology and notation introduced in Chapter 2, Subsection 2.2.3, will be used.

Let (X, d_X) and (Y, d_Y) be complete metric spaces, and let $\mathcal{X} := \{X_i \,|\, i = 1, \ldots, N\}$ be a partition of X. Let $\mathcal{X}' := \{X'_k \,|\, k = 1, \ldots, N'\}$ be a partition of X with the property that each X'_k is a union of elements of $\mathcal{X}$. Define functions $u_i : X \to X$ so that the associated set-valued map $\mathfrak{u}_i : \mathfrak{H}(X) \to \mathfrak{H}(X)$, $\mathfrak{u}_i E = \{u_i(x) \,|\, x \in E\}$, maps elements of $\mathcal{X}'$ onto elements of $\mathcal{X}$. Clearly, the u_i are then contractive with constants $|a_i| < 1$, $i = 1, \ldots, N$.

Now let $v(x, \cdot) : Y \to Y$ by defined as in (5.3) and require that conditions (B^0) and (C) hold. As in Section 5.1, the maximum of the a_i is denoted by a and the uniform Lipschitz constant of the $v_i(x, \cdot)$ by L. Endowing $X \times Y$ with the metric d_θ ((5.10)), it is seen that the mappings $w_i : X \times Y \to$

$X \times Y$, $w_i(x,y) := (u_i(x), v_i(x,y))$, become contractive with contractivity $\max\{(1+a)/2, s\} < 1$. To obtain a recurrent structure, a connection matrix $C = (c_{ki})$ is defined by

$$c_{ki} := \begin{cases} 1 & \text{if } X_i \subset X'_k \\ 0 & \text{otherwise.} \end{cases}$$

Then, defining a mapping $X\mathfrak{w} : X\mathfrak{H}(X) \to X\mathfrak{H}(X)$ as in Section 2.2.3 (Eq. (2.45)) and applying Corollary 2.1, the existence of a set XA such that $X\mathfrak{w}(XA) = XA$ is inferred. Arguments analogous to those given in Section 5.1 show that XA is the graph of a continuous function $f_\Phi : X \to Y$. This function is called a *recurrent fractal function*. The set of all such recurrent fractal functions is denoted by $\mathcal{RF}(X, Y)$.

Next it is shown that recurrent fractal functions also possess the interpolation property. This is done for $X = [0,1]$ and $Y = \mathbb{R}$ and a special class of recurrent fractal functions. The general case can then easily be inferred from this special one.

Definition 5.10 Let X and Y be metric spaces. A mapping $\alpha : X \to Y$ is called affine iff $\alpha = L + v$, where $L \in \mathcal{L}(X, Y)$ and $v \in Y$. The linear space of all affine mappings from X into Y is denoted by $\mathcal{A}(X, Y)$.

Definition 5.11 A fractal function $f_\Phi \in \mathcal{F}(X, Y)$ is called affine iff $u_i \in \mathcal{A}(X, X)$ and $v_i \in \mathcal{A}(X \times Y, Y)$ for all $i = 1, \dots, N$.

Given an interpolation set $\Delta = \{(x_j, y_j) \mid j = 0, 1, \dots, N\} \subseteq [0,1] \times \mathbb{R}$, note that $\{x_j \mid j = 0, 1, \dots, N\}$ induces a partition on $[0,1]$. Let $X_i := [x_{i-1}, x_i]$ and let $X'_k := [x_{k-1}, x_k]$ be the union of some of the X_i. Define mappings $w_i \in \mathcal{A}(X \times Y, X \times Y)$ by

$$w_i(x,y) := \begin{pmatrix} a_i & 0 \\ c_i & s_i \end{pmatrix} \begin{pmatrix} x \\ y \end{pmatrix} + \begin{pmatrix} b_i \\ d_i \end{pmatrix}, \tag{5.31}$$

where the a_i, b_i, c_i, and d_i are uniquely determined by the conditions

$$w_i(x_{k-1}, y_{k-1}) = (x_{i-1}, y_{i-1}), \quad \text{and} \quad w_i(x_k, y_k) = (x_i, y_i),$$

(or $w_i(x_{k-1}, y_{k-1}) = (x_i, y_i)$, and $w_i(x_k, y_k) = (x_{i-1}, y_{i-1})$) and the $s_i \in (-1, 1)$ are free parameters. The recurrent fractal function f_Φ then interpolates Δ, i.e., it is an element of $\mathcal{RF}|\Delta([0,1], \mathbb{R})$.

Example 5.9 Let $\Delta := \{(x_0, y_0), (x_1, y_1), (x_2, y_2), (x_3, y_3), (x_4, y_4)\}$.   Let $X_1' := [0, 1/2]$ and let $X_2' := [1/2, 1]$. Define

$$
\begin{aligned}
w_1(x_0, y_0) &= (x_0, y_0), & w_1(x_2, y_2) &= (x_1, y_1), \\
w_2(x_2, y_2) &= (x_1, y_1), & w_2(x_4, y_4) &= (x_2, y_2), \\
w_3(x_0, y_0) &= (x_3, y_3), & w_3(x_2, y_2) &= (x_4, y_4), \\
w_4(x_2, y_2) &= (x_3, y_3), & w_4(x_4, y_4) &= (x_4, y_4).
\end{aligned}
$$

The connection matrix is then given by

$$
C = \begin{pmatrix}
1 & 1 & 0 & 0 \\
0 & 0 & 1 & 1 \\
1 & 1 & 0 & 0 \\
0 & 0 & 1 & 1
\end{pmatrix}.
$$

The attractor $\mathsf{X}A = (A_1, A_2, A_3, A_4)$ then satisfies (cf. also Corollary 2.1):

$$
\begin{aligned}
A_1 &= w_1(A_1) \cup w_1(A_2), & A_2 &= w_2(A_3) \cup w_2(A_4) \\
A_3 &= w_3(A_1) \cup w_3(A_2), & A_4 &= w_4(A_3) \cup w_4(A_4).
\end{aligned}
$$

In Fig. 5.7 it is shown how the recurrent fractal function interpolating Δ is constructed (also [10]).

5.5   Hidden Variable Fractal Functions

In this section a new class of fractal functions called *hidden variable fractal functions* is introduced. These fractal functions are the projections of a continuous function F whose graph is the attractor of a hyperbolic IFS. The class of hidden variable fractal functions is more diverse than $\mathcal{F}(X, Y)$, as their values depend continuously on all the "hidden" variables determining F, thus making them more appealing as interpolants.

Suppose that (X, d_X) and (Y, d_Y) are complete metric spaces. On the cartesian product $X \times Y$ the metric $d_\theta = d_X + \theta \, d_Y$ is introduced, where $\theta > 0$ has yet to be specified. Let $u : X \to Y$ and $v : X \times Y \to Y$ be mappings required to satisfy

$$u \in \mathrm{Lip}^{(\leq s_1)}(X), \text{ where } s_1 \in [0, 1), \tag{5.32}$$

$$\forall \, y \in Y : v(\,\cdot\,, y) \in \mathrm{Lip}(X), \tag{5.33}$$

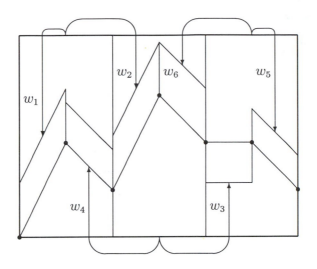

Figure 5.7: A recurrent fractal function with interpolation property.

and

$$\forall\, x \in X : \; v(x, \cdot\,) \in \mathrm{Lip}^{(\leq s_2)}(Y), \; \text{where } s_2 \in [0,1). \qquad (5.34)$$

If $w : X \times Y \to X \times Y$ is defined by $w(x,y) := (u(x), v(x,y))$, then w is a contraction on the complete metric space $(X \times Y, d_\theta)$ with $\theta := (1 - s_1)/2L$, where L denotes the uniform Lipschitz constant of $\{v(\,\cdot\,, y)\,|\, y \in Y\}$ (cf. the proof of Theorem 5.3).

Now let $N \in \mathbb{N} \setminus \{1\}$, and let $w_i : X \times Y \to X \times Y$ be maps of the structure $w_i(x,y) = (u_i(x), v_i(x,y))$, where u_i and v_i satisfy Eqs. (5.32), (5.33), and (5.34), for all $i = 1, \ldots, N$. If θ is chosen as before, it is easily seen that $(X \times Y, \mathbf{w})$ is a hyperbolic IFS on $X \times Y$. Its attractor is denoted by $A(X \times Y)$. It follows directly from Eq. (5.32) that $(X, \mathbf{u})$ is also a hyperbolic IFS; its attractor is denoted by $A(X)$. Note that $\mathrm{proj}_X A(X \times Y) = A(X)$. The idea is to impose conditions on w_i so that $A(X \times Y)$ is the graph of a continuous function $F : A(X) \subseteq X \to Y$.

Theorem 5.7 *Let $(X \times Y, \mathbf{w})$ be the IFS defined earlier. Let $\Delta := \{(x_j, y_j) : j = 0, 1, \ldots, N\}$ be an interpolation set in $X \times Y$, and suppose that*

$$\forall\, i = 1, \ldots, N : \; w_i(x_0, y_0) = (x_{i-1}, y_{i-1}) = w_{i-1}(x_N, y_N). \qquad (5.35)$$

Furthermore, assume that

$$\forall\, i = 1, \ldots, N :\ u_i \text{ is invertible on } u_i A(X), \tag{5.36}$$

$$\forall\, i = 1, \ldots, N :\ u_i A(X) \cap u_{i+1} A(X) = \{x_i\}, \tag{5.37}$$

and

$$\forall\, |i - k| \notin \{0, 1\} :\ u_i A(X) \cap u_k A(X) = \emptyset. \tag{5.38}$$

Then $A(X \times Y) = \mathrm{graph}(F_\Phi)$, *where* $F_\Phi \in \mathcal{F}|\Delta(A(X), Y)$.

Proof. First it is shown that F_Φ is the unique fixed point of an appropriate Read-Bajrakarević operator Φ. Let $C^*(A(X), Y) := \{f \in C(A(X), Y) | \forall\, j = 0, 1, \ldots, N :\ f(x_j) = y_j\}$. Then $C^*(A(X), Y)$ is a complete metric space in the sup-norm. (Note that $A(X)$ is a complete metric space in the relative strong topology.) Define $\Phi : C^*(A(X), Y) \to Y^{A(X)}$ by

$$\Phi f\,(x) := \sum_{i=1}^{N} v_i(u_i^{-1}(x), f \circ u_i^{-1}(x)) \chi_{X_i}(x),$$

where $X_i := u_i X$, $i = 1, \ldots, N$. Clearly, by Eqs. (5.37) and (5.38), Φf is well-defined on $A(X) \setminus \{x_0, \ldots, x_N\}$ and continuous on X_i. Moreover,

$$\lim_{x \to x_i^-} \Phi f\,(x) \;=\; v_i(u_i^{-1}(x_i), f \circ u_i^{-1}(x_i))$$

$$=\; v_i(x_N, f(x_N)) = v_i(x_N, y_N)$$

$$=\; v_{i+1}(x_0, y_0) = v_{i+1}(x_0, f(x_0))$$

$$=\; v_{i+1}(u_{i+1}^{-1}(x_i), f \circ u_{i+1}^{-1}(x_i)) = \lim_{x \to x_i^+} \Phi f\,(x).$$

A similar calculation shows that $\Phi f\,(x_0) = y_0$ and $\Phi f\,(x_N) = y_N$. Hence, $\Phi f \in C^*(A(X), Y)$.

The contractivity of Φ follows as in the proof of Theorem 5.2. Thus, the unique fixed point of Φ is a continuous fractal function $F_\Phi : A(X) \to Y$, i.e., an element of $\mathcal{F}(A(X), Y)$. To show that F_Φ also belongs to $\mathcal{F}|\Delta(A(X), Y)$, note that $\Delta \subset A(X)$, since (x_0, y_0) is the fixed point of w_1 and (x_N, y_N) the fixed point of w_N, and $(x_i, y_i) \in A(X)$, for $i = 2, 3, \ldots, N-1$, as $w_i(x_N, y_N) = (x_i, y_i)$. ∎

Now suppose that X is also a linear space and a finite direct sum of linear subspaces: $X = \bigoplus_{k=1}^{K} X_k$. The orthogonal projection of $\mathrm{graph}(F_\Phi)$ onto $X_k \times Y$, $k = 1, 2$, yields the graph of a continuous function $f_{\Phi,k} : X_k \to Y$.

Definition 5.12 The function $f_{\Phi,k} : X_k \to Y$ whose graph is the orthogonal projection of $\text{graph}(F_\Phi) \subset X \times Y$ onto $X_k \times Y$, $k = 1, 2$, is called a hidden variable fractal (interpolation) function.

Remarks.

1. The term "hidden variable" was first introduced in [11] and [130]. It has its origin in the fact that $\text{graph}(f_{\Phi,k})$ depends continuously on all the parameters that generate F. This then implies that, in general, $\text{graph}(f_{\Phi,k})$ is *not* made up of finitely many copies of itself.

2. The function $f_{\Phi,k}$ interpolates the data set $\Delta_k := \{(x_j^{(k)}, y_j) \mid j = 0, 1, \ldots, N\}$, where $X \ni x = (x^{(1)}, x^{(2)}, \ldots, x^{(K)})$. The set of all hidden variable fractal functions $f_{\Phi,k} : X_k \to Y$ is denoted by $\mathcal{HF}(X_k, Y)$ and, if the interpolation property of $f_{\Phi,k}$ needs to be emphasized, by $f_{\Phi,k} \in \mathcal{HF}|\Delta_k(X_k, Y)$.

There exists a natural way of defining hidden variable fractal functions: Begin with an attractor $A(X)$ of a certain hyperbolic IFS $(X, \mathbf{w})$ and choose Y homeomorphic to the code space associated with $(X, \mathbf{w})$. More precisely, let X be again a complete metric space and let $(X, \mathbf{w})$ be a hyperbolic IFS whose maps w_i, $i = 1, \ldots, N$, satisfy the following conditions:

(C1) $\forall\, i = 1, \ldots, N : w_i \in \text{Lip}^{(\leq 1)}(X)$.

(C2) If $x_0, \ldots, x_N$ are $N + 1$ distinct points in X, then $w_1(x_0) = x_0$, $w_N(x_N) = x_N$, and $\forall\, k = 1, \ldots, N - 1 : w_{k+1}(x_0) = x_k = w_k(x_N)$.

(C3) The polygon joining the points $x_0, \ldots, x_N$ is the image of $[0, 1]$ under the homeomorphism h with $h(0) = x_0$ and $h(1) = x_N$.

(C4) There exists a closed ball $B \subseteq X$ such that

    1. B contains the attractor A of $(X, \mathbf{w})$.
    2. $B \subseteq \bigcup_{i=1}^N w_i B$.
    3. $x_0, x_N \in \partial B \wedge \forall\, k = 1, \ldots, N - 1 : w_k B \cap w_{k+1} B = \{x_{k+1}\}$.

Let (Σ, d_F) be the code space associated with the IFS $(X, \mathbf{w})$. Let I denote the unit interval $[0, 1]$. Define a partition $\mathcal{P}$ on I by $\mathcal{P} := \{t_0, \ldots, t_N\}$, with $t_j := j/N$, $j = 0, 1, \ldots, N$, and its iterates by $\mathcal{P}^2 := \mathcal{P} \vee \mathcal{P} := \{t_{00}, \ldots, t_{0,N-1}, \ldots, t_{N-1,0}, \ldots, t_{N-1,N}, t_{N,N}\}$, where $t_{j_1 j_2} := j_1/N + j_2/N^2$, $j_1, j_2 \in \{0, 1, \ldots, N - 1\}$, and $\mathcal{P}^n := \bigvee_{m=1}^n \mathcal{P}$, with $t_{j_1 \ldots j_n} = j_1/N + j_2/N^2 + \ldots j_n/N^n$, for all

$j_1, j_2, \ldots, j_n \in \{0, 1, \ldots, N-1\}$, $n \in \mathbb{N}$. The partitions $\{\mathcal{P}^n\}_{n \in \mathbb{N}}$ define the N-ary expansion of a point $t \in I$ which is unique except when $t = q/N^n$, for $q \in \{1, \ldots, N-1\}$, $n \in \mathbb{N}$, in which case there are exactly two representations. The code $\mathbf{j} := (j_1\, j_2\, \cdots\, j_n\,)$ is called the representation of $t = t_{j_1 j_2 \ldots j_n \ldots} \in I$. On the collection $\mathfrak{J}$ of all such codes $\mathbf{j}$ an operator $\sigma_+ : \mathfrak{J} \to \mathfrak{J}$ is defined by $\sigma_+(j_1\, j_2\, \cdots\, j_n\,) := (j_1 + 1\, j_2 + 1\, \ldots\, j_n + 1\,)$. The next result is straightforward to prove.

Proposition 5.6 *The mapping $\xi : I \to \Sigma$ defined by*

$$\xi(t) := \begin{cases} \sigma_+ \mathbf{j} & \text{if } 1 \neq t \in [t_{j_1\, j_2\, \ldots\, j_n\, \ldots}, t_{j_1\, j_2\, \ldots\, j_n + 1\, \ldots}] \\ \sigma_+(\mathbf{N}-1) & \text{if } t = 1, \end{cases}$$

where $\mathbf{N} - 1 := (N-1\, N-1\, \ldots\, N-1\, \ldots)$, is a homeomorphism. ∎

Recall that there exists a continuous surjection $\gamma : \Sigma \twoheadrightarrow \beth$, given by $\gamma(\mathbf{i}) = \lim_{n \to \infty} w_{\mathbf{i}(n)}(x)$, where the limit is independent of $x \in A$ (Theorem 2.4). Define a map $\psi : I \to A$ by $\psi := \gamma \circ \xi$. This then induces a mapping $F : I \to X$ via $F(t) := x$, where $x = \psi(t) \in A$. Hence, the following result holds:

Proposition 5.7 *The map F as defined earlier is an element of $\mathcal{F}|\Delta(I, X)$, where $\Delta = \{(t_j, x_j)| j = 0, 1, \ldots, N\}$. Furthermore, $\mathrm{proj}_X(\mathrm{graph}(F)) = A$.*

Proof. Since every code in Σ determines exactly one point on A, the mapping F is well-defined and a function. It is easy to show that, if $\mathbf{j}_1$ and $\mathbf{j}_2$ are the codes giving the two representations of $t_j = j/N \in I$, $F(t_j) = x_j$ with no ambiguity. Moreover, conditions (C1) — (C4) imply that the set $G := \{(t, x)| F(t) = x \wedge t \in I\}$ is connected and that $G\big|_{[t_{j-1}, t_j]} \cap G\big|_{[t_j, t_{j+1}]} = \{(t_j, x_j)\}$, for all $j = 0, 1, \ldots, N$. The proof of continuity of F is essentially given in Theorem 2.4. ∎

The foregoing arguments also indicate that $\mathrm{graph}(F)$ is the attractor of a hyperbolic IFS on $I \times X$.

Proposition 5.8 *Let $(X, \mathbf{w})$ be the IFS and $F \in \mathcal{F}|\Delta(I, X)$ the fractal function as defined earlier. Then $\mathrm{graph}(F)$ is the attractor of the hyperbolic IFS $(I \times X, \mathbf{W})$, where $\mathbf{W} := \{W_i : I \times X \to I \times X | i = 1, \ldots, N\}$, with*

$$W_i(t, x) := \begin{pmatrix} 1/N(t + i - 1) \\ w_i(x) \end{pmatrix}.$$

Proof. Clearly, $(I \times X, \mathbf{W})$ is an IFS. To show hyperbolicity, we introduce a metric $\mathfrak{L} : (I \times X) \times (I \times X) \to \mathbb{R}_0^+$ by $\mathfrak{L}((t, x), (t', x')) := |t - t'| + d_X(x, x')$. Then,

$$\mathfrak{L}(W_i(t, x), W_i(t', x')) \;=\; \left| \frac{1}{N}(t - t') \right| + d_X(w_i(x), w_i(x'))$$

$$\leq \; s\, \mathfrak{L}((t, x), (t', x')),$$

with $s := \max\{1/N, s_1, \dots, s_N\}$, where s_i denotes the Lipschitz constant of w_i. The mappings W_i induce a Read-Bajraktarević operator $\mathfrak{W}$ on $C^*(I, X)$ via

$$\mathfrak{W}g\,(t) := \sum_{i=1}^{N} w_i \circ g(Nt - i + 1)\chi_{[t_{i-1}, t_i]}.$$

Note that Condition (C2) implies that $\mathfrak{W}$ is well-defined. Also, since x_0 is the fixed point of w_1 and x_N the fixed point of w_N, we have that $\mathfrak{W}g \in C^*(I, X)$. Let us now show that $\mathfrak{W}$ is contractive on $C^*(I, X)$: Let $i \in \{1, \dots, N\}$, choose $t \in [t_{i-1}, t_i]$ and $g_1, g_2 \in C^*(I, X)$. Then

$$d_X(\mathfrak{W}g_1\,(t), \mathfrak{W}g_2\,(t)) \;\leq\; s_i\, d_X(g_1(Nt - i + 1), g_2(Nt - i + 1))$$

$$\leq \; s'\, \|g_1 - g_2\|_\infty,$$

where $\|\cdot\|_\infty$ denotes the sup-norm on $C^*(I, X)$ induced by d_X, and $s' := \max\{s_i \,|\, i = 1, \dots, N\}$. Hence, $\mathfrak{W}$ has a unique fixed point $\hat{g} \in C^*(I, X)$. Denote the graph of $\hat{g}$ by $\hat{G}$. The results in the previous section imply that $\mathfrak{w}(\hat{G}) = \hat{G}$, where $\mathfrak{w}$ is the set-valued map associated with the collection $\mathbf{W}$. Thus, $\hat{G}$ is the unique attractor of the hyperbolic IFS $(I \times X, \mathbf{W})$. It remains to be proven that $\hat{G} = \text{graph}(F)$. To this end, notice that F and $\hat{g}$ agree on the set $E := \left\{ \bigcup_{m=0}^{N^n} \{m/N^n\} \,|\, n \in \mathbb{N} \right\} \subseteq I$. As E is dense in I and both F and $\hat{g}$ map into a Hausdorff space, $F \equiv \hat{g}$. ∎

The function F can now be used to construct hidden variable fractal functions. Suppose that X is a linear space and a finite direct sum of linear subspaces: $X = \bigoplus_{k=1}^{K} X_k$. Then the orthogonal projection of graph F induces a function-valued projection $F \mapsto f_k$, where $f_k : I \to X_k$ is a hidden variable fractal function. The next two examples show such hidden variable fractal functions.

Example 5.10 Let $X := \mathbb{R}^2$ and choose as maps w_i those that generate the Sierpiński triangle S. Fig. 5.8 shows the hidden variable functions f_1 and f_2 (also [11] and [130]).

Example 5.11 Let $N = 2$, let $X := I \times I$ and let

$$w_i(x) := \begin{pmatrix} 1/2 & 1/2(-1)^{i-1} \\ 1/2(-1)^{i-1} & -1/2 \end{pmatrix} x + \begin{pmatrix} 1/2(i-1) \\ 1/2(i-1) \end{pmatrix}.$$

The attractor A of the IFS $(X, \mathbf{w})$ is a so-called Peano curve. Fig. 5.9 shows A and the graphs of the hidden variable fractal functions (also [11]).

Remark. The proof of Proposition 5.8 indicates that a slightly more general setup can be used. Instead of using $u_i(t) = 1/N(t + i - 1)$ in $W_i(t, x) = (u_i(t), w_i(x))$, any collection of contractive bijections $u_i : I \to I$ with $\mathbf{u}I = I$ can be considered. This then also defines a slightly more general class of hidden variable fractal functions.

5.6   Properties of Fractal Functions

The objective in this section is the investigation of some of the properties of fractal functions. Most of these properties are a direct consequence of their method of construction. This allows us, for instance, to derive recursive formulae for the moments and transforms of fractal functions. It is also shown that affine fractal functions belong to a certain Lipschitz class. The location and uniqueness of the extreme values of fractal functions is also studied.

The notation and terminology that is used for the remainder of this section is introduced first. It is assumed that all fractal functions considered in this section are the unique fixed points of a Read-Bajraktarević operator Φ associated with functions $b : [0, 1] \to [0, 1]$ and $v(x, \cdot) : \mathbb{R} \to \mathbb{R}$ as defined in Eqs. (5.2) and (5.3). Furthermore, it is assumed that conditions (A), (B⁰), and (C) are satisfied, and that the u_i are contractive with contractivity constants $a_i < 1$. As usual, the constant of contractivity of $v_i(x, \cdot)$ is denoted by s_i, $i = 1, \ldots, N$.

5.6.1   Moment theory of fractal functions

The moment theory of IFSs was considered in Subsection 2.2.2 and the notation introduced there will be used again here. Let $([0, 1] \times \mathbb{R}, \mathbf{w})$ be the

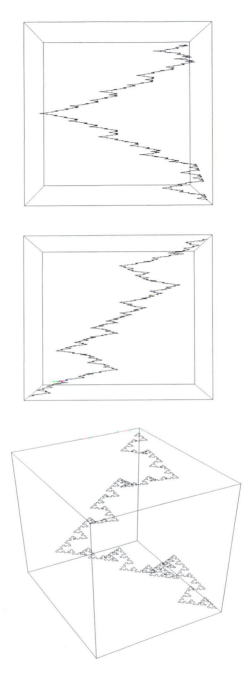

Figure 5.8: The graphs of the hidden variable fractal functions defined in Example 5.10.

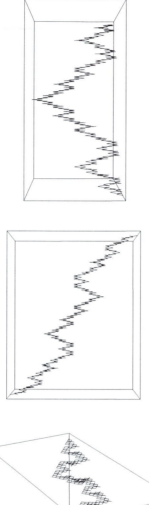

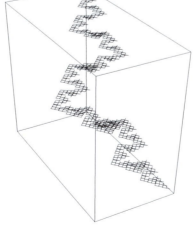

Figure 5.9: A Peano curve and the graphs of the hidden variable fractal functions.

hyperbolic IFS associated with a fractal function $f : [0,1] \to \mathbb{R}$ interpolating the set $\Delta := \{(x_j, y_j) | 0 = x_0 < x_1 < \cdots < x_N = 1 \wedge \forall\, j = 0, 1, \ldots, N : y_j \in \mathbb{R}\}$. Denote the **p**-balanced measure of $([0,1], \mathbf{w})$ by μ. Note that the IFS $([0,1], \mathbf{u})$ whose attractor is $[0,1]$ also admits a unique **p**-balanced measure $\hat{\mu}$ with supp $\hat{\mu} = [0,1]$. If the measure spaces of $[0,1]$ and $G := \operatorname{graph}(f)$ are denoted by $\mathcal{M}([0,1])$ and $\mathcal{M}(G)$, respectively, and a homeomorphism $h : [0,1] \to G$ is defined by $h(x) := (x, f(x))$, then the following relation between measures on $[0,1]$ and on G holds.

Theorem 5.8 *The homeomorphism $h : [0,1] \to G$ as defined above induces a contravariant homeomorphism $\mathcal{M}(h) : \mathcal{M}(G) \to \mathcal{M}([0,1])$. Moreover, for all $\hat{E} \in \mathcal{B}^1([0,1])$, $\hat{\mu}\hat{E} = \mu h(\hat{E})$. Also, if $g \in L^1(G, \mu)$ then*

$$\int_G g \, d\mu = \int_{[0,1]} g \circ h \, d\hat{\mu} = \int_{h^{-1}[0,1]} g \, d\hat{\mu}.$$

Proof. Define $\mathcal{M}(h) : \mathcal{M}(G) \to \mathcal{M}([0,1])$ by

$$\mathcal{M}(h)(\mu)(\hat{E}) := \mu h(\hat{E}),$$

for all $\hat{E} \in \mathcal{B}^1([0,1])$ and all $\mu \in \mathcal{M}(G)$. Clearly, $\mathcal{M}(h)$ is a contravariant homeomorphism. Now suppose that μ is any **p**-balanced measure in $\mathcal{M}(G)$. Denoting the probabilities in the IFS by $\{p_i | i = 1, \ldots, N\}$, and using the properties of **p**-balanced measures, one has $\mu(E) = \sum_i p_i \, \mu(w_i^{-1} \circ h)\hat{E} = \sum_i p_i \, \mu(h \circ u_i^{-1})\hat{E} = \sum_i p_i \, \hat{\mu}(u_i^{-1}\hat{E})$. However, since $\hat{\mu}$ is the stationary measure for the IFS $([0,1], \mathbf{u})$, $\hat{\mu}\hat{E} = \sum_i p_i \, \hat{\mu}(u_i^{-1}\hat{E})$, for all $\hat{E} \in \mathcal{M}([0,1])$. The uniqueness of the **p**-balanced measure now implies that $\mathcal{M}(\mu) = \hat{\mu}$. The integral identity now follows easily. ∎

If one-dimensional Lebesgue measure on $\mathbb{R}$ is denoted by λ, and $p_i := a_i$, then the preceding integral identity gives Corollary 5.2:

Corollary 5.2 *If the probabilities are chosen as above, then, for all $g \in L^1(G, \mu)$,*

$$\int_G g \, d\mu = \int_{[0,1]} g \circ h \, d\lambda. \tag{5.39}$$

Eq. (5.39), together with the stationarity of the **p**-balanced measure μ, can be used to obtain a recursion relation for the moments of fractal functions. To this end, let $g \in L^1(G, \mu)$ and let $p_i := a_i$, $i = 1, \ldots, N$. Then

$$\int_G g(x, y) \, d\mu(x, y) = \sum_i a_i \int_G g \circ w_i(x, y) \, d\mu(x, y)$$

$$= \sum_i a_i \int_G g(u_i(x), v_i(x,y))\, d\mu(x,y)$$

$$= \int_{[0,1]} g(x, f(x))\, d\lambda.$$

However,

$$\int_G g(x,y)\, d\mu(x,y) = \sum_i a_i \int_{[0,1]} g(u_i(x), v_i(x, f(x)))\, d\lambda,$$

and thus,

$$\int_{[0,1]} g(x, f(x))\, d\lambda = \sum_i a_i \int_{[0,1]} g(u_i(x), v_i(x, f(x)))\, d\lambda. \tag{5.40}$$

Definition 5.13 Let $f : X \subset \mathbb{R}^n \to \mathbb{R}$ be an integrable function. Let $\alpha \in \mathbb{N}_0^n$ be a multi-index and let $\beta \in \mathbb{N}_0$. The αth moment of f with respect to n-dimensional Lebesgue measure λ is defined as

$$f_{\lambda\alpha} := \int_X x^\alpha f(x)\, d\lambda, \tag{5.41}$$

and the generalized moment of f with respect to λ as

$$f_{\lambda;\alpha,\beta} := \int_X x^\alpha f(x)^\beta\, d\lambda. \tag{5.42}$$

Now the moment theorem for affine fractal functions can be stated.

Theorem 5.9 *Let G be the graph of an affine fractal function $f : [0,1] \to \mathbb{R}$. Then the moments $f_{\lambda;\alpha}$ and $f_{\lambda;\alpha,\beta}$ are uniquely and recursively determined by the lower order moments, the interpolation set Δ, and the contractivity constants $\{s_i \mid i = 1, \dots, N\}$.*

Proof. Note that since f is an affine fractal function, $u_i(x) = a_i x + b_i$, and $v_i(x,y) = c_i x + s_i y + d_i$, where the a_i, b_i, c_i, and d_i are uniquely determined by Δ. Equation (5.40) implies

$$f_{\lambda;\alpha} = \sum_i a_i \int_{[0,1]} (a_i x + b_i)^\alpha (c_i x + s_i f(x) + d_i)\, d\lambda$$

$$= \sum_i a_i \int_{[0,1]} \left[\sum_{k=0}^{\alpha} \binom{\alpha}{k} a_i^k x^k b_i^{\alpha-k} s_i f(x) + u_i^\alpha(x)(c_i x + d_i) \right] d\lambda$$

$$= \sum_i \sum_{k=0}^{\alpha} \binom{\alpha}{k} a_i^{k+1} b_i^{\alpha-k} s_i \int_{[0,1]} x^k f(x)\, d\lambda + Q(\alpha),$$

where $Q(\alpha) := \sum_i a_i \int_{[0,1]} (a_i x + b_i)^\alpha (c_i x + d_i)\, d\lambda$. Transposing the αth term gives

$$(f_{\lambda;\alpha})(1 - \sum_i a_i^{\alpha+1} s_i) = \sum_{k=0}^{\alpha-1} \binom{\alpha}{k} f_{\lambda;k} \sum_i a_i^{k+1} b_i^{\alpha-k} s_i + Q(\alpha).$$

Hence, since $\sum_i a_i = 1$ and $|s_i| < 1$,

$$f_{\lambda;\alpha} = \frac{\sum_{k=0}^{\alpha-1} \binom{\alpha}{k} f_{\lambda;k} \sum_i a_i^{k+1} b_i^{\alpha-k} s_i + Q(\alpha)}{1 - \sum_i a_i^{\alpha+1} s_i}. \tag{5.43}$$

Similarly, one has

$$\begin{aligned}
f_{\lambda;\alpha,\beta} &= \sum_i a_i \int_{[0,1]} (a_i x + b_i)^\alpha (c_i x + s_i f(x) + d_i)^\beta\, d\lambda \\
&= \sum_i \sum_{k=0}^{\alpha} \sum_{\ell=0}^{\beta} \binom{\alpha}{k} \binom{\beta}{\ell} a_i^k b_i^{\alpha-k} s_i^\ell \int_{[0,1]} x^k (c_i x + d_i)^{\beta-\ell} f^\ell(x)\, d\lambda \\
&= \sum_{\ell=0}^{\beta} \sum_{n=0}^{\alpha+\beta-\ell} Q_{\alpha,\beta,\ell,n} \int_{[0,1]} x^n f^\ell(x)\, d\lambda,
\end{aligned}$$

for an appropriate polynomial $Q_{\alpha,\beta,\ell,n}$. If $\ell = \alpha$ and $n = \beta$, then $|Q_{\alpha,\beta,\alpha,\beta}| = |\sum_i a_i^{\alpha+1} s_i^\beta| < 1$, and so

$$f_{\lambda;\alpha,\beta} = \frac{\sum_{n=0}^{\alpha-1} \binom{\alpha}{n} f_{\lambda;n,\beta} \sum_i a_i^{n+1} s_i b_i^{\alpha-n} + \sum_{\ell=0}^{\beta-1} \sum_{n=0}^{\alpha+\beta-\ell} Q_{\alpha,\beta,\ell,n} f_{\lambda;\ell,n}}{1 - \sum_i a_i^{\alpha+1} s_i^\beta}.$$

∎

5.6.2   Integral transforms of fractal functions

The moments $f_{\lambda;\alpha}$ are examples of a general integral transform of the form

$$f \longmapsto \int_{[0,1]} K(x,y) f(x)\, d\mu(x),$$

where the *kernel* $K : \mathbb{K} \times \mathbb{K} \to \mathbb{K}$ is assumed to be continuous on non-empty compact subsets of $\mathbb{K} \times \mathbb{K}$ (recall that $\mathbb{K}$ is a subfield of $\mathbb{C}$ invariant under the involuntary automorphism $z \mapsto \bar{z}$). Such an integral transform of a function f is denoted by $\hat{f}$. In case f is a fractal function of the form considered earlier, Eq. (5.40) gives

$$\hat{f}(y) = \sum_i a_i \int_{[0,1]} K(u_i(x), y) v_i(x, f(x))\, d\lambda(x). \tag{5.44}$$

In particular, if f is affine, then

$$\hat{f}(y) = \sum_i a_i s_i \int_{[0,1]} K(u_i(x), y) f(x)\, d\lambda(x) + \hat{R}(y), \qquad (5.45)$$

with $R(x) := c_i u_i^{-1}(x) + d_i$.

As an example the Fourier transform of an affine fractal function $f :$ $[0,1] \to \mathbb{K}$ is considered. The kernel K is the *Fourier kernel* $K(x,y) := e^{-ixy}$. One quickly obtains

$$\hat{f}(y) = \sum_{k=1}^{N} a_k s_k e^{-ib_k y} \hat{f}(a_k y) + \hat{R}(y),$$

where

$$\hat{R}(y) = \sum_{k=1}^{N} \frac{2 e^{-i((a_k/2) - b_k)y}}{iy} \left(\left(\frac{c_k}{a_k y} - 1 \right) \sin(a_k y / 2) + c_k e^{-ia_k/2} \right).$$

It is not hard to see that any kernel of the form $K(x,y) = K(x - y)$ or $K(x,y) = K(xy)$ allows us to express $\hat{f}(\,\cdot\,)$ is terms of $\hat{f}(a_k \cdot\,)$.

The moments $f_{\mu;\alpha}$ also play a role in the calculation of the inner product between two affine fractal functions. So, suppose that f and $\tilde{f}$ are two affine fractal functions interpolating the sets $\Delta := \{(x_j, y_j) \,|\, 0 = x_0 < x_1 < \cdots < x_N = 1 \land \forall\, j = 0, 1, \ldots, N : y_j \in \mathbb{R}\}$ and $\tilde{\Delta} := \{(x_j, \tilde{y}_j) \,|\, 0 = x_0 < x_1 < \cdots < x_N = 1 \land \forall j = 0, 1, \ldots, N : \tilde{y}_j \in \mathbb{R}\}$, respectively. It should be noted that the fixed-point property of f and $\tilde{f}$ under the correspnding Read-Bajraktarević operators can be expressed as follows:

$$f(x) = c_i u_i^{-1}(x) + d_i + s_i f(u_i^{-1}(x)), \qquad x \in u_i[0,1],$$

and similarly for $\tilde{f}$.

Proposition 5.9 *Let f and $\tilde{f}$ be affine fractal functions interpolating Δ and $\tilde{\Delta}$, respectively. Then*

$$\int_{[0,1]} f(x)\tilde{f}(x)\, d\lambda = \frac{\left[\begin{array}{c} \sum_{i=1}^{N} a_i [s_i \tilde{c}_i\, f_1 + s_i \tilde{d}_i f_0 + \tilde{s}_i c_i\, \tilde{f}_1 + \tilde{s}_i d_i \tilde{f}_0 \\ + (c_i + \tilde{c}_i)/3 + (c_i \tilde{d}_i + d_i \tilde{c}_i)/2 + d_i \tilde{d}_i] \end{array} \right]}{1 - \sum_i a_i s_i \tilde{s}_i},$$

where

$$f_0 = \frac{\sum_i a_i((c_i/2) + d_i)}{1 - \sum_i a_i s_i},$$

$$f_1 = \frac{\sum_i a_i(b_1 s_i f_0 + (a_i c_i)/3 + (b_i c_i + a_i d_i)/2 + b_i d_i)}{1 - \sum_i a_i^2 s_i},$$

and similarly for $\tilde{f}_k$, $k = 0, 1$.

Proof. The remark immediately preceding the proposition implies that

$$\int_{[0,1]} f(x)\tilde{f}(x)\,d\lambda(x)$$

$$= \sum_i \int_{x_{i-1}}^{x_i} f(x)\tilde{f}(x)\,d\lambda(x)$$

$$= \sum_i \int_{x_{i-1}}^{x_i} v_i(u_i^{-1}(x), f \circ u_i^{-1}(x))\,\tilde{v}_i(u_i^{-1}(x), \tilde{f} \circ u_i^{-1}(x))\,d\lambda(x).$$

Now, letting $\xi_i := u_i^{-1}(x)$, the preceding integral reduces to

$$\sum_i \int_{[0,1]} v_i(\xi_i, f(\xi_i))\,\tilde{v}_i(\xi_i, \tilde{f}(\xi_i))a_i\,d\xi_i.$$

After some straightforward — though considerable — algebra, the given formula is obtained. ∎

5.6.3 Fractal functions and Dirichlet splines

There is an interesting relationship between the moments of affine fractal functions and the moments of *Dirichlet splines*. This relationship is obtained by expressing the binomial coefficient in Eq. (5.43) as a hypergeometric function. Since the graph of an affine fractal function f is obtained by infinite iteration of a Read-Bajraktarević operator applied to, say a piecewise linear or polynomial function, f can be thought of — in nonstandard analytical terms — as a *hyperfinite spline*. Using this interpretation it was therefore natural to ask whether or not it is possible to compute the moments of certain (standard) splines in terms of the moments of affine fractal functions. As it turns out, this can be done for Dirichlet splines. These splines were introduced by Dahmen and Micchelli ([38]) and are a special case of the family of simplex splines defined by de Boor [43]. In [145], a recurrence formula for moments of Dirichlet splines was derived, and applications of these moments to various hypergeometric functions and classical orthogonal polynomials were reported.

The nice relationship between the moments of affine fractal functions and the moments of certain Dirichlet splines gives an algorithm for the computation of the moments of these splines. The motivation for obtaining these moments is given in [144], where it is shown that some special functions can be expressed as the moments of Dirichlet splines.

In order to define the Dirichlet spline, some notation needs to be introduced. The *Euclidean n-simplex* S^n is given by

$$S^n := \left\{ (t_1, \ldots, t_n) \in \mathbb{R}_+ \middle| t_j \geq 0 \wedge j = 1, \ldots, n \wedge \sum_{j=1}^{n} t_j \leq 1 \right\}. \qquad (5.46)$$

The residual t_0 is given by $t_0 = 1 - t_1 - \cdots t_n$. Let $b = (b_0, \ldots, b_n) \in \mathbb{R}^{n+} := \{x \in \mathbb{R}^n : x > 0\}$ with $c = b_0 + \cdots + b_n$. Then the *Dirichlet density function on S^n* is denoted by

$$\phi_b(t) = \Gamma(c) \prod_{j=0}^{n} \frac{t_j^{b_j - 1}}{\Gamma(b_j)}, \qquad (5.47)$$

where Γ is the Gamma function:

$$\Gamma(z) := \int_0^\infty t^{z-1} e^{-t} dt, \qquad \mathrm{Re}\ z > 0.$$

Next, the definition of two important special functions is presented. *Gauß's $_2F_1$ function* is given by

$$_2F_1(a, b; c; z) = \sum_{m=0}^{\infty} \frac{(a, m)(b, m)}{(c, m)} \frac{z^m}{m!}, \qquad |z| < 1,$$

where (z, k) is the *Appell symbol*

$$(z, k) = \begin{cases} 1 & k = 0 \\ z(z+1) \cdots (z + k - 1) & k > 0, \end{cases}$$

for $z \in \mathbb{C}$. The *R-hypergeometric function* is also utilized in what follows:

$$R_\alpha(b, Z) = \int_{S^n} \langle Z, u \rangle^\alpha \phi_b(u) du, \qquad (5.48)$$

where $Z = \{z_0, \ldots, z_n\} \subset \mathbb{C}$. If $\alpha \notin \mathbb{N}$, then one must insist that 0 is not contained in the convex hull $[Z]$ of Z. The interested reader is referred to [33] for more details on the special functions listed above.

Now the univariate Dirichlet spline as introduced in [38] can be defined. For this purpose, let $\Xi = \{\xi_0, \ldots, \xi_n\} \in \mathbb{R}$, $n \geq 1$. Assume that the convex hull $[\Xi]$ of the points of Ξ has non-zero length. The Dirichlet spline, $M(x \mid b; \Xi)$, is defined by requiring that

$$\int_{\mathbb{R}} g(x) M(x \mid b; \Xi) dx = \int_{S^n} g(\langle \Xi, t \rangle) \phi_b(t) dt \qquad (5.49)$$

holds for all $g \in \mathcal{K}_{\mathbb{R}}(\mathbb{R})$. (Recall that $\mathcal{K}_{\mathbb{R}}(\mathbb{R})$ denotes the space of all continuous functions on $\mathbb{R}$ with compact support.)

It can be shown ([38]) that $M(x \mid b; \Xi)$ is positive on its support $[\Xi]$, and using the defining relation (5.49), it is readily verified that

$$\int_{\mathbb{R}} M(x \mid b; \Xi)dx = 1 \tag{5.50}$$

Fig. 5.10 shows the graph of a Dirichlet spline with $n = 2$, $\Xi = \{1, 2, 3\}$, and $b = (1/2, 1/2, 1/2)$. For $\beta \in \mathbb{R}$, the moment of the Dirichlet spline is given by

$$m_\beta(b; \Xi) = \int_{\mathbb{R}} x^\beta M(x \mid b; \Xi)\,dx. \tag{5.51}$$

In the case when $\beta < 0$, we must choose the points of Ξ so that $0 \notin [\Xi]$.

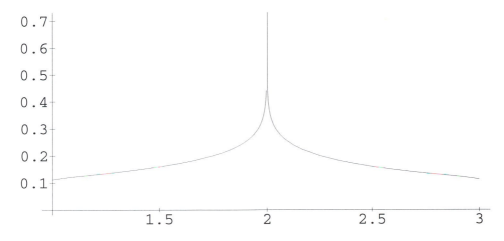

Figure 5.10: The graph of a Dirichlet spline function.

The following simple lemma provides a bridge between the moments of affine fractal functions and the moments of Dirichlet splines.

Lemma 5.6 *Let $0 \leq r \leq \rho$ be integers. Then*

$$\binom{\rho}{r} = \binom{\frac{1}{2}\rho}{\frac{1}{2}r} {}_2F_1\left(\rho - r, r; \frac{1}{2}(\rho + 1); \frac{1}{2}\right). \tag{5.52}$$

Remark. In the case when r is odd, the binomial coefficient on the right-hand side of (5.52) can be evaluated using the Gamma function (cf. 6.1.21 in [1]).

Proof. Using the duplication formula for the Gamma function (see, for instance, [1]) gives

$$
\begin{pmatrix} \rho \\ r \end{pmatrix} = \frac{\Gamma(\rho + 1)}{\Gamma(r+1)\Gamma(\rho - r + 1)}
$$

$$
= \pi^{\frac{1}{2}} \frac{\Gamma(\frac{1}{2}\rho + 1)}{\Gamma(\frac{1}{2}r + 1)\Gamma(\frac{1}{2}\rho - \frac{1}{2}r + 1)} \cdot \frac{\Gamma(\frac{1}{2}\rho + \frac{1}{2})}{\Gamma(\frac{1}{2}r + \frac{1}{2})\Gamma(\frac{1}{2}\rho - \frac{1}{2}r + \frac{1}{2})}
$$

$$
= \begin{pmatrix} \frac{1}{2}\rho \\ \frac{1}{2}r \end{pmatrix} {}_2F_1\left(\rho - r, r; \frac{1}{2}(\rho + 1); \frac{1}{2}\right). \tag{5.53}
$$

∎

Using the preceding lemma, Eq. (5.43) can be rewritten as

$$
f_\rho := f_{\mu;\rho} = \gamma \left[\sum_{i=1}^{N} \sum_{r=0}^{\rho-1} {}_2F_1\left(\rho - r, r; \frac{1}{2}(\rho + 1); \frac{1}{2}\right) K_i \, f_r + Q(\rho) \right], \tag{5.54}
$$

where $\gamma(\rho) := \left(1 - \sum_{i=1}^{N} a_i^{\rho+1} s_i\right)^{-1}$, and $K_i = \begin{pmatrix} \frac{\rho}{2} \\ \frac{r}{2} \end{pmatrix} a_i^{r+1} b_i^{\rho-r} s_i$.

Example 5.12 This example shows how easy it is to calculate the moments of affine fractal functions. For this purpose let $N := 2$, $s_1 = s_2 =: s$, and $\Delta := \{(0,0), (1/2, 1/2), (1,0)\}$. Then

$$
u_1(x) = \frac{1}{2}x, \qquad u_2(x) = \frac{1}{2}(x + 1),
$$

$$
v_1(x, y) = \frac{1}{2}x + sy, \qquad v_2(x, y) = -\frac{1}{2}(x - 1) + sy.
$$

Hence,

$$
Q(\rho) = \sum_{i=1}^{2} \sum_{r=0}^{\rho-1} \begin{pmatrix} \rho \\ r \end{pmatrix} \left(\frac{1}{2}\right)^{r+1} \left(\frac{i-1}{2}\right)^{\rho-r} \int_0^1 x^r q_i(x)\,dx
$$

$$
= \frac{1}{2} \sum_{r=0}^{\rho} \begin{pmatrix} \rho \\ r \end{pmatrix} \left(\frac{1}{2}\right)^{\rho-1} \int_0^1 x^r (1 - x)\,dx
$$

$$
= \left(\frac{1}{2}\right)^{\rho+2} \sum_{r=0}^{\rho} \begin{pmatrix} \rho \\ r \end{pmatrix} \frac{1}{(r+1)(r+2)},
$$

$$\gamma(\rho) = (1 - 2^{-\rho} s)^{-1}, \quad \text{and} \quad K_i = \begin{cases} s\left(\tfrac{1}{2}\right)^{\rho+1} \begin{pmatrix} \frac{\rho}{2} \\ \frac{r}{2} \end{pmatrix} & i = 2 \\ 0 & i = 1. \end{cases}$$

Thus, the following recursive formula for the moments holds f_ρ, $\rho \in \mathbb{N}_0$:

$$f_\rho = \frac{1}{2^\rho - s}\left[\frac{s}{2}\sum_{r=0}^{\rho}\begin{pmatrix} \frac{\rho}{2} \\ \frac{r}{2} \end{pmatrix} {}_2F_1\left(\rho - r, r; \frac{1}{2}(\rho+1), \frac{1}{2}\right)f_r\right.$$

$$\left. + \frac{1}{4}\sum_{r=0}^{\rho}\begin{pmatrix} \rho \\ r \end{pmatrix}\frac{1}{(r+1)(r+2)}\right]. \tag{5.55}$$

If one specializes even further and takes $s = 1/4$, then it is straightforward to show that $f(x) = 2x(1-x)$. Therefore,

$$f_0 = \int_0^1 2x(1-x)\,dx = \frac{1}{3}.$$

And thus,

$$f_1 = \frac{1}{42}\left[{}_2F_1\left(1, 0; 1, \frac{1}{2}\right) + 4\right] = \frac{5}{42},$$

etc.

Now the main result of this subsection can be stated. The theorem that follows uses the ${}_2F_1$ function to illustrate how the moment of an affine fractal function can be expressed as a linear combination of moments of Dirichlet splines.

Theorem 5.10 *Assume $\rho \geq 1$ is an integer. Let $\Xi = \{\xi, 2\xi\}$, $\xi > 0$, and $b = (r, \frac{1}{2}(\rho+1) - r)$, $r = 1, \ldots, [\![\rho]\!]$. (Here $[\![\cdot]\!] : \mathbb{R} \to \mathbb{Z}$ denotes the greatest integer function.) Then*

$$f_\rho = \gamma(\rho)\left[\sum_{i=1}^{N}\{a_i b_i^\rho s_i\, f_0 + Q(\rho)\right.$$

$$\left. + s_i \sum_{r=1}^{\frac{1}{2}\rho}{}'(2\xi)^{\rho-r}[a_i^{r+1}b_i^{\rho-r}f_r + a_i^{\rho-r+1}b_i^r f_{\rho-r}]m_{r-\rho}(b;\Xi)\}\right]$$

for ρ even, and

$$f_\rho = \gamma(\rho)\left[\sum_{i=1}^{N}\{a_i b_i^\rho s_i\, f_0 + a_i^{\frac{1}{2}(\rho+1)}b_i^{\frac{1}{2}(\rho-1)}s_i f_{\frac{1}{2}(\rho+1)} + Q(\rho)\right.$$

$$\left. + s_i \sum_{r=1}^{\frac{1}{2}(\rho-1)}{}'(2\xi)^{\rho-r}[a_i^{r+1}b_i^{\rho-r}f_r + a_i^{\rho-r+1}b_i^r f_{\rho-r}]m_{r-\rho}(b;\Xi)\}\right]$$

when ρ is odd. (The prime on the summation means to divide the final term in the sum by 2.)

Proof. Consider the case of even ρ first. Then the defining relation (5.51) along with Eq. (5.48) yields

$$m_t(b; Z) = R_t(b, Z).$$

Now for $0 < r < \frac{1}{2}(\rho + 1)$, Proposition 5.9–7 in [33] yields

$$_2F_1\left(\rho - r, r; \frac{1}{2}(\rho + 1); \frac{1}{2}\right) = (2\xi)^{\rho - r} m_{r-\rho}(b; \Xi). \tag{5.56}$$

The condition $0 < r < \frac{1}{2}(\rho + 1)$ is necessary to utilize the aforementioned proposition in [33]. For $\frac{1}{2}(\rho + 1) < r < \rho - 1$, the symmetry in the first two arguments of the $_2F_1$ function is exploited to give

$$_2F_1\left(r, \rho - r; \frac{1}{2}(\rho + 1); \frac{1}{2}\right) = (2\xi)^{-r} m_{-r}(b'; \Xi), \tag{5.57}$$

where $b' := (\rho - r, \frac{1}{2} + r - \frac{1}{2}\rho)$.
Inserting Eqs. (5.56) and (5.57) into Eq. (5.54) gives

$$f_\rho = \gamma(\rho) \left[\sum_{i=1}^{N} \{a_i b_i^\rho s_i f_0 + Q(\rho) + i^{r+1} b_i^{\rho - r} s_i f_r \right.$$

$$\left. \times \left(\sum_{r=1}^{\frac{1}{2}\rho} (2\xi)^{\rho - r} m_{r-\rho}(b; \Xi) + \sum_{r=\frac{1}{2}\rho+1}^{\rho-1} (2\xi)^{-r} m_{-r}(b'; \Xi) \right) \right].$$

A change of indices yields the desired result.

Since ρ is even, $r \neq \frac{1}{2}(\rho + 1)$, but this is not the case when ρ is odd. Indeed, if $r = \frac{1}{2}(\rho + 1)$, Proposition 5.9–7 cannot be employed. In this case, the binomial coefficient in the $r = \frac{1}{2}(\rho + 1)$ term in Eq. (5.54) is retained. Thus,

$$f_\rho = \gamma(\rho) \left[\sum_{i=1}^{N} \{a_i b_i^\rho s_i f_0 + a_i^{\frac{1}{2}(\rho+1)} b_i^{\frac{1}{2}(\rho-1)} s_i f_{\frac{1}{2}(\rho+1)} + Q(\rho) \right.$$

$$\left. + s_i \sum_{r=1}^{\frac{1}{2}(\rho-1)} {}'(2\xi)^{\rho - r} [a_i^{r+1} b_i^{\rho - r} f_r + a_i^{\rho - r + 1} b_i^r f_{\rho - r}] m_{r-\rho}(b; \Xi) \} \right].$$

■

The proof of Theorem 5.10 gives some insight into constructing an algorithm for obtaining the moments $m_{r-\rho}(b, \Xi)$, $r = 1, \ldots, \llbracket \rho \rrbracket$.

One begins by choosing *any* affine fractal function and computes the moments $f_0, \ldots, f_\rho$. Now instead of using Lemma 5.6 at each step in the summation in Theorem 5.10, it is only used once. Thus one obtains an equation involving the known moments $f_0, \ldots, f_\rho$, and the one unknown moment of the Dirichlet spline. This process is repeated for each $r = 1, \ldots, \llbracket \rho \rrbracket$ to obtain the desired moments of the Dirichlet spline functions.

5.6.4 Lipschitz continuity of fractal functions

Next it is shown that affine fractal functions belong to a certain Lipschitz class $\mathrm{Lip}^\alpha([0,1], \mathbb{R})$. In order to do this, a dimension result that will be proven in the next chapter is needed.

Theorem 5.11 *Let* $f : [0,1] \to \mathbb{R}$ *be an affine fractal function and* G *its graph. Suppose that the set* $\mathfrak{w}[0,1]$ *is not collinear and that* $\sum_{i=1}^N |s_i| > 1$. *Then the box dimension* $\dim_\beta G$ *is the unique positive solution of*

$$\sum_i^N |s_i| a_i^{d-1} = 1; \tag{5.58}$$

otherwise, $\dim_\beta G = 1$.

The following theorem relates the box dimension of G to the Hölder or Lipschitz exponent α of f.

Theorem 5.12 *Let* f *be an affine fractal function and* G *its graph. Then, for any* $0 < h < 1$ *and all* $x \in [0,1]$,

$$|f(x+h) - f(x)| \le c\, h^\alpha, \tag{5.59}$$

where $c > 0$ *and* $\alpha = 2 - \dim_\beta G$. *That is,* $f \in \mathrm{Lip}^{2-\dim_\beta G}([0,1], \mathbb{R})$.

Proof. Let $h \in (0,1)$ be given, and let $0 < h \le \varepsilon$. Then, if G is covered by squares of side ε, the proof of Theorem 5.11 implies that $|f(x+h) - f(x)| \le \mathcal{N}_\varepsilon(G)\varepsilon^2$, where $\mathcal{N}_\varepsilon(G)$ denotes the minimum number of such squares needed to cover G. It can also be shown that there exists a positive constant c such that $\mathcal{N}_\varepsilon(G) \ge c\varepsilon^{-d}$, where $d := \dim_\beta G$. Hence,

$$\frac{\log |f(x+h) - f(x)|}{\log h} \ge \frac{\log \mathcal{N}_\varepsilon(G)\varepsilon^2}{\log h} \ge \frac{\log c\varepsilon^{2-d}}{\log h} \ge \frac{\log c}{\log h} + (2 - d),$$

since $h \le \varepsilon$. ∎

A converse of the preceding theorem can also be proven. But first a definition.

Definition 5.14 Let $X \in \mathfrak{H}(\mathbb{R}^n)$ and let $Y \in \mathfrak{H}(\mathbb{R}^m)$. The oscillation of a function $f : X \to Y$ over X is defined by

$$\operatorname{osc}(f; X) := \sup\{f(x) - f(x') \mid x, x' \in X\}. \qquad (5.60)$$

Theorem 5.13 *Let $f : [0,1] \to \mathbb{R}$ be a continuous function and G its graph. If there exists a $1 \le d \le 2$ such that $f \in \operatorname{Lip}^{2-d}([0,1], \mathbb{R})$, then $\dim_H G \le \dim_B G \le d$. Moreover, $\mathcal{H}^d(G) < \infty$.*

Proof. Let $0 < \varepsilon < 1$ be given, and let $C_\varepsilon(G)$ be a minimal cover of G by squares of side ε. Denote the cardinality of such a minimal cover by $\mathcal{N}_\varepsilon(G)$. Let n be the smallest integer greater than or equal to $1/\varepsilon$. Then, by the continuity of f, we have

$$\mathcal{N}_\varepsilon(G) \le \sum_{k=0}^{n-1} (2 + \varepsilon^{-1} \operatorname{osc}(f; [k\varepsilon, (k+1)\varepsilon])).$$

Since $f \in \operatorname{Lip}^{2-d}([0,1], \mathbb{R})$, $\operatorname{osc}(f; [k\varepsilon, (k+1)\varepsilon]) \le c\varepsilon^{2-d}$, for all $k = 0, 1, \ldots,$ $n-1$ and some $c > 0$ independent of k. Observing that $n < 1 + \varepsilon^{-1}$, one thus obtains

$$\mathcal{N}_\varepsilon(G) \le (1 + \varepsilon^{-1})(c\varepsilon^{-1}\varepsilon^{2-d}) \le c'\varepsilon^{-d},$$

with $c' > 0$ independent of ε. Therefore, using the definitions of Hausdorff and box dimension, the first statement of the theorem is obtained. That $\mathcal{H}^d(G) < \infty$ follows from the fact that $\mathcal{H}^d_\varepsilon(G) \le \mathcal{N}_\varepsilon(G)\varepsilon^{-d}$. ∎

5.6.5   Extrema of fractal functions

This subsection deals with the location of the extrema of a class of affine fractal functions. This class consists of all non-constant affine fractal functions interpolating $\Delta := \{(j/N, y_j) \mid \forall\, j = 0, \ldots, N : \; y_j \in \mathbb{R}\}$ and satisfying the fixed-point equation

$$f(x) = y_{i-1} + s_i f \circ u_i^{-1}(x), \qquad x \in [(i-1)/N, i/N],$$

where $s_i := y_i - y_{i-1} \in (-1, 1)$, and $i = 1, \ldots, N$. This class of affine fractal functions will be denoted by $\mathcal{E}$. Observe that these conditions imply that all a_i are equal to $1/N$. Most of the results presented here can also be found in [57].

Let $f \in \mathcal{E}$, let $m := \min\{f(x) \mid x \in [0,1]\}$, and let $M := \max\{f(x) \mid x \in [0,1]\}$. The following lemma relates the quantities m and M to the y_j and the differences s_i.

Lemma 5.7 *Let* $f \in \mathcal{E}$ *and assume that* f *takes its minimum value* m *on* $[(i-1)/N, i/N]$, *where* $i \in \{1, \ldots, N\}$. *If* $s_i \geq 0$, *then* $m = y_{i-1} + s_i m$. *If* $s_i < 0$, *then* $m = y_{i-1} + s_i M$.

Proof. Assume first that $s_i \geq 0$. As $m = \min_{x \in [(i-1)/N, i/N]}\{f(x)\}$ and since on $[(i-1)/N, i/N]$, $f(x) = s_i f \circ u_i^{-1}(x) + y_{i-1}$, it is easily seen that $m = y_{i-1} + s_i \min\{f \circ u_i^{-1}(x) \mid x \in [(i-1)/N, i/N]\}$. But this last minimum equals m. This proves the first statement.

Now suppose $s_i < 0$. Then, using similar reasoning, $m = y_{i-1} + s_i \max\{f \circ u_i^{-1}(x) \mid x \in [(i-1)/N, i/N]\} = y_{i-1} + s_i M$. ∎

It should be clear that Lemma 5.7 also applies to the maximum value M. Now suppose that m is attained on $[(i-1)/N, i/N]$ and M is attained on $[(k-1)/N, k/N]$. It is an immediate consequence of Lemma 5.7 that $i \neq k$. For, if $i = k$ and $s_i \geq 0$, then $m = y_{i-1} + s_i m$ and $M = y_{i-1} + s_i M$, which implies that $m = M$. On the other hand, if $s_i < 0$ then, again by Lemma 5.7, $m = y_{i-1} + s_i M$ and $M = y_{i-1} + s_i m$, also implying that $m = M$.

According to the sign of s_i and s_k, there are three possible ways of calculating m:

1. $s_i \geq 0$: Then $m = y_{i-1} + s_i m$, i.e., $m = y_{i-1}/(1 - s_i)$.

2. $s_i < 0$ and $s_k \geq 0$: Then m and M satisfy $m = y_{i-1} + s_i M$ and $M = y_{k-1} + s_k m$. Thus, $m = y_{i-1} + s_i y_{k-1}/(1 - s_k)$.

3. $s_i \geq 0$ and $s_k < 0$: Then $m = y_{i-1} + s_i M$ and $M = y_{k-1} + s_k m$ imply $m = (y_{i-1} + s_i y_{k-1})/(1 - s_i s_k)$.

For $i = 1, \ldots, N$ let $x_i^{(1)} := (i-1)/(N-1)$, $x_{i,k}^{(2)} := [(N-1)(i-1) + (k-1)]/N(N-1)$, and $x_{i,k}^{(3)} := [N(i-1) + (k-1)]/(N^2 - 1)$. The straightforward proof of the next lemma is left to the reader (also [57]).

Lemma 5.8 *Let* $f \in \mathcal{E}$. *Then* $f(x_i^{(1)}) = y_{i-1}/(1 - s_i)$, $f(x_{i,k}^{(2)}) = y_{i-1} + s_i y_{k-1}/(1 - s_k)$, *and* $f(x_{i,k}^{(3)}) = (y_{i-1} + s_i y_{k-1})/(1 - s_i s_k)$.

The lemma and the arguments preceding it suggest the introduction of an $N \times N$ matrix Y whose rows are given by

$$\begin{pmatrix} m_i \\ 0 \\ \vdots \\ 0 \end{pmatrix} \quad \text{and} \quad \begin{pmatrix} m_{i,1} \\ m_{i,2} \\ \vdots \\ m_{i,N} \end{pmatrix},$$

according to $s_i \geq 0$ or $s_i < 0$. In the former case, let $m_i := f(x_i^{(1)})$, and in the latter, $m_{i,k} := f(x_{i,k}^{(2)})$ if $s_k \geq 0$, and $m_{i,k} := f(x_{i,k}^{(3)})$ if $s_k < 0$. In a similar fashion one defines an $N \times N$ matrix X whose rows are given by

according to $s_i \geq 0$ or $s_i < 0$. As before, if $s_k \geq 0$, then $\ell := 2$, and if $s_k < 0$ then $\ell := 3$. Extending f to a function on $N \times N$ matrices $A = (a_{ik})$ via $f((a_{ik})) := (f(a_{ik}))$, the results in Lemma 5.8 can be expressed as $f(X) = Y$.

Combining Lemmas 5.7 and 5.8 and the preceding characterization in terms of the matrices X and Y, gives the next theorem.

Theorem 5.14 *Let $f \in \mathcal{E}$. The minimum value of f is the smallest entry in the matrix Y, and the corresponding entry in X is one of the locations where f attains this minimum value. Analogously, the maximum value of f is the largest entry in X, and the corresponding entry in X is one of the locations where f attains this maximum value.*

Proof. Lemma 5.7 shows that m is an entry in X, and Lemma 5.8 gives $m_i = f(x_i^{(1)}) \geq m$ and $m_{i,k} = f(x_{i,k}^{(\ell)}) \geq m$. This proves the result for the minimum value. The statement for the maximum value is proven similarly. ∎

After having established the numerical value of the extremum of a fractal function $f \in \mathcal{E}$, the location of this extremum and its uniqueness need to be investigated. For this purpose, the following two sets are introduced:

$$I_m := \{i \in \{1, \dots, N\} | f(x) = m \text{ on } [(i-1)/N, i/N]\}$$

and

$$I_M := \{i \in \{1, \dots, N\} | f(x) = M \text{ on } [(i-1)/N, i/N]\}.$$

Lemma 5.7 allows us to characterize the sets I_m and I_M even further:

$$I_m = \{i | (s_i \geq 0 \wedge m = y_{i-1} + s_i m) \vee (s_i < 0 \wedge m = y_{i-1} + s_i M)\}$$

and

$$I_M = \{i | (s_i \geq 0 \wedge M = y_{i-1} + s_i M) \vee (s_i < 0 \wedge M = y_{i-1} + s_i m)\}.$$

With these sets a directed graph $\mathsf{G} = (\mathsf{V}, \mathsf{E})$ is associated in the following manner: The vertex set $\mathsf{V} := \{1, \ldots, N\}$ and the edge set $\mathsf{E} := \{(i, k) \in \mathsf{V} \times \mathsf{V} \mid (i \in I_m \wedge s_i \geq 0 \Rightarrow k \in I_m) \vee (i \in I_m \wedge s_i < 0 \Rightarrow k \in I_M) \vee (i \in I_M \wedge s_i \geq 0 \Rightarrow k \in I_M) \vee (i \in I_M \wedge s_i < 0 \Rightarrow k \in I_m)\}$. Note that, since f cannot attain its minimum and maximum value on the same interval, the $\vee$ symbols appearing in the definition of the edge set are exclusive.

In order to describe the location of an extremum of a fractal function $f \in \mathcal{E}$, every $x \in [0, 1]$ needs to be expressed in N-ary expansion: $x = (i_1 - 1)/N + (i_2 - 1)/N^2 + \cdots + (i_n - 1)/N^n + \cdots$, with $i_\ell \in \{1, \ldots, N\}$. A more compact and succinct notation for the N-ary expansion of $x \in [0, 1]$ is given by $x = \sigma_-\mathbf{i}$. Here the operator $\sigma_- : \{1, \ldots, N\}^{\mathbb{N}} \to \{0, 1, \ldots, N-1\}^{\mathbb{N}}$ is defined by $\sigma_-(i_1\, i_2\, \ldots\, i_n\, \ldots) := (i_1 - 1\, i_2 - 1\, \ldots\, i_n - 1\, \ldots)$. The next theorem gives necessary and sufficient conditions for $x \in [0, 1]$ to yield the minimum value of f.

Theorem 5.15 *A number $x \in [0, 1)$ yields the minimum value of a fractal function $f \in \mathcal{E}$ iff its N-ary expansion $\sigma_-(i_1 i_2 \ldots i_n \ldots)$ satisfies the following two conditions:*

1. $i_1 \in I_m$;

2. *If i_n follows i_m in $\sigma_-(i_1 i_2 \ldots i_n \ldots)$ and if $s_{i_m} \neq 0$, then $(i_m, i_n) \in \mathsf{E}$.*

Proof. Necessity is shown first. For this purpose, assume that f attains its minimum value at $x \in [0, 1)$. Then write $Nx = [\![Nx]\!] + \xi$, where $[\![\cdot]\!] : \mathbb{R} \to \mathbb{Z}$ denotes the greatest integer function. Let $i_1 := [\![Nx]\!] + 1$. Then $x \in [(i-1)/N, i/N)$, and thus f attains its minimum on $[(i_1 - 1)/N, i_1/N]$. Therefore, $i_1 \in I_m$. To prove that the second condition also holds, note that the fixed-point property of f implies that $m = f(x) = y_{i-1} + s_i f(\xi)$. Hence, $f(\xi) = m$ if $s_i \geq 0$ and $f(\xi) = M$ is $s_i < 0$. Defining $i_2 := [\![N\xi]\!] + 1$, it is easy to see that $i_2 \in I_m$ if $s_{i_1} \geq 0$ and $i_2 \in I_M$ if $s_{i_1} < 0$. Thus, condition 2 is satisfied for the first two consecutive integers in the N-ary expansion of x. Applying the preceding argumentation to ξ instead of x, one easily verifies condition 2 for the next pair of consecutive integers in the N-ary expansion of x. Proceeding inductively yields the validity of the second condition in general.

The sufficiency part of the theorem is proved next. Suppose then that the N-ary expansion of $x \in [0, 1)$ satisfies conditions 1 and 2. As before, denote by $i_1 - 1$ and ξ_1 the integer and fractional part of x, respectively. Again, by the fixed-point property of f, $f(x) = y_{i_1-1} + s_{i_1} f(\xi_1)$. Now assume that $s_{i_1} \geq 0$.

Then, using Lemma 5.7 and condition 1, the preceding fixed-point property can be rewritten so that $f(x) - m = s_{i_1}(f(\xi_1) - m)$. In the same manner one obtains $f(x) - m = s_{i_1}(f(\xi_1) - M)$ if $s_{i_1} < 0$. Applying the same reasoning to ξ_1, it is seen — depending on the sign of s_{i_2} — that the integer part $i_2 - 1$ and the fractional part ξ_2 of $N\xi_1$ satisfy either $f(\xi_1) - m = s_{i_2}(f(\xi_2) - m)$ or $f(\xi_1) - m = s_{i_2}(f(\xi_2) - M)$. However, $f(x) - m = s_{i_1}s_{i_2}(f(\xi_2) - m)$ if $s_{i_1}s_{i_2} \geq 0$, or $f(x) - m = s_{i_1}s_{i_2}(f(\xi_2) - M)$ if $s_{i_1}s_{i_2} < 0$. Induction yields after n steps that either $f(x) - m = s_{i_1}s_{i_2}\cdots s_{i_n}(f(\xi_n) - m)$, if $s_{i_1}s_{i_2}\cdots s_{i_n} \geq 0$, or $f(x) - m = s_{i_1}s_{i_2}\cdots s_{i_n}(f(\xi_n) - M)$ if $s_{i_1}s_{i_2}\cdots s_{i_n} < 0$. Since, by the continuity of f the set $\{f(\xi_1), f(\xi_2), \ldots, f(\xi_n), \ldots\}$ is uniformly bounded and since $s_{i_1}s_{i_2}\cdots s_{i_n} \to 0$ as $n \to \infty$, $f(x) - m \to 0$.   ∎

Now necessary and sufficient conditions for the existence of a unique extremum value for $f \in \mathcal{E}$ can be given.

Theorem 5.16 *The function $f \in \mathcal{E}$ attains a unique minimum iff the following two conditions hold:*

1. $\mathrm{card}(I_m) = 1$.

2. *If i in the unique element in I_m and if $s_i < 0$, then $\mathrm{card}(I_M) = 1$.*

Proof. NECESSITY: If $i \in I_m$, the graph G can be traversed along an infinite path starting at vertex i. This infinite path yields the N-ary expansion of a point $x \in [0, 1]$ which, by the preceding theorem, satisfies $f(x) = m$. Hence, there exists a point x yielding the minimum of f whose N-ary expansion begins with i. Now suppose that f has a unique minimum value. Then, by the preceding arguments, I_m consists of only one element, say i. If it happens that $s_i < 0$, then the uniqeness of the minimum also forces the set I_M to have cardinality one.
SUFFICIENCY: Assuming that conditions 1 and 2 hold, one immediately concludes the existence of a unique path that originates at the vertex of I_m. The preceding theorem gives then desired the result.   ∎

To illustrate the foregoing theoretical considerations, consider a specific example, namely, Kiesswetter's fractal function f_K (Section 5.2).

Example 5.13 Clearly, $f_K \in \mathcal{E}$. Recall that the y-values of the interpolating set are given by $y_0 = 0 = y_2$, $y_1 = -1/2$, $y_3 = 1/2$, and $y_4 = 1$. This then gives $s_1 = -1/2$ and $s_2 = s_3 = s_4 = 1/2$. The matrices X and Y are as follows:

$$X = \begin{pmatrix} 5/24 & 1/3 & 2/3 & 1 \\ 1/4 & 0 & 0 & 0 \\ 3/10 & 0 & 0 & 0 \\ 7/20 & 0 & 0 & 0 \end{pmatrix}, \quad Y = \begin{pmatrix} 0 & -1 & 0 & 1 \\ 1/2 & 0 & 0 & 0 \\ 0 & 0 & 0 & 0 \\ -1/2 & 0 & 0 & 0 \end{pmatrix}.$$

The minimum and maximum values of f_K are then -1 and 1, respectively. Also, $I_m = \{2\}$ and $I_M = \{4\}$. Thus, by Theorem 5.16 the unique minimum and the unique maximum are attained at $x = 1/3$ and $x = 1$, respectively.

5.7 Peano Curves

Space-filling curves have been known in mathematics since the latter part of the 19th century, when Peano ([148]) showed that it is possible to map the interval $[0, 1]$ continuously onto a compact subset of $X \subset \mathbb{R}^2$. It came as a surprise that such a continuous vector-valued function $p : [0, 1] \to X$ exists, and it showed that continuity in the component functions of p is not enough to ensure bijectivity. In [91], Hilbert replaced Peano's arithmetic definition of p by a simple geometric construction. Several other geometric constructions appeared in the literature, until Knopp showed in [106] that all Peano curves can be generated using a simple geometric principle.

In this section it is shown that all the known Peano curves can indeed be generated using IFSs, and that they are projections of iso-dimensional fractal functions. For a list of Peano curves the interested reader is referred to [106, 127, 159, 160, 182]. Some of the material presented here can also be found in [159].

Let I denote the unit interval $[0, 1] \subset \mathbb{R}$. It should be clear that if in the construction of univariate fractal functions on I having the interpolation property with respect to $\Delta := \{(x_j, yj) \in I \times \mathbb{R} | j = 0, 1, \ldots, N\}$ one does *not* insist on ordering the x_j linearly, a *fractal curve* instead of a fractal function is obtained.

Definition 5.15 If the attractor A of a hyperbolic IFS $(I \times \mathbb{R}, \mathbf{w})$ is the graph of a continuous fractal curve $p : I \to \mathbb{R}^2$ such that the $p(I)$ has positive two-dimensional Jordan content in $\mathbb{R}^2$, then p is called a Peano or space-filling curve.

Now suppose that $\Delta := \{(x_j, y_j) \in I \times \mathbb{R} | \forall j = 0, 1, \ldots, N : x_j = j/N \wedge y_0 = y_N = 0\}$ is an interpolation set and $f_\Phi : I \to \mathbb{R}$ the unique fractal function interpolating Δ. It is also assumed that graph(f_Φ) is the unique attractor of the hyperbolic IFS $(I \times \mathbb{R}, \mathbf{w})$, where $w_i \in \mathcal{A}(I \times \mathbb{R}, I \times \mathbb{R})$ or $w_i \in \mathcal{S}^*(I \times \mathbb{R}, I \times \mathbb{R})$, for all $i = 1, \ldots, N$. More precisely, for all $i = 1, \ldots, N$ and some $s \in (-1, 1)$,

$$w_i(x, y) = \begin{pmatrix} 1/N & 0 \\ y_i - y_{i-1} & s \end{pmatrix} \begin{pmatrix} x \\ y \end{pmatrix} + \begin{pmatrix} (i-1)/N \\ y_{i-1} \end{pmatrix},$$

or

$$w_i(x, y) = H_s \circ \tau_v \circ O \begin{pmatrix} x \\ y \end{pmatrix},$$

where $H_s : \mathbb{R}^2 \to \mathbb{R}^2$, $H_s(x) := sx$, is a homothety, $\tau_v : \mathbb{R}^2 \to \mathbb{R}^2$, $\tau_v x := x + v$ a translation, and $O : \mathbb{R}^2 \to \mathbb{R}^2$ an orthonormal operator on $\mathbb{R}^2$ (also the remark on page 102). Denote by $\mathfrak{w}$ the associated set-valued mapping $\mathfrak{w} : \mathfrak{H}(I \times \mathbb{R}) \to \mathfrak{H}(I \times \mathbb{R})$. Note that $\mathfrak{w}[0, 1]$ is a piecewise linear function $\tilde{\theta} : I \to \mathbb{R}$ interpolating Δ. The function $\tilde{\theta}$ is extended to a piecewise linear function $\theta : \mathbb{R} \to \mathbb{R}$ by setting

$$\theta(x) := \begin{cases} \tilde{\theta}(x) & x \in I \\ \tilde{\theta}(-x) = \tilde{\theta}(x+1) & \text{otherwise.} \end{cases} \tag{5.61}$$

This function is called the *periodic extension of* $\mathfrak{w}[0, 1]$ *to* $\mathbb{R}$. Using θ, the fractal function $f_\Phi \in \mathcal{F}|\Delta(I, \mathbb{R})$ can be expressed in terms of the following infinite series:

$$f_\Phi(x) = \sum_{k=0}^{\infty} s^k \theta(N^k x), \qquad x \in [0, 1]. \tag{5.62}$$

The partial sums $\mathfrak{p}_n(x) := \sum_{k=0}^n s^k \theta(N^k x)$, $x \in [0, 1]$ and $n \in \mathbb{N}$, are called the *nth polygonal approximations* to graph(f_Φ). Clearly, $\mathfrak{p}_n \to f_\Phi$ in the sup-norm as $n \to \infty$.

In what follows, without loss of generality it is assumed that the considered Peano curves are contained in $Q := [0, 1] \times [0, 1] \subset \mathbb{R}^2$. (This can always by achieved be an appropriate retraction.) The first example is Hilbert's construction of Peano's original space-filling curve (cf. [91]). In many ways this is an archetypical example exhibiting the general charcteristics of Peano curves and their relation to hyperbolic IFSs.

Let $\Delta := \{(0, 1/4), (1/4, 1/4), (1/4, 3/4), (3/4, 3/4)\}$ and define contractive mappings on Q by

$$w_1(x,y) := \begin{pmatrix} 0 & 1/2 \\ -1/2 & 0 \end{pmatrix} \begin{pmatrix} x \\ y \end{pmatrix} + \begin{pmatrix} 0 \\ 1/4 \end{pmatrix}$$

$$w_2(x,y) := \begin{pmatrix} -1/2 & 0 \\ 0 & 1/2 \end{pmatrix} \begin{pmatrix} x \\ y \end{pmatrix} + \begin{pmatrix} 1/2 \\ 1/2 \end{pmatrix}$$

$$w_3(x,y) := \begin{pmatrix} 1/2 & 0 \\ 0 & -1/2 \end{pmatrix} \begin{pmatrix} x \\ y \end{pmatrix} + \begin{pmatrix} 1/2 \\ 1/2 \end{pmatrix}$$

$$w_4(x,y) := \begin{pmatrix} 0 & -1/2 \\ 1/2 & 0 \end{pmatrix} \begin{pmatrix} x \\ y \end{pmatrix} + \begin{pmatrix} 1 \\ 1/2 \end{pmatrix}.$$

It is easy to see that $(Q, \mathbf{w})$ is a hyperbolic IFS satisfying (C1) — (C4) in Section 5.5. Hence, the attractor A is the graph of a continuous function $p : I \to Q$ interpolating Δ. By Propositions 5.7 and 5.8 graph(p) can be viewed as the orthogonal projection of a fractal function $\hat{p} \in \mathcal{F}|\Delta(I, \mathbb{R}^2)$ onto Q. In the next chapter it is not only shown that graph(p) has two-dimensional Jordan content, but that indeed $\dim_H \mathrm{graph}(\hat{p}) = \dim_H \mathrm{graph}(p) = 2$. Hence, graph$(p)$ is the *iso-dimensional orthogonal projection* of graph$(\hat{p})$.

In Fig. 5.11 the *generator* $\mathfrak{w}[0,1]$ of Peano's curve and the third approximating polygon $\mathfrak{p}_3$ is given.

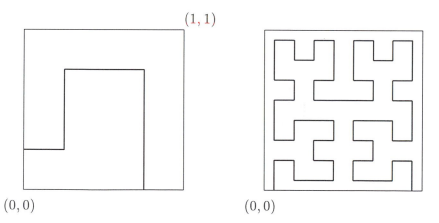

$(1,1)$

$(0,0)$ $(0,0)$

Figure 5.11: The generator and $\mathfrak{p}_3$ of Peano's space-filling curve.

Remark. Comparing our generator to Peano's, the careful reader may have noticed that ours has additional "legs": $w_1[0,1]$ and $w_4[0,1]$. These were introduced to guarantee the connectedness of the attractor.

All the Peano curves investigated in [182] can be obtained by using the preceding construction (slight modifications should be made in some cases to guarantee that $\mathfrak{w}[0,1]$ is connected). One more such example, namely a triangular Peano curve, is given: Define contractive mappings $w_i : Q \to Q$, $i = 1, 2$, by

$$w_i(x, y) := 1/2 \begin{pmatrix} 1 & (-1)^{i-1} \\ (-1)^{i-1} & 1 \end{pmatrix} \begin{pmatrix} x \\ y \end{pmatrix} + \begin{pmatrix} 1/2(i-1) \\ 1/2(i-1) \end{pmatrix}.$$

The attractor of the hyperbolic IFS $(Q, \mathbf{w})$ is an isosceles triangle with vertices at $(0,0)$, $(1,0)$, and $(1/2, 1/2)$. The generator of the associated Peano curve is depicted in Fig. 5.12.

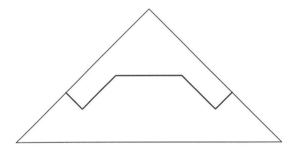

Figure 5.12: The generator of a triangular Peano curve.

In 1938, I. J. Schoenberg introduced a space-filling curve that is similar to the curve introduced earlier by Lebesgue. Next it will be seen how both curves can be obtained using IFSs. The construction of Lebesgue's space-filling curve proceeds as follows: Let C be the classical Cantor set and let $x \in \mathsf{C}$. It is wellknown that every point on the Cantor set has a ternary representation of the form

$$x = \sum_{n \in \mathbb{N}} \frac{2\omega_n}{3^n}, \qquad \omega_n \in \{0, 1\}.$$

Define continuous surjections $\beta, \delta : \mathsf{C} \twoheadrightarrow \mathsf{C}$, by

$$\beta(x) := \sum_{n \in \mathbb{N}} \frac{2\omega_{2n-1}}{3^n} \qquad \text{and} \qquad \delta(x) := \sum_{n \in \mathbb{N}} \frac{2\omega_{2n}}{3^n}.$$

These surjections are then used to define continuous fractal functions $f_j :$

$[0, 1] \to \mathbb{R}$, $j = 1, 2$, by

$$f_1(x) := \sum_{n \in \mathbb{N}} \frac{\omega_n}{2^n} \qquad \text{for } x \in \beta\mathsf{C},$$

$$f_2(x) := \sum_{n \in \mathbb{N}} \frac{\omega_n}{2^n} \qquad \text{for } x \in \delta\mathsf{C},$$

and for $x \in [0, 1] \setminus \beta\mathsf{C}$, respectively $x \in [0, 1] \setminus \delta\mathsf{C}$, inductively by

$$f_1(x) := 1/2 \qquad \text{and} \qquad f_2(x) := 1 - 3x, \qquad x \in (1/3, 2/3),$$

and continued in this manner on the remaining intervals. Lebesgue's Peano curve is then the curve $p_L : I \to Q$, $p_L := (f_1, f_2)$.

To see how the nth-approximating polygon is the image under a set-valued map associated with a hyperbolic IFS, functions $w_i^{(1)} : [0, 1/2] \to \mathbb{R}$ and $w_i^{(2)} : [0, 1/2] \to \mathbb{R}$, $i = 1, 2, 3, 4$, have to be defined first: For $i = 1, 2$, $w_i^{(1)}$ maps $J^{(1)} := [0, 1/2]$ homeomorphically onto $\{\frac{1}{2}J^{(1)} + \frac{1}{2}(i - 1)\} \times 0$, and for $i = 3, 4$, homeomorphically onto $\{\frac{1}{2}J^{(1)} + \frac{1}{2}(i - 3)\} \times \{1/2\}$. Here the notation $aX + b$ is used to denote the set $\{ax + b \mid x \in X \subseteq \mathbb{R}^n\}$, $a \in \mathbb{R}$, $b \in \mathbb{R}^n$. Let $J^{(2)} := J^{(1)} \times \{0\} \cup J^{(1)} \times \{1/2\} \cup \{(x, y) \mid 2x + 2y - 1 = 0\}$. The maps $w_i^{(2)}$ are to map $J^{(2)}$ homeomorphically onto $\{\frac{1}{2}J^{(2)}, \frac{1}{2}J^{(2)} + (0, 1/2), \frac{1}{2}J^{(2)} + (1/2, 0)$, and $\frac{1}{2}J^{(2)} + (1/2, 1/2)$, respectively.

The line segment connecting the initial point of $w_{i+1}^{(j)}J^{(j)}$ to the terminal point of $w_i^{(j)}J^{(j)}$ will be denoted by $L_i^{(j)}(J^{(j)})$, $j = 1, 2$.

The set-valued map associated with $w_i^{(j)}$, $j = 1, 2$, is defined in the usual way: $\mathfrak{w}^{(j)}(E) := \bigcup_{i=1}^{4} w_i^{(j)}(E)$, for $E \in \mathfrak{H}(Q)$. Finally, let $\widetilde{\mathfrak{w}^{(j)}} : \mathfrak{H}(Q) \to \mathfrak{H}(Q)$ be defined by

$$\widetilde{\mathfrak{w}^{(j)}} := \mathfrak{w}^{(j)} \cup \bigcup_{k=1}^{3} L_k^{(j)},$$

$$\widetilde{\mathfrak{w}^{(j)}}^n := \mathfrak{w}^{(j)} \circ \widetilde{\mathfrak{w}^{(j)}}^{n-1} \cup \bigcup_{k=1}^{3} L_k^{(j)} \circ \widetilde{\mathfrak{w}^{(j)}}^{n-1}.$$

The nth-approximating polygon $\mathfrak{p}_n^L$ to p_L is then given by

$$\mathfrak{p}_n^L = \begin{cases} \widetilde{\mathfrak{w}^{(1)}}^m (J^1) & \text{for } n = 2m - 1 \\ \widetilde{\mathfrak{w}^{(2)}}^m (J^2) & \text{for } n = 2m. \end{cases} \tag{5.63}$$

The sixth and seventh approximating polygon for p_L are shown in Fig. 5.13.

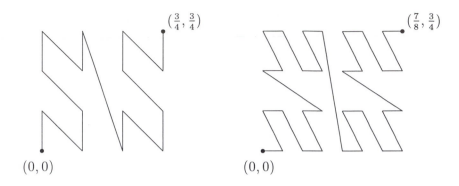

Figure 5.13: $\mathfrak{p}_4$ and $\mathfrak{p}_5$ for p_L.

Schoenberg's space-filling curve p_S is an example of a more elaborate construction. It is explicitly given by $p_S = (f, g) : [0, 1] \to 2Q$, where

$$f(x) := \sum_{k \in \mathbb{N}} \frac{1}{2^k} \theta(3^{2k-2}x),$$

and

$$g(x) := \sum_{k \in \mathbb{N}} \frac{1}{2^k} \theta(3^{2k-1}x),$$

with $\theta : \mathbb{R} \to \mathbb{R}$ defined by

$$\theta(x) := \begin{cases} 0 & \text{for } x \in [0, 1/3) \\ 3x - 1 & \text{for } x \in [1/3, 2/3) \\ 1 & \text{for } x \in [2/3, 1] \\ \theta(-x) & \text{otherwise.} \end{cases}$$

Using the periodic function θ, f and g can be expressed in the following way:

$$f = \frac{1}{2} \sum_{m=0}^{\infty} \left(\frac{1}{\sqrt{2}}\right)^{2m} \theta(3^{2m} \cdot) \text{ and } g = \frac{1}{\sqrt{2}} \sum_{m=1}^{\infty} \left(\frac{1}{\sqrt{2}}\right)^{2m-1} \theta(3^{2m-1} \cdot).$$

Note that the preceding representations of f and g imply that $2f + \sqrt{2}g$ is a fractal function $h : [0, 1] \to \mathbb{R}$, expressed in the form (5.62):

$$h = \sum_{m=0}^{\infty} \left(\frac{1}{\sqrt{2}}\right)^{m} \theta(3^{m} \cdot).$$

It is also easy to see that graph(h) is the unique attractor of the hyperbolic IFS $(2Q, \mathbf{w})$ with

$$w_1(x, y) = \begin{pmatrix} 1/3 & 0 \\ -1/\sqrt{2} & 1/\sqrt{2} \end{pmatrix} \begin{pmatrix} x \\ y \end{pmatrix}$$

$$w_2(x, y) = \begin{pmatrix} 1/3 & 0 \\ 1 - 1/\sqrt{2} & 1/\sqrt{2} \end{pmatrix} \begin{pmatrix} x \\ y \end{pmatrix} + \begin{pmatrix} 1/3 \\ 0 \end{pmatrix}$$

$$w_3(x, y) = \begin{pmatrix} 1/3 & 0 \\ -1/\sqrt{2} & 1/\sqrt{2} \end{pmatrix} \begin{pmatrix} x \\ y \end{pmatrix} + \begin{pmatrix} 2/3 \\ 1 \end{pmatrix}$$

$$w_4(x, y) = \begin{pmatrix} 1/3 & 0 \\ -1/\sqrt{2} & 1/\sqrt{2} \end{pmatrix} \begin{pmatrix} x \\ y \end{pmatrix} + \begin{pmatrix} 1 \\ 1 \end{pmatrix}$$

$$w_5(x, y) = \begin{pmatrix} 1/3 & 0 \\ -1 - 1/\sqrt{2} & 1/\sqrt{2} \end{pmatrix} \begin{pmatrix} x \\ y \end{pmatrix} + \begin{pmatrix} 4/3 \\ 1 \end{pmatrix}$$

$$w_6(x, y) = \begin{pmatrix} 1/3 & 0 \\ -1/\sqrt{2} & 1/\sqrt{2} \end{pmatrix} \begin{pmatrix} x \\ y \end{pmatrix} + \begin{pmatrix} 5/3 \\ 0 \end{pmatrix}.$$

The set Δ of interpolation points is given by $\{(0, 0), (1/3, 0), (2/3, 1), (1, 1), (4/3, 1), (5/3, 0), (2, 0)\}$. Fig. 5.14 shows how p_S can be constructed from f and g. The nth-approximating polygon $\mathfrak{p}_n^S$ of p_S can be obtained from the

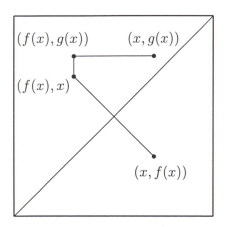

Figure 5.14: The construction of p_S from f and g.

nth-partial sums f_n and g_n of f and g, respectively. The vertices of $\mathfrak{p}_n^S$ which are given by $(f_n(q/3^n), g_n(q/3^n))$, with $q = 0, 1, \ldots, 3^n - 1$, are the images of $[0, 2]$ under the set-valued map $\mathfrak{u} : \mathfrak{H}([0, 2]) \to \mathfrak{H}([0, 2])$ associated with $u_i = 1/3(\cdot + i - 1)$, $i = 1, \ldots, 6$. Note that the u_i are the "horizontal" components of the w_i.

Peano curves can also be constructed using the recurrent set formalism. This is done by means of a specific example which can also be found in [45]. There, several more examples of space-filling curves are listed.

Let $X := \{a, b, c, d\}$ and let $S[X]$ be the free semigroup generated by X. Define a free semigroup endomorphism $\vartheta : S[X] \to S[X]$ by

$$\vartheta(a) := a\,b\,a\,d, \qquad \vartheta(b) := a\,b\,a\,b,$$
$$\vartheta(c) := c\,d\,c\,b, \qquad \vartheta(d) := c\,d\,c\,d.$$

Let $V := \{(x, y, x, y) \,|\, x, y \in \mathbb{R}\} \subseteq \mathbb{R}^4$. Then it is readily seen that V is ϑ^*-invariant. Let $f : S[X] \to \mathbb{R}^2$ be defined by

$$f(a) := (1, 0) =: -f(c), \qquad f(b) := (0, 1) =: -f(d).$$

Thus, L_ϑ is given by

$$L_\vartheta = \begin{pmatrix} 2 & 2 \\ 0 & 2 \end{pmatrix}.$$

Finally, define $K : S[X] \to \mathbb{R}^2$ by

$$K(x) := \{t f(x) \,|\, t \in [0, 1]\}.$$

The recurrent set $K_\vartheta(a)$ is then a space-filling curve whose second approximation is depicted in Fig. 5.15.

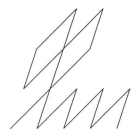

Figure 5.15: The second approximation to $K_\vartheta(a)$.

5.8 Fractal Functions of Class C^k

In Section 5.1 it was shown how to construct continuous fractal functions as unique fixed points of a Read-Bajraktarević operator Φ acting on a set of continuous functions. It is natural to ask whether it is possible to obtain fixed points of class C^k starting with a Read-Bajraktarević operator Φ acting on C^k functions. Here in this section the answer to this question is given, and a method is introduced that allows the construction of C^k-fractal functions via integration.

Throughout this section $X := [0, 1]$ and $Y := \mathbb{R}$, although a more general setup could be considered. It is also assumed that the maps u_i are affine contractions on $[0, 1]$ satisfying $u_i^{-1}(x) = 1$ and $u_{i+1}^{-1}(x) = 0$, for $x \in u_i[0, 1] \cap u_{i+1}[0, 1]$, and that the mappings $v_i(x, \cdot)$ are of the form $v_i(x, \cdot) = \lambda_i(x) + s_i(\cdot)$, with $\lambda_i \in \text{Lip}(X)$ and $s_i \in (-1, 1)$, $i = 1, \ldots, N$. Again, these are certainly not the most general assumptions; however, for our purposes they will prove more than sufficient. For the sake of convenience, let $u_i[0, 1] =: I_i$ and denote compositions of the form $f \circ u_i^{-1}$ by $^\#f_i$. Let k always denote a non-negative integer. In order to proceed, two definitions are needed.

Definition 5.16 Let X and Y be subsets of $\mathbb{R}^n$ endowed with the sup-norm $\| \cdot \|_\infty$, and let $C^k(X, Y)$ denote the linear space of all functions $f : X \to Y$ that possess continuous and bounded derivatives up to order k on X. The C^k-topology on $C^k(X, Y)$ is the topology induced by the norm

$$\|f\|_{C^k} := \sup\{\|f^{(\ell)}\|_\infty | \ell = 0, 1, \ldots k\}, \qquad f \in C^k(X, Y).$$

Remark. Note that convergence in $C^k(X, Y)$ means uniform convergence in X not only of the sequence of functions itself, but also of the sequence of derivatives up to order k.

Definition 5.17 Let X and Y be subsets of $\mathbb{R}^n$, and let $f : X \to Y$ be an element of $C^k(X, Y)$. The function f is said to belong to the class $\text{Lip}^{\alpha+k}(X, Y)$ iff $f^{(k)} \in \text{Lip}^\alpha(X, Y)$, $0 < \alpha < 1$.

As in Eqs. (5.2) and (5.3), define $b : [0, 1] \to [0, 1]$ by $b(x) = \sum_i u_i^{-1}(x) \chi_{I_i}(x)$ and $v(x, \cdot) : \mathbb{R} \to \mathbb{R}$ by $v(x, y) = \sum_i (\lambda_i(x) + s_i y) \chi_{I_i}(x)$. Instead of requiring condition (B^0) to hold, the stronger condition (B^k) is required to hold:

(B^k) There exists a family $\mathfrak{F}^k([0,1],\mathbb{R})$ of C^k-functions such that for all $\mathfrak{f} \in \mathfrak{F}^k([0,1],\mathbb{R})$, the functions $\lambda_i + \mathfrak{f} \circ b$ are elements of $\mathfrak{F}^k([0,1],\mathbb{R})$ and

$$\frac{d^\ell}{dx^\ell}\left({}^\#\lambda_{i,i}(x) + s_i \, {}^\#\mathfrak{f}_i(x)\right) = \frac{d^\ell}{dx^\ell}\left({}^\#\lambda_{i+1,i+1}(x) + s_{i+1} \, {}^\#\mathfrak{f}_{i+1}(x)\right) \quad (5.64)$$

for all $\ell \in \{0,1,\ldots,k\}$ and $x \in I_i \cap I_{i+1}$.

Suppose that conditions (A) and (B^k) are satisfied. Let $u_i(x) := a_i x + b_i$, $a_i \neq 0$, $i = 1,\ldots,N$, and define a linear space $\Lambda^k := \{g_i \in C^k([0,1]) | \forall \ell \in \{0,1,\ldots,k\} \, \forall \, i \in \{2,\ldots,N-1\} : a_i^{-\ell} g_i^{(\ell)}(1)(1 + s_i \, (a_1^\ell - s_1)^{-1}) = a_{i+1}^{-\ell} g_{i+1}^{(\ell)}(0)(1 + s_{i+1}(a_N^\ell - s_N)^{-1})\}$. The Read-Bajraktarević operator associated with the functions b and $v(x,\cdot)$ defined earlier is denoted by Ψ. Let f_Ψ be the unique fixed point of this operator. Below it is shown that f_Ψ is of class C^k.

Theorem 5.17 *Let Ψ be the Read-Baraktarević operator defined earlier, and let $\lambda_i \in \Lambda^k$ for all $i = 1,\ldots,N$. Suppose that $(\min_i\{a_i\})^{-k} \max_i\{|s_i|\} < 1$. Then the unique fixed point f_Ψ of Ψ is of class C^k.*

Proof. It is necessary to express the dependence of Ψ upon the λ_i and the s_i explicitly. Therefore, Ψ is written as $\Psi(\lambda_i, s_i)$. Define $\mathfrak{F}^k([0,1],\mathbb{R}) := C^k([0,1],\mathbb{R})$, and let f_Ψ be the unique fixed point of Ψ. Let $\ell \in \{0,1,\ldots,k\}$. Then, for all $i = 1,\ldots,N$,

$$f_\Psi^{(\ell)}(x) = a_i^{-\ell} \lambda_i^{(\ell)} \circ u_i^{-1}(x) + a_i^{-\ell} s_i f_\Psi^{(\ell)} \circ u_i^{-1}(x), \qquad x \in I_i.$$

Using the fact that all $\lambda_i \in \Lambda^k$ and that $(\min_i\{a_i\})^{-k} \max_i\{|s_i|\} < 1$, it is readily seen that $\Psi(\lambda_i, s_i)$ is a contraction in the C^ℓ-topology with contractivity $(\min_i\{a_i\})^{-k} \max_i\{|s_i|\}$. Hence, its unique fixed point f_Ψ is of class C^k. Moreover, by the preceding equation, notice that $f_{\Psi(\lambda_i,s_i)}^{(\ell)}$ is the unique fixed point of the Read-Bajraktarević operator $\Psi(a_i^{-\ell} \lambda_i^{(\ell)}, a_i^{-\ell} s_i)$, i.e.,

$$f_{\Psi(\lambda_i,s_i)}^{(\ell)} = f_{\Psi(a_i^{-\ell} \lambda_i^{(\ell)}, a_i^{-\ell} s_i)}.$$

(Here the uniqueness of the fixed point was used.)   ■

This theorem gives rise to a definition.

Definition 5.18 The unique fixed point of the Read-Bajraktarević operator Ψ is called a fractal function of class C^k. The collection of all such fractal functions is denoted by $\mathcal{F}^k([0,1],\mathbb{R})$; if $k = 0$, then $\mathcal{F}^0([0,1],\mathbb{R}) := \mathcal{F}([0,1],\mathbb{R})$.

Remark. Notice that the proof of Theorem 5.17 implies that if f is a fractal function of class C^k, then its kth derivative is a fractal function as introduced in Section 5.1. In particular, if f is an affine fractal function and of class C^k, then $f \in \mathrm{Lip}^{\alpha+k}([0,1], \mathbb{R})$, where $\alpha = 2 - \dim_\beta \mathrm{graph}(f)$.

As an example, the case $a_i := 1/N$ and $s_i := s$, for all $i = 1, \ldots, N$, is considered, where $1 \neq N \in \mathbb{N}$ and $|s| < 1$. Then $\Lambda^{(k)}$ consists of all those functions $g_i \in C^k([0,1])$ that satisfy

$$(1 - sN^\ell)\left(g_{i+1}^{(\ell)}(0) - g_i^{(\ell)}(1)\right) + (sN^\ell)\left(g_1^{(\ell)}(0) - g_N^{(\ell)}(1)\right) = 0, \qquad (5.65)$$

for all $i = 2, \ldots, N-1$. Equation (5.65) has the following — later very important — interpretation: Let $\lambda := (\lambda_1, \ldots, \lambda_N)^t$, where ^{t} denotes "transpose." Then λ is an element of the space $\prod_{i=1}^N C^k([0,1])$. Here $\prod$ denotes the direct product of linear spaces; it is the extension of the direct product of the underlying abelian groups. The outer or scalar multiplication is defined in the usual componentwise way. For $i = 2, \ldots, N-1$, let $L_i^{(\ell)} : \prod_{i=1}^N C^k([0,1]) \to \mathbb{R}$ be the operator defined by

$$L_i^{(\ell)}\lambda := \mathfrak{D}_i^\ell (\, \lambda_1 \quad \cdots \quad \lambda_i \quad \lambda_{i+1} \quad \cdots \quad \lambda_N\,)^t, \qquad (5.66)$$

where

$$\mathfrak{D}^\ell := \mathfrak{n}_i^t \frac{d^\ell}{dx^\ell},$$

and

$$\mathfrak{n}_i := (\, sN^\ell \quad 0 \quad \cdots \quad 0 \quad -(1 - sN^\ell) \quad (1 - sN^\ell) \quad 0 \quad 0 \quad \cdots \quad -sN^\ell\,)^t.$$

This characterization implies that

$$\lambda \in \prod_{i=1}^N \Lambda^k \iff \lambda \in \bigcap_{\ell=0}^{k} \bigcap_{i=2}^{N-1} \ker(L_i^{(\ell)}).$$

Hence, a fractal function f generated by the previously defined special maps is of class C^k only if $N^k|s| < 1$ and $\lambda \in \bigcap_{\ell=0}^{k} \bigcap_{i=2}^{N-1} \ker(L_i^{(\ell)})$. In Chapter 7 this description of a fractal function of class C^k in terms of the λ_i will be reconsidered.

Example 5.14 Reconsider Example 5.3; there $N = 2$, $a_1 = a_2 = 1/2$, $s = 1/4$, and $\lambda_1(x) = x = 1 - \lambda_2(x)$. It is not hard to verify that the hypotheses of Theorem 5.17 are satisfied for $k = 1$: Clearly, $|s|N < 1$. The operators $L_i^{(\ell)}$ are given by

$$\ell = 0: \quad L_1^{(0)}\lambda = (1 - 3/4)(\lambda_2(0) - \lambda_1(1)) + 3/4(\lambda_1(0) - \lambda_2(1))$$
$$\ell = 1: \quad L_1^{(1)}\lambda = (1 - 1/2)(\lambda_2'(0) - \lambda_1'(1)) + 1/2(\lambda_1'(0) - \lambda_2'(1)),$$

with $\lambda = (\lambda_1, \lambda_2)$. One easily verifies that $\lambda \in \ker(L_1^{(0)}) \cap \ker(L_1^{(1)})$. Thus, the unique fixed point f_Ψ of the Read-Bajrakarević operator Ψ (Example 5.3) is a fractal function of class C^1, i.e., an element of $\mathcal{F}^1([0,1], \mathbb{R})$. Furthermore, it is seen that the value for s, namely $1/4$, is unique.

Next, another approach to construct fractal functions of higher regularity is presented. This is done by introducing a class of admissible functions $v_i(x, \cdot)$ that is different from that considered at the beginning of this section. Begin with an interpolation set $\Delta := \{(x_j, y_j) \mid 0 = x_0 < x_1 < \cdots < x_N = 1 \wedge \forall\, j = 0, 1, \ldots, N: \; y_j \in \mathbb{R} \wedge 1 < N \in \mathbb{N}\}$. The objective is the construction of a fractal function of class C^k, $k = 1, 2$, interpolating Δ whose graph is the attractor of a hyperbolic IFS on $[0, 1] \times \mathbb{R}$.

In this setup, the affine contractions $u_i : [0, 1] \to [0, 1]$ can explicitly be written as $u_i(x) = a_i x + x_{i-1}$, where $a_i = x_i - x_{i-1}$, $i = 1, \ldots, N$. For $i \in \{1, \ldots, N\}$, let $K_i(\xi, \eta)$ be a symmetric bilinear form on $\mathbb{R}^2$. For $\xi := (x, y) \in \mathbb{R}^2$, define maps $v_i(x, \cdot) : [0, 1] \times \mathbb{R} \to \mathbb{R}$ by $v_i(\xi) := K_i(\xi, \xi) + c_i$, $c_i \in \mathbb{R}$ and $i = 1, \ldots, N$. Note that v_i may also be written in the form

$$v_i(x, y) = b_i x^2 + s_i xy + t_i y^2 + c_i,$$

for constants b_i, c_i, s_i, and t_i. Requiring that

$$v_i(0, y_0) = y_{i-1} = v_{i-1}(1, y_N)$$

determines the b_i and c_i uniquely:

$$b_i = y_i - y_{i-1} - 2s_i y_N - t_i(y_0^2 + y_N^2), \qquad c_i = y_{i-1} - t_i y_0^2.$$

In order for the maps $v_i(x, \cdot)$ to satisfy condition (A), the following must hold:

$$\max\{|s_i| + |t_i| \mid i = 1, \ldots, N\} \le r < 1/2. \tag{5.67}$$

The constants s_i and t_i, as long as they satisfy Eq. (5.67), are free parameters. Let $\alpha, \beta \in \mathbb{R}$ be arbitrary. Define $\mathfrak{F}^1([0,1], \mathbb{R}) := \{f \in C^1([0,1], \mathbb{R}) \mid \forall\, j =$

$0, 1, \ldots, N : \mathfrak{f}(x_j) = y_j \wedge \mathfrak{f}'(0) = \alpha \wedge \mathfrak{f}'(1) = \beta\}$. Let $\Psi : \mathfrak{F}^1([0,1], \mathbb{R}) \to \mathbb{R}^{[0,1]}$ be the operator given by

$$\Psi f(x) := v_i(u_i^{-1}(x), {}^{\#}f_i(x)), \qquad x \in I_i,$$

$i = 1, \ldots, N$. Then

$$(\Psi f)' = \frac{2}{a_i} \left(b_i u_i^{-1} + s_i {}^{\#}f_i + s_i u_i^{-1} {}^{\#}f_i' + t_i {}^{\#}f_i {}^{\#}f_i' \right).$$

If it is required that

$$(\Psi f)'(0) = f'(0), \quad (\Psi f)'(1) = f'(1), \quad \lim_{x \to x_i^-} (\Psi f)'(x) = \lim_{x \to x_i^+} (\Psi f)'(x), \quad (5.68)$$

for $i = 1, \ldots, N-1$, then it is easy to see that $\Psi f \in C^1([0,1], \mathbb{R})$, and that Ψ is contractive in the C^1-topology with contractivity constant $2 \max_i\{|s_i| + |t_i|\} < 1$. Hence, the unique fixed point f_Ψ of Ψ is an element of $\mathcal{F}^1([0,1], \mathbb{R})$. Moreover, by construction, f_Ψ interpolates Δ, i.e., $f_\Psi \in \mathcal{F}^1|\Delta([0,1], \mathbb{R})$.

Example 5.15 As an example, consider the special case $y_0 = 0 = y_N$. Then the preceding requirements yield

$$s_i = -\frac{b_i}{f'(1)} \quad (i = 1, \ldots, N-1), \quad s_N = \frac{a_N}{2} - \frac{b_N}{f'(1)}, \quad f'(0) = \alpha = 0.$$

The coefficients t_i, as long as they satisfy (5.67), are free parameters. Fig. 5.16 shows a C^1-interpolating fractal function for $\Delta := \{(0,0), (1/2, 1/2), (1, 0)\}$, $s_1 = 3/10$, $s_2 = 1/5$, $t_1 = 1/10$, $t_2 = 1/5$, and $\beta = -10$.

The class of $v_i(x, \cdot)$ also allows us to impose conditions so that we obtain C^2-interpolating fractal functions. The additional requirements are

$$\lim_{x \to x_i^-} (\Psi f)''(x) = \lim_{x \to x_i^+} (\Psi f)''(x) \quad (i = 1, \ldots, N-1),$$

and

$$(\Psi f)''(1) = f''(1).$$

Then the Read-Bajraktarević operator Ψ when acting on $\mathfrak{F}^2([0,1], \mathbb{R}) := \{f \in \mathfrak{F}^1([0,1], \mathbb{R}) | f''(1) = \gamma\}$, where γ is a given arbitrary real number, is contractive in the C^2-topology. Thus, its unique fixed point f_Ψ is an element of $\mathcal{F}^2([0,1], \mathbb{R})$, and also interpolates Δ.

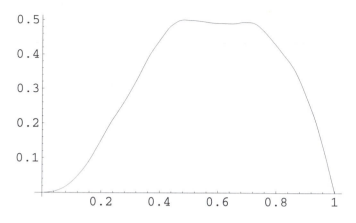

Figure 5.16: A fractal function in $\mathcal{F}^1|\Delta([0,1], \mathbb{R})$.

***Example* 5.16** For the preeding example, i.e., $y_0 = 0 = y_N$,

$$(\Psi f)''(0) = \frac{2b_i}{a_i^2},$$

$$t_i = \frac{b_{i+1}(a_i/a_{i+1})^2 + b_i(1 + f''(1)/f'(1))}{f'(1)[f'(1) + f''(1)]} \quad (i = 1, \ldots, N-1),$$

$$t_N = \frac{a_N^2 f''(1) - 2b_N - 2s_N(2f'(1) - f''(1))}{2f'(1)[f'(1) + f''(1)]}.$$

Such a fractal function of class C^2 is depicted in Fig. 5.17. There Δ is chosen as $\{(0,0), (1/3, 1/3), (2/3, -1/3), (1,0)\}$, $s_1 = -2/9$, $s_2 = 4/9$, $s_3 = 1/9$, $t_1 = -t_2 = t_3 = 1/10$, and $\beta = 18$.

Comparing the approach given at the beginning of this section to that presented earlier gives the following conclusions. In the former case, fractal functions of arbitrary regularity can be constructed by choosing for the λ_i polynomials of increasing degree. Since the only free parameter is s, it is, in general, not possible to obtain a fractal function of class greater than C^k using polynomial λ_i of degree less than or equal to $k+1$. In the latter case, more than one free parameter is available, and thus there is more flexibility: Fractal functions in, for instance, $\mathcal{F}^2([0,1], \mathbb{R})$ are obtained by choosing the correct values for the s_i and the t_i. Clearly, the specific application will determine which of these two methods is more appropriate.

A somewhat different approach to construct fractal functions of class C^k was undertaken in [12]. The basic idea is to take the indefinite integral of

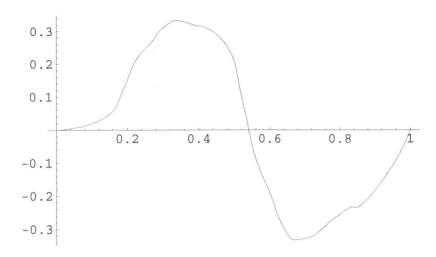

Figure 5.17: A fractal function in $\mathcal{F}^2|\Delta([0,1], \mathbb{R})$.

a C^0-fractal function, thus obtaining a more regular fractal function. This method will be investigated in the remainder of this section.

Let f be a fractal function generated by a Read-Bajraktarević operator Φ associated with maps u_i and $v_i(x, \cdot) := \lambda_i(x) + s_i(\cdot)$ as defined at the beginning of this section. Denote by Δ the $N+1$ distinct points $(x_j, y_j) \in \mathbb{R}^2$ determined by the maps $u_i^{-1}(0)$, $u_i^{-1}(1)$ and $f(x_j)$, respectively. We define

$$F(x) := \tilde{y}_0 + \int_0^x f(t)\, dt, \qquad \tilde{y}_0 \in \mathbb{R}. \tag{5.69}$$

Recall that the fixed-point equation for f may be written in the form

$$f \circ u_i(x) = \lambda_i(x) + s_i f(x), \qquad x \in [0,1],$$

$i = 1, \ldots, N$. Hence,

$$
\begin{aligned}
F \circ u_i(x) &= \tilde{y}_0 + \int_0^{u_i(x)} f(t)\, dt \\[2mm]
&= \left(\tilde{y}_0 + \int_0^{x_{i-1}} f(t)\, dt \right) + \int_{x_{i-1}}^{u_i(x)} f(t)\, dt \\[2mm]
&= \left(\tilde{y}_0 + \int_0^{x_{i-1}} f(t)\, dt \right) + a_i \int_0^x f \circ u_i(t)\, dt.
\end{aligned}
$$

Setting $\tilde{y}_{i-1} := \tilde{y}_0 + \int_0^{x_{i-1}} f(t)\, dt$, and using the fixed-point equation for f, yields

$$F \circ u_i(x) = \left(\tilde{y}_{i-1} - a_i s_i \tilde{y}_0 + a_i \int_0^x \lambda_i(t)\, dt \right) + a_i s_i F(x).$$

Now let $\tilde{\lambda}_i := \tilde{y}_{i-1} - a_i s_i \tilde{y}_0 + a_i \int_0^x \lambda_i(t)\, dt$ and $\tilde{s}_i := a_i s_i$, for $i = 1, \ldots, N$. One then easily verifies that $\tilde{\Phi} := \tilde{\lambda}_i \circ u_i^{-1} + \tilde{s}_i(\,\cdot\,) \circ u_i^{-1}$ is a contractive Read-Bajraktarević operator on $\mathfrak{F}([0,1], \mathbb{R}) := \{\mathfrak{f} \in C([0,1], \mathbb{R}) \mid \mathfrak{f}(0) = y_0 \wedge \mathfrak{f}(1) = y_N\}$ and that F is its unique fixed point.

Let $\tilde{y}_i := F(x_i)$, $i = 1, \ldots, N$. Then, letting $x = 1$ in the functional equation for F gives

$$y_i = F \circ u_i(1) = \tilde{y}_{i-1} - a_i s_i \tilde{y}_0 + a_i \int_0^1 \lambda_i(t)\, dt + a_i s_i \tilde{y}_N.$$

Therefore, since $\tilde{y}_i = \tilde{y}_0 + \sum_{k=1}^i (\tilde{y}_k - \tilde{y}_{k-1})$,

$$\tilde{y}_i = \tilde{y}_0 + \sum_{k=1}^i a_k \left(s_k[\tilde{y}_N - \tilde{y}_0] + \int_0^1 \lambda_k(t)\, dt \right). \tag{5.70}$$

Setting $i = N$ in the preceding equation, the following expression for $\tilde{y}_N$ is derived:

$$\tilde{y}_N = \tilde{y}_0 + \frac{\sum_{i=1}^N a_i \int_0^1 \lambda_i(t)\, dt}{1 - \sum_{i=1}^N a_i s_i}. \tag{5.71}$$

These results are summarized in the following theorem.

Theorem 5.18 *Let $f \in \mathcal{F}|\Delta([0,1], \mathbb{R})$ for a Read-Bajraktarević operator of the form $\Phi := \lambda_i \circ u_i^{-1} + s_i(\,\cdot\,) \circ u_i^{-1}$. Then the indefinite integral*

$$F(x) := \tilde{y}_0 + \int_0^x f(t)\, dt, \qquad \tilde{y}_0 \in \mathbb{R}$$

is an element of $\mathcal{F}^1|\tilde{\Delta}([0,1], \mathbb{R})$, with $\tilde{\Delta} := \{(x_j, \tilde{y}_j) \mid j = 0, 1, \ldots, N\}$ where

$$\tilde{y}_i = \tilde{y}_0 + \sum_{k=1}^i a_k \left(s_k[\tilde{y}_N - \tilde{y}_0] + \int_0^1 \lambda_k(t)\, dt \right) \qquad (i = 1, \ldots, N-1),$$

and

$$\tilde{y}_N = \tilde{y}_0 + \frac{\sum_{i=1}^N a_i \int_0^1 \lambda_i(t)\, dt}{1 - \sum_{i=1}^N a_i s_i}.$$

Proof. It suffices to show that $F \in \mathcal{F}^1([0,1], \mathbb{R})$. This, however, is immediate. ∎

It is clear that the foregoing procedure can be iterated indefinitely, generating fractal functions of higher and higher regularity. This is most easily done when the λ_i are polynomials. It is worthwhile mentioning that in the case of polynomial λ_i and $s_i = 0$, for all $i = 1, \ldots, N$, the fractal functions are the classical splines. (The reader who is knowledgeable in nonstandard analysis may have noticed that interpolating fractal functions are indeed *hyperfinite* splines.)

An example ends this section.

***Example* 5.17** Let f be the unique affine fractal function interpolating $\Delta :=$ $\{(0,0),(2/3,3/4),(1,1/2)\}$. Let $s_1 := 1/4$ and $s_2 := -1/2$. Then $\lambda_1(x) = (5/8)\,x$ and $\lambda_2(x) = 3/4$. The graph of f is shown in Fig. 5.18. Let $\tilde{y}_0 := 0$. The indefinite integral F of f interpolates the set $\tilde{\Delta} := \{(0,0),(2/3,31/144),$ $(1,11/24)\}$. Furthermore, $\tilde{\lambda}_1(x) = (5/24)\,x^2$ and $\tilde{\lambda}_2(x) = (1/4)\,x + 31/144$. Fig. 5.19 displays the graph of F.

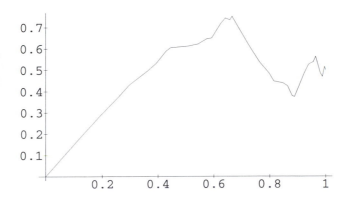

Figure 5.18: The fractal function f in Example 5.16.

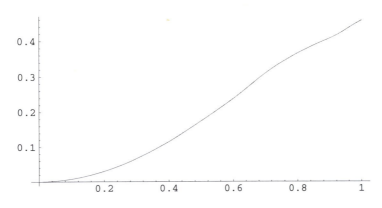

Figure 5.19: The indefinite integral of f.

Chapter 6

Dimension of Fractal Functions

The present chapter is devoted to dimension calculations for univariate fractal functions. First, the dimension theorems for those functions considered in Sections 5.1, 5.4, and 5.5 are proven. Then a dimension result for $\mathbb{R}^m$-valued fractal functions is stated without proof. The last section deals with the characterization of the box dimension of a class of fractal functions in terms of certain smoothness classes of functions, so-called *Besov spaces*.

6.1 Dimension Calculations

First a dimension theorem for affine fractal functions (Example 5.4) is proven. The methods used in the proof of this theorem then lead the way to dimension formulae for recurrent, hidden variable, and $\mathbb{R}^m$-valued fractal functions.

6.1.1 Affine fractal functions

Let $X := [0,1]$ and let $Y := \mathbb{R}$. Let $f : [0,1] \to \mathbb{R}$ be an affine fractal function and let G denote its graph. Using the notation introduced in Section 5.1, in particular Example 5.4, the following two assumptions are made:

$$(\mathrm{I}): \ \sum_{i=1}^{N} |s_i| > 1 \qquad \text{and} \qquad (\mathrm{II}): \ \Delta \text{ is not collinear.}$$

The dimension theorems stated later involve expressions of the form $\sum_{i=1}^{N} |s_i| a_i^{d-1}$, where the s_i satisfy (I) and $\sum_{i=1}^{N} a_i = 1$. Note that the function

205

$h : \mathbb{R} \to \mathbb{R}$ defined by $h(d) := \sum_{i=1}^{N} |s_i| a_i^{d-1}$ is strictly increasing in d and satisfies $\lim_{d \to -\infty} h(d) = \infty$ and $h(2) \leq \max\{|s_i| \, | \, i = 1, \ldots, N\} < 1$. Hence, there exists a unique $d^* \in (-\infty, 2)$ such that $h(d^*) = 1$. Furthermore, $d^* > 1$ iff (I) is satisfied.

The main idea behind all the dimension calculations is to define the right covers for the graphs G of the fractal functions. Now a class of such covers is defined which allows one to relate covers of different sizes.

Definition 6.1 Let $0 < \varepsilon < 1$. The set $\{\mathfrak{P}_\varkappa \, | \, \varkappa = 0, 1, \ldots, K\}$ is called an ε-partition iff

1. $\forall \, \varkappa \in \{0, 1, \ldots, K\} : \; -\varepsilon/2 < \mathfrak{P}_\varkappa < 1$.

2. $\forall \, \varkappa \in \{0, 1, \ldots, K-1\} : \; \varepsilon/2 < \mathfrak{P}_{\varkappa+1} - \mathfrak{P}_\varkappa \leq \varepsilon$.

A cover C of G is called an ε-column cover with associated ε-partition $\{\mathfrak{P}_\varkappa\}$ iff there exist $n_\varkappa \in \mathbb{N}_0$ and $\xi_\varkappa \in \mathbb{R}$, $\varkappa \in \{0, 1, \ldots, K\}$, such that

$$
\begin{aligned}
C \;=\; & \{[\mathfrak{P}_\varkappa, \mathfrak{P}_\varkappa + \varepsilon] \times [\xi_\varkappa + (j_\varkappa - 1)\varepsilon, \xi_\varkappa + j_\varkappa \varepsilon] \, | \, j_\varkappa = 1, \ldots, n_\varkappa; \\
& \varkappa = 0, 1, \ldots, K\}.
\end{aligned}
$$

The class of all such covers of G is denoted by $\mathcal{C}^*(\varepsilon)$. A non-overlapping ε-cover of G is a cover of G consisting of $\varepsilon \times \varepsilon$-squares with disjoint interiors of the form $[\varkappa\varepsilon, (\varkappa+1)\varepsilon] \times [y, y+\varepsilon]$, where $\varkappa \in \{0, 1, \ldots, [\![\varepsilon^{-1}]\!]\}$ and $y \in \mathbb{R}$. Let $\mathcal{C}^{**}(\varepsilon)$ denote the class of all non-overlapping ε-covers of G.

Clearly, a cover $C \in \mathcal{C}^*(\varepsilon)$ consists of $\sum_{\varkappa=0}^{K} n_\varkappa$ closed $\varepsilon \times \varepsilon$-squares arranged in $K + 1$ columns. Let $\mathcal{N}^*(\varepsilon) := \min\{\|C\|_c \, | \, C \in \mathcal{C}^*(\varepsilon)\}$, let $\mathcal{N}^{**}(\varepsilon) := \min\{\|C\|_c \, | \, C \in \mathcal{C}^{**}(\varepsilon)\}$, and let $\mathcal{N}(\varepsilon)$ denote the minimum number of $\varepsilon \times \varepsilon$-squares $[x, x+\varepsilon] \times [y, y+\varepsilon]$, $x, y \in \mathbb{R}$, necessary to cover G. The next result shows that for the calculation of the box dimension of G it suffices to consider covers from $\mathcal{C}^*(\varepsilon)$.

Proposition 6.1 $\forall \, 0 < \varepsilon < 1 : \; \mathcal{N}(\varepsilon) \leq \mathcal{N}^*(\varepsilon) \leq 2 \mathcal{N}(\varepsilon)$.

Proof. The definition of non-overlapping ε-cover implies $\mathcal{N}^{**}(\varepsilon) \leq 2 \mathcal{N}(\varepsilon)$. Also, if $C \in \mathcal{C}^{**}(\varepsilon)$ then, since f is continuous, $C \in \mathcal{C}^*(\varepsilon)$. Hence, $\mathcal{N}^*(\varepsilon) \leq \mathcal{N}^{**}(\varepsilon) \leq 2 \mathcal{N}(\varepsilon)$. ∎

For the proof of Theorem 6.1 the following lemmas are needed.

Lemma 6.1 *If conditions (I) and (II) are satisfied, then*

$$
\lim_{\varepsilon \to 0} \varepsilon \mathcal{N}^*(\varepsilon) = \infty.
$$

Proof. Since Δ is not collinear, there exists an $n \in \{1, \ldots, N\}$ such that $V := |y_n - y_0 - (y_N - y_0)x_n| > 0$. Let $\underline{a} := \min\{a_i \mid i = 1, \ldots, N\}$. The continuity of f implies

$$\mathcal{N}^*(\varepsilon) \geq \sum_{i_1=1}^{N} \cdots \sum_{i_k=1}^{N} \left(|s_{i_1} \cdots s_{i_k}| V \varepsilon^{-1} \right)$$

$$\geq \left(\sum_{i=1}^{N} |s_i| \right)^k V \varepsilon^{-1},$$

for all $0 < \varepsilon < \underline{a}^k$ and all $k \in \mathbb{N}$. Thus,

$$\lim_{\varepsilon \to \infty} \varepsilon \mathcal{N}^*(\varepsilon) \geq \lim_{k \to \infty} \left(\sum_{i=1}^{N} |s_i| \right)^k V = \infty,$$

by condition (I). ∎

Lemma 6.2 *Let $x_0 \in (1, \infty)$ and let $\Xi : (0, x_0] \to \mathbb{R}$ be a decreasing function. Assume that $\lim_{x \to \infty} x \Xi(x) = \infty$. Furthermore, suppose that for all $0 \leq x \leq x_0$, Ξ satisfies the functional inequalities*

$$\sum_{i=1}^{N} \frac{|s_i|}{a_i} \Xi \left(\frac{x}{a_i} \right) - \frac{\alpha}{x} \leq \Xi(x) \leq \sum_{i=1}^{N} \frac{|s_i|}{a_i} \Xi \left(\frac{x}{a_i} \right) + \frac{\beta}{x},$$

where $0 < a_i, |s_i| < 1$, $\sum_{i=1}^{N} a_i = 1$, and $\alpha, \beta > 0$. If $\sum_{i=1}^{N} |s_i| > 1$, then there exist positive constants A and B such that

$$A x^{-d^*} \leq \Xi(x) \leq B x^{-d^*}, \tag{6.1}$$

where d^* is the unique positive solution of

$$\sum_{i=1}^{N} |s_i| a_i^{d-1} = 1.$$

If $\sum_{i=1}^{N} |s_i| \leq 1$, then there exists a $B > 0$ so that $\Xi(x) \leq B x^{-1}$.

Proof. First it is assumed that $\sum_{i=1}^{N} |s_i| > 1$. Let $\underline{a} := \min\{a_i \mid i = 1, \ldots, N\}$ and let $\bar{a} := \max\{a_i \mid i = 1, \ldots, N\}$. As $\lim_{x \to \infty} x \Xi(x) = \infty$, there exists an $\tilde{x} > 0$ small enough so that

$$\forall x \in (0, \tilde{x}/\underline{a}] : \; \Xi(x) \geq 2 \left(\frac{\alpha}{\sum_i |s_i| - 1} \right) x^{-1}.$$

Now select a $k_1 > 0$ small enough such that

$$k_1 \tilde{x}^{-d^*} \le \left(\frac{\alpha}{\sum_i |s_i| - 1} \right) x^{-1},$$

and choose $k_2 > 0$ large enough so that

$$\Xi(x) \le \left(\frac{\beta}{1 - \sum_i |s_i|} \right) x^{-1} + k_2 x^{-d^*},$$

for $\tilde{x} \le x \le \tilde{x}/\underline{a}$. Recall that $d^* > 1$ iff $\sum_i |s_i| > 1$.

Now define upper and lower functions $\overline{\Xi}, \underline{\Xi} : (0, x_0] \to \mathbb{R}$ by

$$\overline{\Xi}(x) := \left(\frac{\beta}{1 - \sum_i |s_i|} \right) x^{-1} + k_2 x^{-d^*},$$

respectively,

$$\underline{\Xi}(x) := \left(\frac{\alpha}{\sum_i |s_i| - 1} \right) x^{-1} + k_1 x^{-d^*}.$$

Then the preceding arguments yield

$$\forall\, x \in [\tilde{x}, \tilde{x}/\underline{a}] : \ \underline{\Xi} \le \Xi \le \overline{\Xi}.$$

Note that by construction the functions $\underline{\Xi}$ and $\overline{\Xi}$ satisfy the following functional equations:

$$\underline{\Xi}(x) = \sum_i \left(\frac{|s_i|}{a_i} \right) \underline{\Xi} \left(\frac{x}{a_i} \right) - \alpha x^{-1}$$

and

$$\overline{\Xi}(x) = \sum_i \left(\frac{|s_i|}{a_i} \right) \overline{\Xi} \left(\frac{x}{a_i} \right) + \beta x^{-1}.$$

If $\overline{a}\tilde{x} \le x \le \tilde{x}$, then $\tilde{x} \le x \le x/\underline{a}$ and thus

$$\Xi(x) \ \le \ \sum_i \left(\frac{|s_i|}{a_i} \right) \Xi \left(\frac{x}{a_i} \right) + \beta x^{-1}$$

$$\le \ \sum_i \left(\frac{|s_i|}{a_i} \right) \overline{\Xi} \left(\frac{x}{a_i} \right) + \beta x^{-1} = \overline{\Xi}(x).$$

A similar argument shows that $\underline{\Xi} \le \Xi$ on $[\overline{a}\tilde{x}, \tilde{x}]$. The preceding argument serves as an induction step: Suppose that for $x \in [\overline{a}^n \tilde{x}, \tilde{x}]$, $n \in \mathbb{N}$, $\underline{\Xi}(x) \le \Xi(x) \le \overline{\Xi}(x)$; then it must also hold for $x \in [\overline{a}^{n+1}\tilde{x}, \tilde{x}]$. Since $\overline{a}^n \to 0$ as $n \to \infty$,

$$\underline{\Xi}(x) \le \Xi(x) \le \overline{\Xi}(x), \qquad \text{for all } x \in (0, \tilde{x}].$$

Furthermore, since $d^* > 1$, there exist two positive constants A and B so that $\Xi(x) \geq A x^{-d^*}$ and $\overline{\Xi}(x) \leq B x^{-d^*}$. This proves the lemma under the assumption that $\sum_{i=1}^{N} |s_i| > 1$.

Assume then that $\sum_{i=1}^{N} |s_i| < 1$. As before, a positive constant k_2 can be found so that

$$\Xi(x) \leq \overline{\Xi}(x) = \left(\frac{\beta}{1 - \sum_i |s_i|}\right) x^{-1} + k_2 x^{-d^*},$$

for all $x \in (0, 1]$. Now, however, $d^* < 1$, and since $\frac{\beta}{1 - \sum_i |s_i|} > 0$, there is some $B > 0$ so that $\Xi(x) \leq B x^{-1}$.

Finally, suppose that $\sum_{i=1}^{N} |s_i| = 1$. Then define an upper function $\overline{\Xi}$ by

$$\overline{\Xi}(x) := \left(\frac{\beta}{\sum_i |s_i| \log a_i}\right) \frac{\log x}{x} + k_2 x^{-1},$$

for any $k_2 \in \mathbb{R}$. This upper function then satisfies

$$\overline{\Xi}(x) = \sum_i \left(\frac{|s_i|}{a_i}\right) \overline{\Xi}\left(\frac{x}{a_i}\right) + \beta x^{-1},$$

for all $x \in (0, 1]$. Hence, as before, a $B > 0$ can be found so that $\Xi(x) \leq \overline{\Xi}(x) \leq B x^{-1}$. This proves the lemma. ∎

Theorem 6.1 *Let $f : [0, 1] \to \mathbb{R}$ be an affine fractal function and G its graph. Suppose that conditions (I) and (II) are satisfied. Then the box dimension $\dim_\beta G$ is the unique positive solution d^* of*

$$\sum_i^N |s_i| a_i^{d-1} = 1; \tag{6.2}$$

otherwise, $\dim_\beta G = 1$.

Proof. Let $([0, 1] \times \mathbb{R}, \mathbf{w})$ be the hyperbolic IFS whose unique attractor is G. Let $\varepsilon \in (0, 1)$ be given. Choose a minimal ε-cover $C(\varepsilon)$ of G. By Proposition 6.1 it may be assumed that $C(\varepsilon) \in \mathcal{C}^*(\varepsilon)$. Let $\{\mathfrak{P}_\varkappa\}$ be the associated ε-partition. For $i \in \{1, \ldots, N\}$, let $[a, b+\varepsilon]$ be the smallest interval of the form $[\mathfrak{P}_{\varkappa_1}, \mathfrak{P}_{\varkappa_2} + \varepsilon]$ that covers $I_i = u_i[0, 1]$. Denote by

$$\mathcal{C}_i^*(\varepsilon) := \{C \in \mathcal{C}^*(\varepsilon) \mid C \subseteq [\mathfrak{P}_{\varkappa_1}, \mathfrak{P}_{\varkappa_2} + \varepsilon] \times \mathbb{R}\}$$

the "restriction" of $\mathcal{C}^*(\varepsilon)$ to I_i and by $\mathcal{N}_i^*(\varepsilon)$ its cardinality. Note that $\|\mathcal{C}_{i+1}^*(\varepsilon) \cap \mathcal{C}_i^*(\varepsilon)\|_c \leq 2$ and that f is uniformly bounded on $[0, 1]$. Therefore, there exists an $\alpha_1 > 0$ such that for all $0 < \varepsilon < 1$,

$$\sum_{i=1}^{N} \mathcal{N}_i^*(\varepsilon) \leq \mathcal{N}^*(\varepsilon) + \alpha_1 \varepsilon^{-1}.$$

Now suppose that $i \in \{1, \ldots, N\}$ is such that $s_i \neq 0$. Let $\mathfrak{R}$ be an element of $\mathcal{C}_i^*(\varepsilon)$ which is the union of n $\varepsilon \times \varepsilon$-squares. Since the map $w_i \in \mathbf{w}$ is invertible, the inverse image $w_i^{-1}\mathfrak{R}$ of $\mathfrak{R}$ is a parallelogram of width $\varepsilon/|a_i|$, height $n\varepsilon/|s_i|$ and shear $c_i/(a_i|s_i|)$. Thus, $w_i^{-1}\mathfrak{R}$ can be covered by

$$\left[\!\left[\frac{na_i}{|s_i|} + \left|\frac{c_i}{a_i}\right| \right]\!\right] + 1$$

$\varepsilon/a_i \times \varepsilon/a_i$ squares. Hence, $w_i^{-1}\mathcal{C}_i^*(\varepsilon)$ is an element of the covering class $\mathcal{C}^*(\varepsilon/a_i)$ of G. Realizing that there are at most $(2a_i)/\varepsilon + 2$ elements $\mathfrak{R}$ in $\mathcal{C}_i^*(\varepsilon)$, a positive constant β_1 can be found such that

$$\sum_i \frac{|s_i|}{a_i} \mathcal{N}^*(\varepsilon/a_i) - \frac{\beta_1}{\varepsilon} \leq \sum_i \mathcal{N}_i^*(\varepsilon) - \frac{\alpha_1}{\varepsilon} \leq \mathcal{N}^*(\varepsilon). \tag{6.3}$$

Next an upper bound for $\mathcal{N}^*(\varepsilon)$ is obtained. Suppose then that $C \in \mathcal{C}^*(\varepsilon/a_i)$ is a minimal cover of G and $\mathfrak{R}$ one of its elements. The direct image of $\mathfrak{R}$ under w_i is also a parallelogram, but of width ε, height $(n|s_i|\varepsilon)/a_i$, and shear $|c_i|\varepsilon$. (Here it is again assumed that $\mathfrak{R}$ is covered by n $\varepsilon/a_i \times \varepsilon/a_i$-squares.) The number of $\varepsilon \times \varepsilon$-squares needed to cover $w_i\mathfrak{R}$ is then given by

$$\left[\!\left[\frac{n|s_i|}{a_i} + \frac{|c_i|}{a_i} \right]\!\right] + 1.$$

This way a cover C_i of w_iG consisting of $\varepsilon \times \varepsilon$-squares is generated. To relate the cardinality of C_i to the cardinality of C, one makes again use of the fact that C contains at most $(2a_i)/\varepsilon + 2$ elements $\mathfrak{R}$ and thus obtains

$$\|C_i\|_c \leq \frac{|s_i|}{a_i} \mathcal{N}^*(\varepsilon/a_i) + \frac{\alpha_2}{\varepsilon},$$

for some positive constant α_2. The union $\bigcup_{i=1}^{N} C_i$ is a cover of G, but in general not an element of $\mathcal{C}^*(\varepsilon)$, since the columns $\mathfrak{R}$ in adjacent covers C_i may not properly join up. However, a cover in $\mathcal{C}^*(\varepsilon)$ can be constructed from

$\bigcup_{i=1}^{N} C_i$ by replacing at most two elements in $C_i \cap C_{i+1}$ with at most two properly spaced ones. This then yields the required upper bound for $\mathcal{N}^*(\varepsilon)$:

$$\mathcal{N}^*(\varepsilon) \leq \| \bigcup_{i=1}^{N} C_i \|_c \leq \sum_i \|C_i\|_c + N \left(\frac{4n}{\varepsilon} + 1 \right).$$

Hence, a positive constant β_2 can be found such that

$$\mathcal{N}^*(\varepsilon) \leq \sum_{i=1}^{N} \frac{|s_i|}{a_i} \mathcal{N}^*(\varepsilon/a_i) + \frac{\beta_2}{\varepsilon}. \tag{6.4}$$

Applying Lemma 6.2 to Eqs. (6.3) and (6.4) yields the required results if condition (II) holds. If condition (II) fails to be true, then it is clear that the attractor G of the IFS $([0,1] \times \mathbb{R}, \mathbf{w})$ is the line segment connecting $(0, y_0)$ and $(1, y_N)$. Hence, $\dim_\beta G = 1$. ∎

6.1.2 Recurrent fractal functions

Next the dimension theorem for recurrent fractal functions is stated (also [10]). At this point the reader may want to review the notation and terminology introduced in Section 5.4. The case where the graph G of the recurrent fractal function is the attractor of a recurrent IFS whose maps w_i are affine will be considered. First some more notation is introduced.

Let $D(d)$ denote the diagonal matrix

$$D(d) := \begin{pmatrix} |s_1| |a_1|^{d-1} & 0 & \cdots & 0 \\ 0 & |s_2| |a_2|^{d-1} & \cdots & 0 \\ 0 & 0 & \ddots & 0 \\ 0 & 0 & \cdots & |s_N| |a_N|^{d-1} \end{pmatrix},$$

and let $r(d) := r(CD(d))$ denote the spectral radius of the product $CD(d)$, where C is the irreducible $N \times N$ connection matrix associated with the recurrent IFS. It is not hard to see that there exists a number $d^* > 0$ such that $r(d^*) = 1$ (use the Perron-Frobenius Theorem and the properties of the function h defined at the beginning of this chapter).

Theorem 6.2 *Let* $f : [0,1] \to \mathbb{R}$ *be a recurrent fractal function and* G *its graph. If* $r(CD(1)) > 1$ *and if* $\{(x_i, y_i) \mid x_i \in X'_k\}$ *is not collinear, then* $\dim_\beta G = d^*$; *otherwise,* $\dim_\beta G = 1$.

Proof. The proof of this theorem relies essentially on the same type of arguments that were used in the proof of Theorem 6.1. The covers needed are those in $\mathcal{C}^*(\varepsilon)$, $0 < \varepsilon < 1$. Since $G = \bigcup_{i=1}^{N} G_i$, where G_i is the portion of G above X_i, one obtains, with only minor modifications — the details of which are left for the reader — the following system of functional inequalities:

$$\sum_{j\in\mathbf{I}(i)} \left|\frac{s_i}{a_i}\right| \mathcal{N}_j^* \left(\frac{\varepsilon}{a_i}\right) - \alpha_i \varepsilon^{-1} \leq \mathcal{N}_i^*(\varepsilon) \leq \sum_{j\in\mathbf{I}(i)} \left|\frac{s_i}{a_i}\right| \mathcal{N}_j^* \left(\frac{\varepsilon}{a_i}\right) + \beta_i \varepsilon^{-1}, \quad (6.5)$$

where $\mathcal{N}_i^*(\varepsilon) := \min\{\|C\|_c \mid C \in \mathcal{C}^*(\varepsilon)\}$ and α_i and β_i are positive constants, $i = 1, \ldots, N$. However, in order to apply Lemma 6.2, Lemma 6.1 needs to be modified.

Lemma 6.3 *If* $\varrho := r(1) > 1$ *and if there exists a* $k \in \{1, \ldots, N\}$ *such that* $\{(x_i, y_i) \mid x_i \in X_k'\}$ *is not collinear, then* $\lim_{\varepsilon\to 0} \varepsilon \mathcal{N}_i^*(\varepsilon) = \infty$, *for all* $i = 1, \ldots, N$.

Proof. As $\{(x_i, y_i) \mid x_i \in X_k'\}$ is not collinear for all $i = 1, \ldots, N$ and C irreducible, the portion G_i of G above X_i is not a line segment for any i. Therefore, one can find three non-collinear points $P_i(\xi_i, \eta_i)$, $P_i'(\xi_i', \eta_i')$, and $P_i''(\xi_i'', \eta_i'')$, on the graph of f that belong to the interior of $X_i \times \mathbb{R}$ and that satisfy $\xi_i < \xi_i' < \xi_i''$. Let V_i denote the *vertical* distance between the point $P_i'(\xi_i', \eta_i')$ and the line segment connecting $P_i(\xi_i, \eta_i)$ and $P_i''(\xi_i'', \eta_i'')$. A simple calculation shows that

$$V_i = \left| \eta_i' - \left(\eta_i + \frac{\eta_i'' - \eta_i}{\xi_i'' - \xi_i}(\xi_i' - \xi_i) \right) \right|.$$

Let $\delta := \min\{|\xi_i'' - \xi_{i+1}| \mid i = 1, \ldots, N-1\}$. The Perron-Frobenius Theorem now guarantees the existence of a positive eigenvector $v = (v_1, \ldots, v_N)$ of $CD(1)$ with eigenvalue ϱ such that $v_i \leq V_i$ for all $i = 1, \ldots, N$. As in Lemma 6.1, the continuity of f implies that $\mathcal{N}_i^*(\varepsilon) \geq V_i \varepsilon^{-1} \geq v_i \varepsilon^{-1}$. Now let $\underline{a} := \min\{|a_i| \mid i = 1, \ldots, N\}$. It follows from the recurrent structure of G, i.e., $G_i = \bigcup_{j\in\mathbf{I}(i)} w_i(A_j)$, that if $\varepsilon < \underline{a}\delta$ then

$$\mathcal{N}_i^*(\varepsilon) \geq |s_i| \sum_{j\in\mathbf{I}(i)} V_j \varepsilon^{-1} \geq |s_i| \sum_{j\in\mathbf{I}(i)} v_j \varepsilon^{-1} = \varepsilon^{-1}(CD(1))_i = \varrho v_i \varepsilon^{-1}.$$

Applying the preceding arguments to the images of $w_i A_j$ under the recurrent IFS mappings, obtain after n steps

$$\varepsilon \mathcal{N}_i^*(\varepsilon) \geq \varrho^n v_i,$$

for $0 < \varepsilon < \delta \underline{a}^n$. This proves the lemma.   ∎

Proof of Theorem 6.2 continued. Since $\varrho = r(1) >$, there exists a unique $d^* > 1$ such that $r(d^*) = 1$. As in the proof of Lemma 6.3, the positive eigenvector of $CD(1)$ is denoted by v and its eigenvalue by ϱ. The positive eigenvector u of $CD(d^*)$ whose eigenvalue is equal to 1 is also needed. Next, real numbers α and β are chosen such that for all $i = 1, \ldots, N$ $\alpha v_i \geq \alpha_i$ and $\beta v_i \geq \beta_i$. With these new additions, the system of functional inequalities (6.5) now reads

$$\sum_{j \in \mathbf{I}(i)} \left| \frac{s_i}{a_i} \right| \mathcal{N}_j^* \left(\frac{\varepsilon}{a_i} \right) - \alpha v_i \varepsilon^{-1} \leq \mathcal{N}_i^*(\varepsilon) \leq \sum_{j \in \mathbf{I}(i)} \left| \frac{s_i}{a_i} \right| \mathcal{N}_j^* \left(\frac{\varepsilon}{a_i} \right) + \beta v_i \varepsilon^{-1}. \quad (6.6)$$

It is easy to verify that the functions

$$\zeta_i(\varepsilon; K, \gamma) := K u_i \varepsilon^{-D^*} + \frac{\gamma}{1 - \varrho} v_i \varepsilon^{-1} \qquad (i = 1, \ldots, N)$$

solve the following system of inhomogenous functional equations associated with (6.6):

$$\zeta_i(\varepsilon) = \sum_{j \in \mathbf{I}(i)} \left| \frac{s_i}{a_i} \right| \zeta_j \left(\frac{\varepsilon}{a_i} \right) + \gamma v_i \varepsilon^{-1} \qquad (i = 1, \ldots, N),$$

where $K \in \mathbb{R}$ is arbitrary and γ denotes either $-\alpha$ or β. Select K_1 large enough so that

$$\forall x \in [\underline{a}, 1] : \mathcal{N}_i^*(\varepsilon) \leq \zeta_i(\varepsilon; K_1, \gamma) \qquad (i = 1, \ldots, N).$$

Let $\overline{a} := \max\{|a_i| \, | \, i = 1, \ldots N\}$. As in the proof of Theorem 6.1 the validity of the preceding inequality for all $x \in (0, \tilde{\varepsilon}]$ needs to be established, where $\tilde{\varepsilon}$ has to be chosen appropriately. For this purpose, observe that if $\overline{a}\,\underline{a} \leq \varepsilon \leq \underline{a}$, then $\underline{a} \leq \varepsilon/|a_i| \leq 1$ and therefore, by system (6.5),

$$\mathcal{N}_i^*(\varepsilon) \leq \sum_{j \in \mathbf{I}(i)} |s_i/a_i| \, \mathcal{N}_j^* \, (\varepsilon/a_i) + \beta v_i \varepsilon^{-1}$$

$$\leq \sum_{j \in \mathbf{I}(i)} |s_i/a_i| \, \zeta_j \, (\varepsilon/a_i; K_1, \beta) + \beta v_i \varepsilon^{-1} \leq \zeta_i(\varepsilon; K_1, \beta),$$

for $i = 1, \ldots, N$. Inductively, one thus obtains

$$\forall x \in [\overline{a}^n \underline{a}, 1] : \mathcal{N}_i^*(\varepsilon) \leq \zeta_i(\varepsilon; K_1, \gamma) \qquad (i = 1, \ldots, N; \, n \in \mathbb{N}),$$

and since $\overline{a}^n \to 0$ as $n \to \infty$,

$$\mathcal{N}_i^*(\varepsilon) \leq \zeta_i(\varepsilon; K_1, \gamma)$$

for all $\varepsilon \in (0,1]$. The definition of box dimension for an attractor of a recurrent IFS gives then an upper bound for $\dim_\beta G$:

$$\dim_\beta G \leq \lim_{\varepsilon \to \infty} \frac{\log \sum_i \zeta_i(\varepsilon; K_1, \beta)}{-\log \varepsilon} = \max\{d^*, 1\}.$$

To obtain a lower bound for $\dim_\beta G$, Lemma 6.3 is used. Choose an $\tilde{\varepsilon} > 0$ small enough so that

$$\forall \varepsilon \in (0, \tilde{\varepsilon}] : \varepsilon \mathcal{N}_i^*(\varepsilon) > \frac{\gamma}{1 - \varrho} v_i \varepsilon^{-1} \qquad (i = 1, \ldots, N).$$

Then it is possible to select a $K_2 > 0$ such that

$$\mathcal{N}_i^*(\varepsilon) \geq \zeta_i(\varepsilon; K_2, \alpha),$$

for $\underline{a}\tilde{\varepsilon} \leq \varepsilon \leq \tilde{\varepsilon}$ and $i = 1, \ldots, N$. Arguments similar to those given earlier then show that

$$\forall \varepsilon \in (0, \tilde{\varepsilon}) : \mathcal{N}_i^*(\varepsilon) \geq \zeta_i(\varepsilon; K_2, \alpha) \qquad (i = 1, \ldots, N).$$

Hence, $\dim_\beta G \geq \max\{d^*, 1\}$ implying then that $\dim_\beta G = d^*$.

If $\varrho \leq 1$, then $d^* \leq 1$, and thus $\dim_\beta G \leq 1$. However, since G is the graph of a continuous function, $\dim_\beta G \geq 1$. Hence, $\dim_\beta G = 1$.

In case $\{(x_i, y_i) \mid x_i \in X_k'\}$ is collinear for all $k = 1, \ldots, N$, then G is the union of a finite number of line segments and hence has box dimension equal to 1.   ∎

6.1.3   Hidden variable fractal functions

In the case of hidden variable fractal functions, there is an interesting relation between the box dimension of $\text{graph}(f_k)$ and the ambient space. However, before this relationship is stated and proved, the necessary background needs to be provided.

The notation and terminology introduced in Section 5.5 is used but will be restricted to the following setup:

(a) The mappings v_i are independent of x: $v_i : Y \to Y$.

(b) $\forall i \in \{1, \ldots, N\} : u_i \in \text{Lip}^{(\leq s_i)}(X) \wedge v_i \in \text{Lip}^{(\leq s_i)}(Y) \ (0 < s_i < 1)$.

Now let $F_\Phi : A(X) \to Y$ be the function given in Theorem 5.7 and let d^* be the unique positive solution of $\sum_{i=1}^N s_i^d = 1$.

Theorem 6.3 *Let $(X \times Y, \mathbf{w})$ be the IFS considered in Theorem 5.7 with $w_i \in \mathcal{S}^*(X \times Y)$. Then $\dim_\beta$ graph $F_\Phi \le d^*$.*

Proof. The proof is rather easy: Let C be a cover of graph F_Φ of minimal cardinality $\mathcal{N}(\varepsilon)$ consisting of $\varepsilon \times \varepsilon$-balls. Denote by $\mathcal{N}_i(\varepsilon)$ the least number of such balls to cover $w_i(\text{graph } F_\Phi)$. Since w_i maps any ε/s_i-ball into an ε-ball, it is seen that $\mathcal{N}_i(\varepsilon) \le \mathcal{N}(\varepsilon/s_i)$. Therefore,

$$\mathcal{N}(\varepsilon) \le \sum_{i=1}^{N} \mathcal{N}_i(\varepsilon) \le \sum_{i=1}^{N} \mathcal{N}(\varepsilon/s_i).$$

This, however, is a special case of Lemma 6.2. Hence, $\mathcal{N}(\varepsilon) \le B\,\varepsilon^{-d^*}$, for some $B > 0$, which proves the theorem. ∎

Since the attractor $A(X)$ of the hyperbolic IFS $(X, \mathbf{u})$ has also box dimension equal to d^*, and since $\text{proj}_X \text{graph}\,(F_\Phi) = \text{proj}_X A(X \times Y) = A(X)$, Corollary 6.1 is obtained:

Corollary 6.1 $\dim_\beta$ graph $F_\Phi = d^*$. ∎

Now suppose that graph F is the attractor of the hyperbolic IFS $(I \times X, \mathbf{W})$ defined in Proposition 5.8, with $0 < |s_i| < 1$ and $w_i \in \mathcal{S}^*(I \times X)$, $i = 1, \dots, N$. If $s_i \ge 1/N$ for all $i = 1, \dots, N$, Corollary 6.1 gives immediately the box dimension of graph F. The goal is to relax this condition on the s_i.

Theorem 6.4 $\dim_\beta$ graph $F = d^*$.

Proof. Let $n \in \mathbb{N}$ and let B_n denote a ball of radius $1/n$. Let $C(n) := \{[(j-1)/N^n, j/N^n] \times B_n \mid j = 1, \dots, N^n\}$ be a minimal cover of $G := \text{graph } F$ of cardinality $\mathcal{N}(n)$. Denote by $\mathcal{N}'(\varepsilon)$ and $\mathcal{N}''(\varepsilon)$ the least number of ε-balls needed to cover G and $A(X)$, respectively. It is easy to see that $\mathcal{N}(n) \ge \mathcal{N}'(2/N^n)$. Let $G_i := G\big|_{[(j-1)/N^n, j/N^n] \times X}$ and let $\mathbf{j}(n)$ be the code corresponding to $[(j-1)/N^n, j/N^n]$. Let C' be a minimal $(N^n s_{\mathbf{j}(n)})^{-1}$-cover of $A(X)$. If the map $W_{\mathbf{j}(n)}$ is applied to $\{[0,1] \times B' \mid B' \in C'\}$, then G_i can be covered by $\mathcal{N}''((N^n s_{\mathbf{j}(n)})^{-1})$ sets from $C(n)$. Hence,

$$\mathcal{N}(n) \le \sum_{j_1=1}^{N} \cdots \sum_{j_n=1}^{N} \mathcal{N}''((N^n s_{\mathbf{j}(n)})^{-1})$$

$$\le B\,N^{nd^*} \sum_{j_1=1}^{N} \cdots \sum_{j_n=1}^{N} s_{\mathbf{j}(n)}^{d^*}$$

$$= B\,N^{nd^*} \left(\sum_{i=1}^{N} s_i^{d^*} \right) = B\,N^{nd^*}.$$

To obtain the second inequality, the fact that $\dim_\beta A(X) = d^*$ was used, i.e., $\mathcal{N}''(\varepsilon) \le B\,\varepsilon^{-d^*}$, for some $B > 0$. If ε is chosen in $[2/N^n, 2/N^{n-1}]$, then

$$\frac{\log \mathcal{N}(\varepsilon)}{-\log \varepsilon} \le \frac{\log B\, N^{nd^*}}{\log N^{n-1}/2},$$

and thus

$$\limsup_{\varepsilon \to 0} \frac{\log \mathcal{N}(\varepsilon)}{-\log \varepsilon} \le d^*.$$

However, as $\dim_\beta A(X) = d^*$, $\dim_\beta G \ge d^*$.   ∎

Next the dimension formula for the hidden variable fractal functions $f_k \in \mathcal{HF}(I, X_k)$ is derived, where X_k is a summand in the finite direct sum decomposition of the linear space X. It is assumed that $X_k = \mathbb{R}$. Following the remark on page 168, maps $W_i(t, x)$ of the form

$$W_i(t, x) = \begin{pmatrix} a_i t + b_i \\ w_i(x) \end{pmatrix} \qquad (i = 1, \dots, N)$$

are considered, where the a_i and b_i are determined by the interpolation property and $w_i \in \mathcal{S}^*(X)$.

Theorem 6.5 *Let $f_k : I \to \mathbb{R}$ be a hidden variable fractal function defined via the preceding maps. Then the box dimension of graph f_k is the unique positive solution d^* of*

$$\sum_{i=1}^{N} s_i a_i^{d-1} = 1. \tag{6.7}$$

Proof. The notation and terminology developed in Theorem 6.1 will be used. Let $\{\mathfrak{P}_\varkappa \mid \varkappa = 0, 1, \dots, K\}$ be the ε-partition associated with a minimal ε-column cover $C(\varepsilon)$ of graph f_k. For a fixed $\varkappa$, consider the union $\mathfrak{U}$ of all $\varepsilon \times \varepsilon$-squares covering graph $f_k \big|_{[\mathfrak{P}_\varkappa, \mathfrak{P}_{\varkappa+1}]}$. The continuity of f_k implies that $\mathfrak{U}$ is a rectangle of width ε and height $h_\varkappa$. Consider the pencil of planes $\Pi := \{\Pi_\theta \mid \theta \in [0, 2\pi)\}$ whose axis coincides with the t-axis. For a plane $\Pi_\theta \in \Pi$, denote by $h_\varkappa(\theta)$ the height of the orthogonal projection of $f_k \big|_{[\mathfrak{P}_\varkappa, \mathfrak{P}_{\varkappa+1}]}$ onto Π_θ. Let $m_\varkappa := \min\{h_\varkappa(\theta) \mid \theta \in [0, 2\pi)\}$ and $M_\varkappa := \max\{h_\varkappa(\theta) \mid \theta \in [0, 2\pi)\}$. Define $\underline{\mathcal{N}}(\varepsilon) := \sum_\varkappa m_\varkappa \varepsilon^{-1}$ and $\overline{\mathcal{N}}(\varepsilon) := \sum_\varkappa M_\varkappa \varepsilon^{-1}$. Then

$$\underline{\mathcal{N}}(\varepsilon) \le \|C(\varepsilon)\|_c \le \overline{\mathcal{N}}(\varepsilon).$$

As in the proof of Theorem 6.1 functional inequalities for $\underline{\mathcal{N}}(\varepsilon)$ and $\overline{\mathcal{N}}(\varepsilon)$ are obtained, yielding the existence of two positive constants A and B so that

$$A\varepsilon^{-d^*} \le \|C(\varepsilon)\|_c \le A\varepsilon^{-d^*},$$

where d^* is the unique positive solution of $\sum_{i=1}^{N} s_i a_i^{d-1} = 1$. Note that the collinearity of Δ_k is equivalent to $\sum_{i=1}^{N} s_i = 1$. ∎

Remark. Since $\dim_\beta A(X) = d^* < \infty$, it might as well be assumed that X has (topological) dimension $n := [\![d^*]\!] + 1$. If X_k is a summand in a finite direct sum decomposition of X, then $\dim X_k =: m < n$. Note that if Δ_k is collinear $(m = 1)$ or coplanar $(m > 1)$, then $\sum_{i=1}^{N} s_i^m = 1$. Also, $\sum_{i=1}^{N} s_i^n \leq 1$.

Theorem 6.5 can be generalized to vector-valued hidden variable fractal functions ([129]). This result is stated without presenting its proof, referring the reader to [86, 129]. The arguments leading to functional inequalities in the vector-valued case are quite similar to those given in Theorem 6.1. The difficulty arises in generalizing Lemma 6.1; its proof becomes rather technical and involved ([86]).

Theorem 6.6 *Let $\boldsymbol{f}_k : I \to X_k$ be a vector-valued hidden variable fractal function. Suppose $\dim X_k = m$ and that $\sum_{i=1}^{N} s_i^m > 1$. Then*

$$\dim_\beta \text{graph } \boldsymbol{f}_k = d^*,$$

where d^* is the unique positive solution of

$$\sum_{i=1}^{N} s_i^m a_i^{d-m} = 1. \tag{6.8}$$

∎

Now the previously mentioned relationship between $\dim_\beta \text{graph } f_k$ and $\dim I \times X_k$ can be proven.

Theorem 6.7 *Suppose that $f_k \in \mathcal{HF}(I, X_k)$ and that $\dim X_k = m < n$. Let $d^* = \dim_\beta \text{graph } f_k$. If $\sum_{i=1}^{N} s_i^m > 1$, then*

$$m \leq d^* \leq (m+1) - \frac{m}{n}. \tag{6.9}$$

Proof. The first inequality is clear. To prove the second, let q be such that $m/n + 1/q = 1$. The Cauchy-Schwarz Inequality implies

$$1 = \sum_i s_i^m a_i^{d^*-m} \leq \left(\sum_i s_i^n\right)^{m/n} \left(\sum_i a_i^{q(d^*-m)}\right)^{1/q}$$

$$\leq \left(\sum_i a_i^{q(d-m)}\right)^{1/q}.$$

But since $\sum_i a_i = 1$, $q(d^* - m) \leq 1$. Thus, $d^* \leq (m+1) - (m/n)$. ∎

For the sake of completeness, a dimension formula for vector-valued fractal functions of the type considered in Example 5.5 is presented. Unfortunately, it is quite impossible to give the very long and involved proof here. The interested reader is referred to [129].

Theorem 6.8 *Let $\boldsymbol{f} : I \to \mathbb{R}^n$ be a vector-valued fractal function. Let $s_i := \prod_{k=1}^n |s_{i,k}|$, and suppose that the interpolation set $\Delta \subset \mathbb{R} \times \mathbb{R}^n$ is not contained in any hyperplane of $\mathbb{R}^{n+1}$.*

1. *If $\sum_{i=1}^N s_i > 1$ then $\dim_\beta \operatorname{graph} \boldsymbol{f} = d^*$, where d^* is the unique positive solution of*

$$\sum_{i=1}^N s_i a_i^{d-n} = 1, \tag{6.10}$$

   where $a_i := x_{i+1} - x_i$, $i = 1, \ldots, N$.

2. *If $\forall k \in \{1, \ldots, n\} : \sum_{i=1}^N |s_{i,k}| \leq 1$, then $\dim_\beta \operatorname{graph} \boldsymbol{f} = 1$.*

3. *Suppose that $\sum_{i=1}^N s_i \leq 1$, and that there exists an $\ell \in \{1, \ldots, n\}$ such that $\sum_{i=1}^N |s_{i,k}| > 1$ for $k \in \{1, \ldots, \ell\}$, and $\sum_{i=1}^N |s_{i,k}| \leq 1$ for $k \in \{\ell + 1, \ldots, n\}$. Denote by $\mathfrak{p}$ any p-tuple of elements from $\{1, \ldots, \ell\}$, and let $s_i^{\mathfrak{p}} := \prod_{k \in \mathfrak{p}} |s_{i,k}|$. Let $\mathfrak{q} := \max\{\mathfrak{p} \,|\, \mathfrak{p} \subseteq \{1, \ldots, \ell\}\}$ such that $\sum_{i=1}^N s_i^{\mathfrak{p}} > 1$. Then*

$$\dim_\beta \operatorname{graph} \boldsymbol{f} = \max\{d^{\mathfrak{q}} \,|\, \mathfrak{q} \subseteq \{1, \ldots, \ell\}\}, \tag{6.11}$$

   where $d^{\mathfrak{q}}$ is the unique positive solution of

$$\sum_{i=1}^N s_i^{\mathfrak{q}} a_i^{d^{\mathfrak{q}} - n} = 1. \tag{6.12}$$

If $\Delta \subseteq H^m$, where H^m is a plane of codimension m in $\mathbb{R}^n$, $1 \leq m \leq n-1$, then conclusions 1 — 3 hold with n replaced by $n - m$. If $\Delta \subseteq H^n$ then $\dim_\beta \operatorname{graph} \boldsymbol{f} = 1$. ∎

6.2 Function Spaces and Dimension

In Subsection 2.1.2 it was seen that the dimension of the graph of a function is a measure of its smoothness; in particular, the Theorem of Besicovitch and Ursell shows that if $f \in \mathrm{Lip}^{\alpha}(\mathbb{R})$, then its Hausdorff (box) dimension $\leq 2 - \alpha$ (also Theorem 5.12). In this present section the previous result is sharpened and an exact characterization of the box dimension of a fractal function in terms of a certain class of smoothness spaces, called *Besov spaces*, is given. Before this result can be presented, however, the reader has to be provided with the necessary background material, especially the classical function spaces and some of their properties. Since the theory of function spaces is a deep and beautiful area of its own, only the most rudimentary and basic concepts can be presented and the interested reader must be referred to the references given in the bibliography, in particular to the excellent book by Stein [165].

6.2.1 Some basic function space theory

Throughout this section the base space will be $\mathbb{R}^n$, $n \geq 1$. In Chapters 1 and 5 two classes of classical function spaces have already been encountered, namely the spaces $C^k = C^k(\mathbb{R}^n)$ and the Lebesgue or L^p spaces $L^p = L^p(\mathbb{R}^n, d\lambda)$, where $d\lambda$ denotes n-dimensional Lebesgue measure on $\mathbb{R}^n$. It is easy to show that under the respective norms (Definitions 5.16 and Eqs. (1.30) and (1.31)) these linear spaces become Banach spaces. The Banach space $C^{k,\alpha} = C^{k,\alpha}(\mathbb{R}^n)$ is defined as the linear space of all bounded functions $f \in \mathrm{Lip}^{\alpha+k}$ endowed with the norm

$$\|f\|_{k,\alpha} := \sup_{\substack{|\ell| \leq k \\ x \in \mathbb{R}^n}} \left\{ \|(\mathfrak{D}^\ell f)(x)\| \right\} + \sup_{\substack{|\ell| = k \\ x,y \in \mathbb{R}^n \\ x \neq y}} \left\{ \frac{(\mathfrak{D}^\ell f)(x) - (\mathfrak{D}^\ell f)(y)}{\|x - y\|^\alpha} \right\}.$$

For the sake of completeness, one also defines

$$C^{0,0} := C, \qquad C^{k,0} := C^k.$$

Definition 6.2 Suppose X and Y are normed spaces. An embedding of X into Y, written $X \hookrightarrow Y$, is such that

1. X is a subvector space of Y;

2. The identity mapping $X \ni x \overset{j}{\mapsto} j(x) := x \in Y$ is continuous, i.e.,

$$\exists\, c > 0 \, \forall\, x \in X : \ \|j(x)\|_Y \leq c\, \|x\|_X.$$

The mapping j is called an embedding (of X into Y).

Using the definition of the respective norms, it is easy to show the validity of the following embeddings: Let $k \in \mathbb{N}_0$ and $0 < \alpha < \beta \leq 1$. Suppose that $\Omega \subset \mathbb{R}^n$. Then

$$C^{k+1}(\overline{\Omega}) \hookrightarrow C^k(\overline{\Omega}),$$
$$C^{k,\alpha}(\overline{\Omega}) \hookrightarrow C^k(\overline{\Omega}),$$
$$C^{k,\beta}(\overline{\Omega}) \hookrightarrow C^{k,\alpha}(\overline{\Omega}).$$

Next properties of functions are related to properties of their partial derivatives. In order to obtain a relation that is widely applicable, a "weak" version of the partial derivative of a function is needed. For $k \in \mathbb{N}_0 \cup \{\infty\}$, let $\mathcal{K}^k = \mathcal{K}_{\mathbb{R}}^k(\mathbb{R}^n)$ denote the collection of all functions in $C^k(\mathbb{R}^n)$ with compact support, and let $\mathcal{D} = \mathcal{D}(\mathbb{R}^n)$ denote the class of all infinitely differentiable functions on $\mathbb{R}^n$ each with compact support. Using the multi-index notation introduced in Subsection 2.2.2, one writes

$$\mathfrak{D}^\alpha := \frac{\partial^{\alpha_1 + \alpha_2 + \dots + \alpha_n}}{\partial \mathfrak{x}_1^{\alpha_1} \partial \mathfrak{x}_2^{\alpha_2} \cdots \partial \mathfrak{x}_n^{\alpha_n}}$$

for the *differential monomial* $\mathfrak{D}^\alpha$ of *order* $|\alpha| := \alpha_1 + \alpha_2 + \dots \alpha_n$. As usual, $\mathfrak{D}^\circ := \mathfrak{I}$, where I denotes the *identity operator on* $\mathbb{R}^n$.

Now suppose two locally integrable functions f and g are given. Then $g = \mathfrak{D}^\alpha f$ *in the weak sense* or f is the *weak derivative of* g iff

$$\int_{\mathbb{R}^n} f(x)(\mathfrak{D}^\alpha \varphi)(x)\, d\lambda = (-1)^\alpha \int_{\mathbb{R}^n} g(x)\varphi(x)\, d\lambda, \qquad \text{for all } \varphi \in \mathcal{D}. \quad (6.13)$$

Note that if f had partial derivatives up to order $|\alpha|$ in the usual (strong) sense, the preceding equation would easily follow from integration by parts.

Definition 6.3 Let $1 \leq q \leq \infty$ and let $s \in \mathbb{N}_0$. The Sobolev space $W_q^s = W_q^s(\mathbb{R}^n)$ is defined as the completion of $\mathcal{K}_{\mathbb{R}}^s(\mathbb{R}^n)$ with respect to the norm

$$\|f\|_{W_q^s} := \sum_{|\alpha| \leq s} \|\mathfrak{D}^\alpha f\|_q, \qquad (6.14)$$

where $\|\cdot\|_q$ denotes the L^q-norm.

Equivalently, W_q^s is the collection of all $\mathbb{R}$-valued functions $f \in L^q$ and where all $\mathfrak{D}^\alpha f$ exist and $\mathfrak{D}^\alpha f \in L^q$ in the weak sense, whenever $|\alpha| \leq s$. It should also be clear that, for $s > 0$, $W_q^s \subset C^{s-1}$ and that W_p^0 can be identified with L^q. If $q = 2$, then the space W_2^s is also denoted by $H^s = H^s(\mathbb{R}^n)$.

The importance of Sobolev spaces is well summarized in the next theorem (cf. also [165]). It shows how restrictions on the partial derivatives of a function give rise to restrictions on the function itself.

Theorem 6.9 *Let s and q be as before, and let p be such that* $1/p = 1/q + s/n$.

1. *If* $p < \infty$, *then* $W_q^s \subset L^p$ *and the inclusion mapping* $j : W_q^s \hookrightarrow L^p$ *is continuous, that is, j is an embedding of* W_q^s *into* L^p.

2. *If* $p = \infty$, *then the restriction of* $f \in W_q^s$ *to* $X \in \mathfrak{H}(\mathbb{R}^n)$ *is an element of* L^r, *for any* $r < \infty$.

3. *If* $q > n/s$, *then every* $f \in W_q^s$ *can be modified on a set of measure zero so that the resulting function is continuous.* ■

A proof of this theorem can be found in [165]. It should be noted that the embedding $W_q^s \hookrightarrow L^p$ is to be interpreted in the following manner: Let f be a representative of the equivalence class $[f] \in W_q^s$. Then, f is equal a.e. (on $\mathbb{R}^n$) to a function $g \in [g] \in L^p$. Before the next class of function spaces will be defined, the concept of the L^p-modulus of continuity ω_p has to be introduced: For any $f \in L^p$, $1 \le p \le \infty$, let

$$\omega_p(x) := \|f(\cdot + x) - f(\cdot)\|_p. \tag{6.15}$$

It is a well-known fact that when $p \in [1, \infty)$, $\lim_{x \to \infty} \omega_p(x) = 0$.

In Section 2.1 the class $\text{Lip}^\alpha(\mathbb{R}^n)$, $0 < \alpha \le 1$, of Lipschitz continuous functions was introduced. Using the earlier definitions of L^p-modulus of continuity, a function space Λ_α associated with $\text{Lip}^\alpha(\mathbb{R}^n)$ can be defined as follows: For $0 < \alpha < 1$, let

$$\dot{\Lambda}_\alpha := \{f \in L^\infty \mid \omega_\infty(x) \le c|x|^\alpha\}, \tag{6.16}$$

for some constant $c \ge 0$. If $\| \cdot \|_{\dot{\Lambda}_\alpha}$ denotes the infimum of all constants c for which Eq. (6.16) holds, then $\| \cdot \|_{\dot{\Lambda}_\alpha}$ is a Banach space *semi-norm* for the *homogeneous Lipschitz space* $\dot{\Lambda}_\alpha$. The function space $\dot{\Lambda}_\alpha$ can be made into a Banach space by identifying functions that differ by a constant: $f \sim g \Longleftrightarrow \exists c \in \mathbb{R} : f = g + c$. It is customary to call $\dot{\Lambda}_\alpha / \sim$ or simply $\dot{\Lambda}_\alpha$ a *Banach space modulo constants*. The *inhomogenous Lipschitz space* Λ_α is obtained when the Banach space norm

$$\|f\|_{\Lambda_\alpha} := \|f\|_\infty + \sup_{|x|>0} \frac{\omega_\infty(x)}{|x|^\alpha}$$

is used. Recall that the *Poisson integral* $u(x, y)$ *of* $f \in L^2$ is defined by

$$u(x, y) := \int_{\mathbb{R}^n} P_y(x) f(x - t) \, dt, \tag{6.17}$$

where $P_y(x)$ denotes the *Poisson kernel*

$$P_y(x) = \frac{\Gamma((n+1)/2)y}{\pi^{(n+1)/2}(\|x\|^2 + y^2)^{(n+1)/2}}.$$

Here Γ represents the Gamma function. The next theorem gives a characterization of functions on Λ_α in terms of their Poisson integrals.

Theorem 6.10 *Let $0 < \alpha < 1$ and suppose that $f \in L^\infty$. Then $f \in \Lambda_\alpha$ iff*

$$\left\| \frac{\partial u(x,y)}{\partial y} \right\|_\infty \leq c\, y^{\alpha-1}.$$

∎

This theorem allows us now to define Lipschitz spaces for all $\alpha > 1$: Let $k := [\![\alpha]\!] + 1$, where $[\![\cdot]\!]$ denotes the greatest integer function. Define

$$\Lambda_\alpha := \left\{ f \in L^\infty \,\middle|\, \left\| \frac{\partial^k}{\partial y^k} u(x,y) \right\|_\infty \leq c\, y^{\alpha-k} \right\}. \tag{6.18}$$

Denote by c_k the infimum of all constants c appearing in the inequality in (6.18). Then, as before, homogeneous and inhomogeneous Lipschitz spaces can be defined by setting $\|f\|_{\dot{\Lambda}_\alpha} := c_k$ and $\|f\|_{\Lambda_\alpha} := c_k + \|f\|_\infty$, respectively. This rather brief discussion of Lipschitz spaces is concluded with a theorem.

Theorem 6.11 *Suppose that $\alpha > 1$. Then $f \in \Lambda_\alpha$ iff $f \in L^\infty$ and $\partial f/\partial x_j \in \Lambda_{\alpha-1}$, $j = 1, \ldots, n$. Moreover, the norms $\| \cdot \|_{\Lambda_\alpha}$ and $\| \cdot \|_\infty + \sum_{j=1}^n \|\partial \cdot /\partial x_j\|_{\Lambda_{\alpha-1}}$ are equivalent.*

∎

As seen earlier, the Lipschitz spaces are defined in terms of their L^p-moduli of continuity and the Sobolev spaces by requiring that the partial derivatives are in some L^p space. It is therefore natural to ask whether or not there exist function spaces that are characterized in terms of both, i.e., whether or not there exist spaces that are intermediate to Sobolev spaces. Two equivalent definitions of these *intermediate Besov spaces* are given. The first follows naturally from our development of function spaces; the second is needed for later purposes.

Define the homogeneous Besov space $\dot{B}_p^{\alpha,q} = \dot{B}_p^{\alpha,q}(\mathbb{R}^n)$ as the linear space consisting of all functions $f \in L^p$ for which the semi-norm

$$\|f\|_{\dot{B}_p^{\alpha,q}} := \begin{cases} \left(\int_{\mathbb{R}^n} \frac{\omega_p^q(x)\, dx}{\|x\|^{n+\alpha q}} \right)^{1/q}, & q < \infty \\[3mm] \sup_{\|x\|>0} \frac{\omega_p(x)}{\|x\|^\alpha}, & q = \infty \end{cases} \tag{6.19}$$

is finite. The inhomogeneous Besov spaces $B_p^{\alpha,q} = B_p^{\alpha,q}(\mathbb{R}^n)$ are defined via the norms

$$\| \cdot \|_{B_p^{\alpha,q}} := \| \cdot \|_p + \| \cdot \|_{\dot{B}_p^{\alpha,q}}.$$

Thus, one may think of the Besov spaces as modified Lipschitz spaces. This is the reason why sometimes in the literature the notation $\Lambda_\alpha^{p,q}$ instead of $B_p^{\alpha,q}$ is encountered. Next some theorems about Besov spaces are stated and some embeddings given.

Theorem 6.12 1. *Let $0 < \alpha < 1$ and suppose $f \in L^p$. Then $f \in B_p^{\alpha,q}$ iff*

$$\left(\int_0^\infty \left(y^{\alpha-1} \left\| \frac{\partial}{\partial y} u(x,y) \right\|_p \right)^q \frac{dy}{y} \right)^{1/q} < \infty.$$

Furthermore, $\| \cdot \|_{B_p^{\alpha,q}}$ is equivalent with the norm

$$\| \cdot \|_p + \left(\int_0^\infty \left(y^{\alpha-1} \left\| \frac{\partial}{\partial y} u(x,y) \right\|_p \right)^q \frac{dy}{y} \right)^{1/q}.$$

2. *Let $\alpha > 1$. Then $f \in B_p^{\alpha,q}$ iff $f \in L^p$ and $\partial f/\partial x_j \in B_p^{\alpha-1,q}$, $j = 1, \dots, n$. Moreover, the norms $\| \cdot \|_{B_p^{\alpha,q}}$ and $\| \cdot \|_\infty + \sum_{j=1}^n \|\partial \cdot /\partial x_j\|_{B_p^{\alpha-1,q}}$ are equivalent.* ∎

The following inclusion relations hold for Besov spaces: If either $\alpha_1 > \alpha_2 > 0$ or $\alpha_1 = \alpha_2 > 0$ and $q_1 \le q_2$, then $B_p^{\alpha_1,q_1} \hookrightarrow B_p^{\alpha_2,q_2}$. If $\alpha \in \mathbb{N}$ and $p \in (1,\infty)$, then

$$W_p^\alpha \sim B_p^{\alpha,p} \quad (p \ge 2), \qquad W_p^\alpha \hookrightarrow B_p^{\alpha,2} \quad (p \le 2),$$

$$B_p^{\alpha,p} \hookrightarrow W_p^\alpha \quad (p \le 2).$$

Remarks.

1. The notation $W_p^\alpha \sim B_p^{\alpha,p}$ has to be interpreted as follows: For $f \in [f] \in W_p^\alpha$, there exists a $g \in [g] \in B_p^{\alpha,p}$ such that $f = g$ a.e. (on $\mathbb{R}^n$).

2. Besov spaces can also be defined using the concept of *interpolation spaces*. The basic idea is as follows: Suppose that $(X_0, \| \cdot \|_0)$ and $(X_1, \| \cdot \|_1)$ are Banach spaces contained in a linear Hausdorff space $\mathcal{X}$, such that the inclusion mappings $j_k : X_k \hookrightarrow \mathcal{X}$, $k = 0, 1$, are continuous.

The linear spaces $X_0 \cap X_1$ and $X_0 + X_1 := \{f \in \mathcal{X} \mid f = f_0 + f_1 \wedge f_k \in X_k \wedge k = 0, 1\}$ are Banach spaces under the norms

$$\|f\|_{X_0 \cap X_1} := \max\{\|f\|_k \mid k = 0, 1\}$$

and

$$\|f\|_{X_0 + X_1} := \inf\{\|f_0\|_0 + \|f_1\|_1 \mid f = f_0 + f_1 \wedge f_k \in X_k \wedge k = 0, 1\},$$

respectively. It is then clear that the inclusions

$$X_0 \cap X_1 \hookrightarrow X_k \hookrightarrow X_0 + X_1 \hookrightarrow \mathcal{X}$$

are continuous. Any Banach space X embedded in $\mathcal{X}$ satisfying

$$X_0 \cap X_1 \hookrightarrow X \hookrightarrow X_0 + X_1 \hookrightarrow \mathcal{X}$$

is called an *intermediate space of X_0 and X_1*. Using the *real interpolation method* one constructs the Besov spaces $B_p^{\alpha,p}$ as real intermediate spaces of two appropriate Sobolev spaces. For more details, the interested reader is referred to [150].

To proceed, an important class of a function space and its dual needs to be defined. The space $\mathcal{S} = \mathcal{S}(\mathbb{R}^n)$ consists of all functions in C^∞ all of whose derivatives remain bounded when multiplied by any polynomial. More precisely, a function $\varphi \in C^\infty$ is an element of $\mathcal{S}$ iff for arbitrary multi-indices α, β,

$$\sup_{x \in \mathbb{R}^n} \|x^\beta (\mathfrak{D}^\alpha \varphi)(x)\| =: c_{\alpha,\beta}(\varphi) < \infty. \tag{6.20}$$

This definition is equivalent to: $\varphi \in C^\infty$ is in $\mathcal{S}$ iff

$$\forall \, \alpha \in \mathbb{N}^n \, \forall \, m \in \mathbb{N} \, \exists \, c \in \mathbb{R} : \|\mathfrak{D}^\alpha \varphi\| \leq \frac{c}{(1 + \|x\|)^m}. \tag{6.21}$$

It follows immediately from (6.20) that $p(x)(\mathfrak{D}^\alpha \varphi)(x)$ is bounded in $\mathbb{R}^n$ for all real-valued polynomials p. Convergence in $\mathcal{S}$ is defined as follows: If $\{\varphi_k\}_{k \in \mathbb{N}} \subset \mathcal{S}$, then $\lim_{k \to \infty} \varphi_k = 0$ in $\mathcal{S}$ means

1. $\forall \, \alpha, \beta \in \mathbb{N}^n \, \exists \, c_{\alpha,\beta} \in \mathbb{R} \, \forall \, k \in \mathbb{N} \, \forall \, x \in \mathbb{R} : \|x^\beta \mathfrak{D}^\alpha \varphi\| \leq c_{\alpha,\beta}.$

2. $\forall \, \alpha \in \mathbb{N}^n : \lim_{k \to \infty} (\mathfrak{D}^\alpha \varphi)(x) = 0$, uniformly in $x \in \mathbb{R}^n$.

A linear continuous functional f defined on $\mathcal{S}$ is called a *tempered distribution*. The set of all tempered distributions forms a linear space denoted by $\mathcal{S}'$. For $f(\varphi) \in \mathbb{R}$ one sometimes writes $\langle f, \varphi \rangle$. If $f : \mathbb{R}^n \to \mathbb{R}$ is a locally integrable function satisfying

$$f(x) \le c(1 + \|x\|)^m,$$

for all $m \in \mathbb{N}$ and some constant $c > 0$, then the integral

$$\langle f, \varphi \rangle = \int_{\mathbb{R}^n} f(x) \varphi(x) \, dx$$

exists and is a continuous linear functional, hence a tempered distribution. In this case, f is called a *regular* tempered distribution. Convergence in $\mathcal{S}'$ is defined as follows:

$$\lim_{k \to \infty} f_k = f \quad \text{in } \mathcal{S}' \iff \forall \varphi \in \mathcal{S} : \lim_{k \to \infty} f_k(\varphi) = f(\varphi). \tag{6.22}$$

For $\varphi \in \mathcal{S}$ and $f \in \mathcal{S}'$, define the *convolution of f with φ* to be the (ordinary) function $f * \varphi$ given pointwise by

$$(f * \varphi)(\xi) := \langle f, \varphi(\xi - \cdot) \rangle. \tag{6.23}$$

If f is a regular tempered distribution, then Eq. (6.23) coincides with the ordinary definition of convolution:

$$f * \varphi(\xi) := \int_{\mathbb{R}^n} f(x) \varphi(\xi - x) \, dx.$$

For a function $\varphi : \mathbb{R}^n \to \mathbb{R}$ the notation $\varphi_t(x)$ is to mean $t^{-n} \varphi(x/t)$, for any $t > 0$. The Fourier transform of a $\varphi \in \mathcal{S}$ is defined by

$$\hat{\varphi}(\xi) := \int_{\mathbb{R}^n} \varphi(x) \, e^{-i \langle x, \xi \rangle} \, dx,$$

where dx denotes n-dimensional Lebesgue measure. The following fact, easy to establish, is well-known:

$$(f * \varphi)\hat{} = \hat{f} \hat{\varphi}.$$

Now a definition of Besov spaces in terms of elements in $\mathcal{S}$ and $\mathcal{S}'$ is given. Using *Calderón's Formula*, one can show that this definition is equivalent to the one given earlier (for more details see [72]). Choose a $\varphi \in \mathcal{S}$ so that

$\operatorname{supp} \hat{\varphi} \subset \{\xi \mid 1/2 \le \|\xi\| \le 2\}$ and $\|\hat{\varphi}\| \ge c > 0$ if $3/5 \le \|\xi\| \le 5/3$. (The choice of these bounds is conventional. There are other choices that work as well.) For $\alpha \in \mathbb{R}$, $p, q \in (0, \infty]$, and $f \in \mathcal{S}'$ we define

$$\|f\|_{\dot{B}_p^{\alpha,q}} := \left\{ \sum_{k \in \mathbb{Z}} \left(2^{k\alpha} \|\varphi_{2^{-k}} * f\|_p \right)^q \right\}^{1/q}. \tag{6.24}$$

For $\varphi \in \mathcal{S}$ and $f \in \mathcal{S}'$ the convolution $\varphi_{2^{-k}} * f$ is a smooth function, and $\|f\|_{\dot{B}_p^{\alpha,q}} = 0$ iff $\varphi_{2^{-k}} * f$ is the zero function for all $k \in \mathbb{Z}$. This, however, is equivalent to $\hat{f}\hat{\varphi}(2^k \xi) = 0$, for all $k \in \mathbb{Z}$. Using the preceding conditions on $\hat{\varphi}$ this is, in turn, equivalent to $\operatorname{supp} \hat{f} = \{0\}$, which implies that $\hat{f}$ is a finite linear combination of the Dirac delta distribution δ and its derivatives, which implies that f is a polynomial (recall that the *Dirac delta distribution* δ is defined by $\delta(\varphi) := \varphi(0)$). Hence, if f is regarded as an element of $\mathcal{S}/\mathcal{P}$, where $\mathcal{P}$ denotes the class of polynomials, then $\|f\|_{\dot{B}_p^{\alpha,q}}$ is a Banach space norm, and $\dot{B}_p^{\alpha,q}$, the homogeneous Besov space, consists of all those functions in $\mathcal{S}/\mathcal{P}$ whose norm $\| \cdot \|_{\dot{B}_p^{\alpha,q}}$ is finite. The corresponding inhomogenous version $B_p^{\alpha,q}$ is obtained by defining a Banach space norm by

$$\|f\|_{B_p^{\alpha,q}} := \|\Phi * f\|_p + \left\{ \sum_{k \in \mathbb{N}} \left(2^{k\alpha} \|\varphi_{2^{-k}} * f\|_p \right)^q \right\}^{1/q}, \tag{6.25}$$

where $\Phi \in \mathcal{S}$ is chosen so that $\operatorname{supp} \hat{\Phi} \subset \{\xi \in \mathbb{R}^n \mid \|\xi\| \le 2\}$ and $\|\hat{\Phi}(\xi)\| \ge c > 0$, for all $\|\xi\| \le 5/3$. Since $\hat{\Phi}(0) \ne 0$, the need to consider distributions modulo polynomials disappears. The definition of the above Banach space norm is independent of the choices of φ and Φ: Different choices lead to equivalent norms.

One can associate a sequence space $\dot{b}_p^{\alpha,q}$ with the Besov space $\dot{B}_p^{\alpha,q}$ in the following way. Let $Q = Q_{k,\ell} := \{x \in \mathbb{R}^n \mid 2^{-k}\ell_i \le x_i \le 2^{-k}(\ell_{i+1} + 1) \wedge i = 1, \ldots, n\}$ denote a *dyadic cube* in $\mathbb{R}^n$, with $k \in \mathbb{Z}$ and $\ell = (\ell_1, \ldots, \ell_n) \in \mathbb{Z}^n$. The collection of all such dyadic cubes is denoted by $\mathcal{Q} := \{Q_{k,\ell} \mid k \in \mathbb{Z} \wedge \ell \in \mathbb{Z}^n\}$. For a dyadic cube $Q \in \mathcal{Q}$, let $l(Q)$ be its side length and $|Q|$ its diameter. Let $s = \{s_Q\}$ be a sequence in $\mathbb{R}^n$ indexed by dyadic cubes in $\mathcal{Q}$. For $\alpha \in \mathbb{R}$ and $p, q \in (0, \infty]$, let $\dot{b}_p^{\alpha,q}$ be the space consisting of all sequences $s = \{s_Q\}$ for which

$$\|s\|_{\dot{b}_p^{\alpha,q}} := \left(\sum_{k \in \mathbb{Z}} \left(\left\| \sum_{l(Q)=2^{-k}} |Q|^{-(\alpha/n+1/2)} \|s_Q\| \chi_Q \right\|_p \right)^q \right)^{1/q} < \infty. \tag{6.26}$$

To introduce what is called the *wavelet transform*, the following lemma, whose proof can be found in [72], needs to be stated.

Lemma 6.4 *Let* $\varphi \in \mathcal{S}$ *satisfy* $\operatorname{supp} \hat{\varphi} \subset \{\xi \in \mathbb{R}^n \mid 1/2 \leq \|\xi\| \leq 2\}$ *and* $\|\hat{\varphi}(\xi)\| \geq c > 0$, *for all* $3/5 \leq \|\xi\| \leq 5/3$. *Then there exists a* $\psi \in \mathcal{S}$ *with the same properties as* φ *so that*

$$\sum_{k \in \mathbb{Z}} \overline{\hat{\varphi}(2^{-k}\xi)}\hat{\psi}(2^{-k}\xi) = 1, \qquad \text{for } \xi \neq 0. \tag{6.27}$$

∎

Using this lemma and Calderón's formula one derives the following *wavelet decomposition of* $f \in \mathcal{S}'$ ([72] for more details):

$$f = \sum_Q \langle f, \varphi_Q \rangle \psi_Q, \tag{6.28}$$

with φ and ψ are as in Lemma 6.4. Here let

$$\varphi_Q := 2^{kn/2}\varphi(2^k \cdot -\ell) \tag{6.29}$$

for $Q = Q_{k,\ell}$ a dyadic cube, and similarly for ψ. The next theorem relates distributions in $\mathcal{S}'/\mathcal{P}$ to sequences in $\dot{b}_p^{\alpha,q}$.

Theorem 6.13 *Let* $\alpha \in \mathbb{R}$ *and let* $p, q \in (0, \infty]$. *Suppose that* φ *and* ψ *are as in Lemma 6.4. Assume that for* $f \in \mathcal{S}'/\mathcal{P}$ *we have the decomposition* $f = \sum_Q \langle f, \varphi_Q \rangle \psi_Q$. *Let* $s_Q := \langle f, \varphi_Q \rangle$, *for each dyadic cube* Q. *Then* $f \in \dot{B}_p^{\alpha,q}$ *iff* $s = \{s_Q\} \in \dot{b}_p^{\alpha,q}$, *and* $\|f\|_{\dot{B}_p^{\alpha,q}} \approx \|s\|_{\dot{b}_p^{\alpha,q}}$.

Proof. See [72]. ∎

Next the preceding theorem is rephrased in terms of retractions and retracts and the φ-transform is introduced. For this purpose, suppose that $f \in \dot{B}_p^{\alpha,q}$ and φ is as in Lemma 6.4. For $Q \in \mathcal{Q}$, let $s_Q := \langle f, \varphi_Q \rangle$, and define a mapping $W_\varphi : \dot{B}_p^{\alpha,q} \to \dot{b}_p^{\alpha,q}$ by

$$W_\varphi f := s = \{s_Q\}. \tag{6.30}$$

This mapping is called the φ-*transform* or the *wavelet transform of* f. The left inverse $\check{W}_\psi : \dot{b}_p^{\alpha,q} \to \dot{B}_p^{\alpha,q}$ of W_φ is defined by

$$\check{W}_\psi s := \sum_{Q \in \mathcal{Q}} s_Q \psi_Q. \tag{6.31}$$

It can be shown that W_φ and $\breve{W}_\psi$ are bounded (the latter follows from the theorem above, the former from estimates involving the so-called *atomic decomposition of f*). Clearly, $\breve{W}_\psi \circ W_\varphi$ is the identity operator on $\dot{B}_p^{\alpha,q}$. Moreover, using Definition 1.34, it is immediately seen that $\breve{W}_\psi$ is a retraction with $\dot{B}_p^{\alpha,q}$ being the retract of $\dot{b}_p^{\alpha,q}$. Therefore, $\dot{B}_p^{\alpha,q}$ is isomorphic to $W_\varphi \dot{B}_p^{\alpha,q} \subset \dot{b}_p^{\alpha,q}$. This observation can be used to characterize bounded linear operators on $\dot{B}_p^{\alpha,q}$ in terms of the lift of W_φ to the space of linear operators on $\dot{B}_p^{\alpha,q}$. For more details the interested reader is referred to [72].

Notice that, by setting $\varphi = \psi$ in Eq. (6.28), the resulting decomposition of f resembles a Fourier series. This resemblance can be further exploited by considering the special case $f \in L^2(\mathbb{R})$. An appropriate $\psi \in \mathcal{S}(\mathbb{R})$ whose dyadic dilates and integer translates ψ_Q form a complete orthonormal basis of $L^2(\mathbb{R})$ needs to be found. (The dyadic cube Q in this case is of course a dyadic interval.) The proof of the following theorem can be found in [72].

Theorem 6.14 *There exists a function $\psi \in \mathcal{S}(\mathbb{R})$ such that the set $\mathcal{B}_\psi :=$ $\{2^{k/2}\psi(2^k \cdot -\ell) \mid k, \ell \in \mathbb{Z}\}$ is an orthonormal basis for $L^2(\mathbb{R})$. Furthermore, $\operatorname{supp} \hat{\psi} \subset [-8\pi/3, -2\pi/3] \cup [2\pi/3, 8\pi/3]$ and thus $\int_{\mathbb{R}} x^m \psi(x)\, dx = 0$, for all $m \in \mathbb{N}_0$.* ∎

The inhomogenous case is now considered briefly. As before, it is assumed that φ satisfies the hypotheses of Lemma 6.4 and that Φ satisfies the conditions leading to the definition of the Banach space norm $\|\cdot\|_{B_p^{\alpha,q}}$. Furthermore, it is assumed that φ also satsfies

$$\hat{\Phi}(\xi) + \sum_{k \in \mathbb{N}} \hat{\varphi}(2^k \xi) = 1.$$

Then, Lemma 6.4 implies the existence of two functions ψ and Ψ satisfying the same conditions as φ and Φ, respectively, as well as

$$\overline{\hat{\Phi}(\xi)}\hat{\Psi}(\xi) + \sum_{k \in \mathbb{Z}} \overline{\hat{\varphi}(2^{-k}\xi)}\hat{\psi}(2^{-k}\xi) = 1, \qquad \text{for } \xi \neq 0. \qquad (6.32)$$

For an $f \in \mathcal{S}'$, the decomposition

$$f = \sum_{l(Q)=1} \langle f, \Phi_Q \rangle \, \Psi_Q + \sum_{l(Q)<1} \langle f, \varphi_Q \rangle \, \psi_Q \qquad (6.33)$$

is obtained. The wavelet transform $W_\varphi : B_p^{\alpha,q} \to b_p^{\alpha,q}$ is then given by

$$W_\varphi f := \begin{cases} \langle f, \Phi_Q \rangle & \text{for } l(Q) = 1 \\ \langle f, \varphi_Q \rangle & \text{for } l(Q) < 1. \end{cases} \qquad (6.34)$$

Finally, the homogeneous Besov spaces for functions on the n-torus $\mathbb{T}^n$ are defined. The torus $\mathbb{T}^n$ is regarded as the unit cube in $\mathbb{R}^n$ with opposite faces identified. As usual, functions on $\mathbb{T}^n$ are identified with n-fold periodic functions on $\mathbb{R}^n$. One possibility of defining the homogeneous Besov space $B_p^{\alpha,q}(\mathbb{T}^n)$ is the following: Let $\mathcal{U}_{\mathbb{T}}^n$ be a fixed bounded neighborhood of $\mathbb{T}^n$, and let $\theta \in C^\infty(\mathbb{R}^n)$ with the property that $\theta = 1$ on $\mathbb{T}^n$ and $\operatorname{supp}\theta \subset \mathcal{U}_{\mathbb{T}^n}$. Then, define a Banach space norm on the linear space of all locally integrable functions $f : \mathbb{T}^n \to \mathbb{T}^n$ by

$$\|f\|_{B_p^{\alpha,q}(\mathbb{T}^n)} := \|\theta f\|_{B_p^{\alpha,q}(\mathbb{R}^n)}. \tag{6.35}$$

The homogeneous Besov space $B_p^{\alpha,q}(\mathbb{T}^n)$ is then the collection of all functions f for which the preceding norm is finite. Note that the norm $\|f\|_{B_p^{\alpha,q}(\mathbb{T}^n)}$ is independent of θ in the sense that a different choice of θ yields an equivalent norm.

In a similar fashion one defines the homogeneous Besov space $b_p^{\alpha,q}(\mathbb{T}^n)$: Let $s = \{s_Q\}$ be a sequence indexed by cubes $Q \subset [0,1]^n$. Then

$$\|s\|_{b_p^{\alpha,q}(\mathbb{T}^n)} := \|\{s_Q\}_{Q \subset [0,1]^n}\|_{b_p^{\alpha,q}(\mathbb{R}^n)} \tag{6.36}$$

is a Banach space norm. The class of all sequences $s = \{s_Q\}_{Q \subset [0,1]^n}$ for which this norm is finite constitutes the homogeneous Besov space $b_p^{\alpha,q}(\mathbb{T}^n)$. As before, one defines the wavelet transform W_φ on $B_p^{\alpha,q}(\mathbb{T}^n)$, yielding an element of $b_p^{\alpha,q}(\mathbb{T}^n)$. It is easy to see that if f is an n-fold periodic function on $\mathbb{R}^n$, then

$$W_\varphi f\big|_{Q_{k,\ell}} = W_\varphi f\big|_{Q_{k,\ell+2^k e}},$$

where $Q_{k,\ell} := 2^{-k}\ell + 2^{-k}[0,1]^n$, $k \in \mathbb{Z}$, $\ell \in \mathbb{Z}^n$, and $e := (1,\dots,1) \in \mathbb{Z}^n$. Hence, the wavelet transform W_φ is determined by all cubes Q contained in $[0,1]^n$. Furthermore, as in Theorem 6.14, one has

$$\|f\|_{B_p^{\alpha,q}(\mathbb{T}^n)} \approx \|W_\varphi f\|_{b_p^{\alpha,q}(\mathbb{T}^n)}.$$

6.2.2 Box dimension and smoothness

Here an exact description of the upper box dimension of a continuous function in terms of its inclusion in a certain range of smoothness spaces is given (also [48]). This description is presented for $n = 1$. In what follows, N will always denote an integer greater than 1 and Q a dyadic subinterval of $[0,1]$, i.e., an interval of the form $[jN^{-k}, (j+1)N^{-k}]$, for $k \in \mathbb{N}_0$ and $j \in \{0, 1, \dots, N^{-k}\}$.

Recall (cf. Definition 5.14) that the oscillation of a measurable function $f : [0, 1] \to \mathbb{R}$ on $Q \subset [0, 1]$ is defined by

$$\text{osc}(f; Q) = \sup\{f(x) - f(x') \mid x, x' \in Q\},$$

or, equivalently, by

$$\text{osc}(f; Q) = \sup_Q f - \inf_Q f.$$

It will prove advantageous to also introduce the *total oscillation* $\text{Osc}(f; k)$ of *order k of f*:

$$\text{Osc}(f; k) := \sum_{|Q|=N^{-k}} \text{osc}(f; Q), \tag{6.37}$$

where the sum extends over all intervals $Q \subset [0, 1]$ of length N^{-k}. As in Section 5.6, the cardinality of a minimal cover of $G = \text{graph } f$ can be expressed in terms of the oscillation of f. For this purpose, let $k \in \mathbb{N}_0$ and let $\mathcal{C}_\varepsilon(G)$ be a minimal cover of graph f consisting of squares of side N^{-k}, where $f \in C([0, 1], \mathbb{R})$. If the cardinality of $\mathcal{C}_\varepsilon(G)$ is denoted by $\mathcal{N}(\varepsilon)$, and the cardinality of $\mathcal{C}_\varepsilon(\text{graph } f |_Q)$ by $\mathcal{N}_Q(\varepsilon)$, then the arguments in the proof of Theorem 6.1 restricted to the case $\varepsilon = N^{-k}$ yield immediately the following inequalities:

$$N^{-k}(\mathcal{N}_Q(\varepsilon) - 2) \leq \text{osc}(f; Q) \leq N^{-k}\mathcal{N}_Q(\varepsilon), \tag{6.38}$$

and thus, after summing over all subintervals Q of length N^{-k},

$$N^{-k}\mathcal{N}(\varepsilon) - 2 \leq \text{Osc}(f; k) \leq N^{-k}\mathcal{N}(\varepsilon). \tag{6.39}$$

It is worthwhile noting that the preceding inequalities are just special cases of those obtained in the proof of Theorem 6.1. Since the N-adic intervals Q form a partition of $[0, 1]$, the extra terms in Eqs. (6.3) and (6.4) are not needed.

Using the definition of upper box dimension (cf. Definition 3.4), one thus obtains

$$\overline{\dim}_\beta G = 1 + \limsup_{k \to \infty} \frac{\log^+ \text{Osc}(f; k)}{\log^+ N^k}. \tag{6.40}$$

Here, $\log^+ x = \max\{\log x, 0\}$ was used with the convention that $\log^+ 0 := 0$. At this point a definition is needed.

Definition 6.4 A function $f : [a, b] \to \mathbb{R}$ is said to be of bounded variation on $[a, b]$ iff there exists a real number $V > 0$ so that

$$\sum_{m=1}^M |f(x_m) - f(x_{m-1})| < V, \tag{6.41}$$

for every finite set of points $\{x_0, x_1, \ldots, x_M\}$ satisfying $a \leq x_0 < x_1 < \ldots < x_M \leq b$. The class of all functions that are of bounded variation on $[a, b]$ is denoted by $\mathrm{BV}([a, b])$.

In view of the preceding considerations, it is not hard to establish that $f \in \mathrm{BV}([0, 1])$ iff $\sup_{k \in \mathbb{N}_0} \mathrm{Osc}(f; k)$ is finite. One can use this characterization to obtain the following class of smoothness spaces: Let $0 \leq \alpha \leq 1$ and let $V^\alpha = V^\alpha(\mathbb{T})$ denote the class of all measurable functions $f : [0, 1] \to \mathbb{R}$ for which

$$\|f\|_{V^\alpha} := \sup_{k \in \mathbb{N}_0} \frac{\mathrm{Osc}(f; k)}{N^{k(1-\alpha)}} < \infty. \tag{6.42}$$

Note that the V^α are only semi-normed spaces; if the collection of all constant functions on $\mathbb{R}$ is denoted by $\mathcal{C}$, then $V^\alpha/\mathcal{C}$ is a normed linear space. Furthermore, $f \in \mathrm{BV}([0, 1])$ iff $f \in V^1$ and $V^\beta \subset V^\alpha$, if $\alpha \leq \beta$. The spaces V^α are related to those studied in [23]. This class of smoothness spaces allows us now to improve the result of Besicovitch and Ursell.

Theorem 6.15 *Let $f \in C([0, 1], \mathbb{R})$ and let $0 < \alpha < 1$. Then*

$$\overline{\dim}_\beta \text{ graph } f = 2 - \alpha \quad \text{iff} \quad f \in \left(\bigcap_{\beta < \alpha} V^\beta \right) \setminus \left(\bigcup_{\alpha < \gamma} V^\gamma \right). \tag{6.43}$$

Proof. Necessity is shown first. Assume then that $\overline{\dim}_\beta \text{ graph } f = 2 - \alpha$. Using the definition of supremum and Eq. (6.40), it is easily seen that for all $\varepsilon > 0$, there exists a $k_0 \in \mathbb{N}_0$ such that

$$\forall k > k_0 : \mathrm{Osc}(f; k) \leq N^{k(1-\alpha+\varepsilon)},$$

and a sequence $\{k_\nu\}_{\nu \in \mathbb{N}}$ with $k_\nu \to \infty$ and

$$\mathrm{Osc}(f; k_\nu) \geq N^{k_\nu(1-\alpha-\varepsilon)}.$$

This last inequality, however, shows that $f \notin \bigcup_{\alpha < \gamma} V^\gamma$. The continuity of f implies its boundedness, and thus, by the first inequality, there exists a constant $c = c(f)$ such that

$$\forall k \in \mathbb{N}_0 : \mathrm{Osc}(f; k) \leq c N^{k(1-\alpha+\varepsilon)},$$

i.e., $f \in \bigcap_{\beta < \alpha} V^\beta$.

To show sufficiency, let $f \in \bigcap_{\beta < \alpha} V^{\beta}$. Then, for all $\varepsilon > 0$, $f \in V^{\alpha - \varepsilon}$ (see the earlier remarks). This, however, is equivalent to

$$\mathrm{Osc}(f; k) \leq c N^{k(1 - \alpha - \varepsilon)},$$

for some $c = c(f) > 0$. Taking logarithms and using Eq. (6.40) yields $\overline{\dim}_{\beta}$ graph $f \leq 2 - \alpha$.

Now suppose that $f \notin \bigcup_{\alpha < \gamma} V^{\gamma}$. Then there exists a small enough $\varepsilon > 0$ so that $f \notin V^{\alpha + \varepsilon}$. Let us fix one of these ε. Then there exists a sequence $\{k_{\nu}\}_{\nu \in \mathbb{N}}$ such that

$$\mathrm{Osc}(f; k_{\nu}) \geq 2^{\nu} N^{k_{\nu}(1 - \alpha - \varepsilon)}.$$

The boundedness of f now implies that the sequence $\{k_{\nu}\}_{\nu \in \mathbb{N}}$ contains a sub-sequence $\{k_{\nu_{\varkappa}}\}_{\varkappa \in \mathbb{N}}$ with $k_{\nu_{\varkappa}} \to \infty$ as $\varkappa \to \infty$. Thus, by Eq. (6.40), $\overline{\dim}_{\beta}$ graph $f \geq 1 - \alpha - \varepsilon$. As $\varepsilon > 0$ was arbitrary, the result follows. ∎

Let us consider an example to illustrate how Theorem 6.15 is an improvement over Theorems 2.1 and 5.12 (also [48]). For $0 < \alpha < 1$, define a one-periodic function $f : [0, 1] \to \mathbb{R}$ by $f(x) := |x - 1/2|^{\alpha}$. Clearly, $\dim_{H}$ graph $f = \dim_{\beta}$ graph $f = 1$. A simple calculation shows that f is an element of $\bigcap_{\beta < 1} V^{\beta}$ and in Lip^{α}. Theorem 6.15 implies then that $\overline{\dim}_{\beta}$graph $f = 1$, whereas the classical Theorems 2.1 and 5.12 only give us $\overline{\dim}_{\beta}$ graph $f \leq 2 - \alpha$.

Next the upper box dimension of a continuous function f is related to the size of the coefficients in its wavelet decomposition (Eq. (6.28). Since these coefficients are elements of a Besov space (Theorem 6.14) we have to express the results in Theorem 6.15 in terms of Besov spaces. For this purpose, the following proposition is needed. Its proof uses methods from function space theory and harmonic analysis not developed in this section and is therefore omitted. The interested reader may want to consult [48]. Also, as the primary interest is in continuous functions, the normed space $V^{\alpha} \cap L^{\infty}(\mathbb{T})$ rather than V^{α} is considered.

Proposition 6.2 *Let $\alpha \in (0, 1)$. Then the following inclusions are embeddings:*

$$B_{1}^{\alpha, 1}(\mathbb{T}) \cap L^{\infty}(\mathbb{T}) \hookrightarrow V^{\alpha} \cap L^{\infty}(\mathbb{T}) \hookrightarrow B_{1}^{\alpha, \infty}(\mathbb{T}) \cap L^{\infty}(\mathbb{T}).$$

∎

Using this proposition, Theorem 6.15 can be rephrased:

Theorem 6.16 *Let $f \in C([0,1], \mathbb{R})$ and let $0 < \alpha < 1$. Then*

$$\overline{\dim}_\beta \text{ graph } f = 2 - \alpha \quad iff \quad f \in \left(\bigcap_{\beta < \alpha} B_1^{\beta, \infty} \right) \setminus \left(\bigcup_{\alpha < \gamma} B_1^{\gamma, \infty} \right). \quad (6.44)$$

■

Now let $N = 2$. Note that for $p = 1$ and $q = \infty$

$$\|f\|_{B_1^{\alpha, \infty}(\mathbb{T})} \approx \sup_{k \in \mathbb{N}_0} \sum_{|Q| = 2^{-k}} |Q|^{1/2 - \alpha} |W_\varphi f|_Q|,$$

where the sum is taken over all dyadic intervals $Q \subset [0,1]$. Replacing $\text{Osc}(f; k)$ in Eq. (6.40) by

$$\sum_{|Q| = 2^{-k}} |Q|^{-1/2} |W_\varphi f|_Q|$$

and using the preceding theorem yields, as a direct consequence of the proof of Theorem 6.15, the following corollary :

Corollary 6.2 *Let $f \in C([0,1], \mathbb{R})$. Then*

$$\overline{\dim}_\beta \text{ graph } f = 1 + \limsup_{\nu \to \infty} \frac{\log^+ \sum_{|Q| = 2^{-k}} |Q|^{-1/2} |W_\varphi f|_Q|}{\log^+ 2^k}. \quad (6.45)$$

■

This section is closed with an application of Theorem 6.13. Suppose that f is a continuous function that vanishes outside $[0,1]$. Setting $p = 1$ and $q = \infty$ in the first statement in Theorem 6.13 yields

$$\|f\|_{B_1^{\alpha, \infty}(\mathbb{R})} \approx \|f\|_{L^1(\mathbb{R})} + \sup_{y > 0} \int_\mathbb{R} y^{1-\alpha} \left| \frac{\partial}{\partial y} u(x, y) \right| dx,$$

where $u(x, y)$ denotes the Poisson integral of f, i.e., $u(x, y) = P_y * f$. Since $\frac{\partial}{\partial y} u(x, y) = f * \frac{\partial}{\partial y} P_y$, one has for all $2^{-k} \le t < 2^{-k+1}$,

$$\int_\mathbb{R} \left| \frac{\partial}{\partial y} u(x, y) \right| dx \approx \int_\mathbb{R} \left| \frac{\partial}{\partial y} u(x, 2^{-k}) \right| dx.$$

Hence, using Corollary 6.2 gives

$$\overline{\dim}_\beta \text{ graph } f = 1 + \limsup_{k \to \infty} \frac{\log^+ \int_\mathbb{R} \left| \frac{\partial}{\partial y} u(x, 2^{-k}) \right| dx}{\log^+ 2^k}$$

$$= 1 + \limsup_{y \to 0} \frac{\log^+ \int_\mathbb{R} \left| \frac{\partial}{\partial y} u(x, y) \right| dx}{\log^+ (1/y)}. \quad (6.46)$$

Chapter 7

Fractal Functions and Wavelets

Wavelet theory has its origin in several disciplines; the types of functions that are now called wavelets were studied in quantum field theory, signal analysis, and function space theory. In all these areas, wavelet-like algorithms replaced the classical Fourier-type expansion of a function. It was not until the mid-1980s that these, at first, seemingly different notions were described in a unified manner. An — albeit incomplete — list of the principal contributors to this unified description is given in the references. The wavelet transform is an alternative to the classical *windowed* Fourier transform. The windowed Fourier transform serves as a means to describe or compare the fine structure of a function at different resolutions. Its basic building blocks are the integer dilates of the sine and cosine functions multiplied by a *window function*, usually a Gaussian. Although quite successful, Fourier analysis is not able to describe highly localized functions. For instance, the Fourier transform of the *Dirac delta distribution* δ supported at a single point x is 1, a function defined on all of $\mathbb{R}$. (Recall that the Dirac delta distribution is that element of $\mathcal{S}'$ satisfying

$$\delta * f = f, \qquad \text{for all } f \in \mathcal{S}.$$

In other words, δ is the identity in the *convolution algebra* $(\mathcal{S}', *)$.) This "spreading-out" of a highly localized function is a direct consequence of the global support of sine and cosine. Using the windowed Fourier transform instead, and allowing the window to "move," yields better localization properties. However, the window size is usually kept fixed, which for certain applications is a disadvantage. To overcome these problems, one replaces sine and cosine by a function that has compact support and is continuous, and whose dilates and translates form an orthonormal basis of $L^2(\mathbb{R})$. (An example of a *discontinuous* orthonormal basis is the *Haar system.*) It can be shown that under rather mild conditions such a *wavelet decomposition of $L^2(\mathbb{R})$* exists. The famous *Daubechies wavelets* ([40]) are a class of such compactly supported, continuous, and orthonormal wavelet bases. The interested reader may consult [40, 41] or, if interested in *spline wavelets,* [35] and the reference given therein, for a more detailed presentation of the construction of wavelets and their usefulness in applications.

Wavelet decompositions are often obtained via a *multiresolution analysis* (see, for instance, [122, 137]). In this chapter it is shown how a class of fractal functions can be used to define a multiresolution analysis yielding a (fractal) wavelet decomposition. Furthermore, the existence of the free parameter s in the construction of fractal functions allows one to obtain a compactly supported, continuous, and orthogonal wavelet basis of $L^2(\mathbb{R})$ that has a higher approximation order than the corresponding Daubechies wavelets.

In the first section a brief introduction to wavelet theory is given, developing the necessary notation and terminology needed later. It is assumed that the reader is familiar with the rudimentary concepts of Fourier analysis. A construction of wavelets based on fractal functions is given in the second section. There a method for constructing stable shift- and dilation-invariant function spaces using the fractal functions defined in Example 5.2 is given. This construction yields a multiresolution analysis on $L^2(\mathbb{R})$. For some special classes of such translation- and dilation-invariant function spaces, exemplary wavelet decompositions of $L^2(\mathbb{R})$ and also $C_b(\mathbb{R})$ are presented. (Here $C_b(\mathbb{R})$ denotes the Banach space of all functions $f \in C(\mathbb{R})$ with the property that $\lim_{|x|\to\infty} f(x) = 0$. In the last section, it is shown how a proper choice of s yields a compactly supported, continuous, and orthonormal wavelet basis of $L^2(\mathbb{R})$.

7.1   Basic Wavelet Theory

Here the basic concepts of wavelet theory are introduced: refinable shift-invariant function spaces, scaling functions, multiresolution analyses, and reconstruction and decomposition algorithms. Because of the limited scope of this book the reader is referred to [35, 40, 41, 39, 90, 121, 122, 137] and the references given therein for a more detailed introduction to this new and fascinating area.

There are two approaches to obtain wavelet decompositions: The first uses the concept of *multiresolution analysis*. This multiresolution analysis yields, in general, a finite set of *scaling functions* which are then used to define the wavelets and the wavelet decompositions. The second method begins with a dilation equation for a given function or set of functions. The non-trivial solutions of this dilation equation then give scaling functions that generate a multiresolution analysis.

Let $F = (F(\mathbb{R}^n, \mathbb{K}), \| \cdot \|_F)$ be a Banach space of functions $f : \mathbb{R}^n \to \mathbb{K} \subseteq \mathbb{C}$ with Schauder basis $\{e_i\}_{i \in \mathbb{N}}$. Here $\| \cdot \|$ denotes a Banach space norm on F and $\mathbb{K}$ is a subfield of $\mathbb{C}$ invariant under the involutary automorphism $z \mapsto \bar{z}$. Suppose that Γ is a lattice in $\mathbb{R}^n$. A linear transformation δ is called a *dilation for* Γ iff δ satisfies the following properties:

1. Γ is δ-invariant, i.e., $\delta\Gamma \subset \Gamma$.

2. $\forall \lambda \in \sigma(\delta) : |\lambda| > 1$.

It should be noted that these properties imply that $2 \le |\det \delta| \in \mathbb{N}$. An example of a dilation δ for any lattice Γ is given by $\delta := N \, id_{\mathbb{R}^n}$, where $2 \le N \in \mathbb{N}$. The dilation δ induces a *dilation operator* U_δ on F via the correspondence

$$U_\delta f := f \circ \delta^{-1}. \tag{7.1}$$

Definition 7.1 Assume that Γ is a lattice in $\mathbb{R}^n$. The operator $\tau_\gamma : \mathbb{K}^{\mathbb{R}^n} \to \mathbb{K}^{\mathbb{R}^n}$ is defined by

$$\tau_\gamma f := f(\cdot - \gamma), \qquad \gamma \in \Gamma. \tag{7.2}$$

A linear subspace V of the function space F is called shift- or translation-invariant iff

$$\forall f \in V \, \forall \gamma \in \Gamma : \ f(\cdot - \gamma) \in V. \tag{7.3}$$

The function $f(\cdot - \gamma)$ is called the shift or translate of f.
The space V is said to be $(\delta$-)refinable or $(\delta$-)dilation-invariant iff

$$f \in V \implies U_\delta f \in V. \tag{7.4}$$

The function $U_\delta f$ is called the δ-dilate of f.

For given functions $\phi^1, \ldots \phi^A \in F$, $A \in \mathbb{N}$, denote by $V[\phi^1, \ldots \phi^A]$ the smallest closed shift-invariant subspace of F containing $\phi^1, \ldots \phi^A$. A function ϕ is called refinable iff $V[\phi]$ is. A refinable function is called a *scaling function*.

For $k \in \mathbb{Z}$, denote by V_k the U_δ^{-k}-dilate of a shift-invariant space V, i.e., $V_k = \{U_\delta^{-k} f \, | \, f \in V\}$. For a given lattice Γ and function space F, let $(\ell(\Gamma), \| \cdot \|_\ell)$ be a Banach space of sequences $c = (c_\gamma) \subset \mathbb{K}^\Gamma$ such that the *coordinate functionals* $\{e_i^*\}_{i\in\mathbb{N}}$ are elements of $\ell(\Gamma)$. (Recall that in a Banach space F with basis $\{e_i\}$, the coordinate functionals are defined by $e_i^* : F \to \mathbb{K}$, $e_i^*(f) := \langle f, e_i \rangle$, for $F \ni f = \sum_{i\in\mathbb{N}} \langle f, e_i \rangle e_i$.) The space of all finitely supported sequences $c \subset \mathbb{K}^\Gamma$ is denoted by $\ell_0(\Gamma)$. For a function f defined on $\mathbb{R}^n$ and a sequence $c \subset \ell(\Gamma)$, the *semi-convolution product* $\circledast$ is defined by

$$f \circledast c := \sum_{\gamma\in\Gamma} c_\gamma f(\,\cdot\, - \ell),$$

whenever this sum makes sense. Note that if $c \in \ell_0(\Gamma)$, then $f \circledast c$ is well-defined for all functions $f : \mathbb{R}^n \to \mathbb{K}$. The shifts of a function $f \in F$ are called *stable* iff there exist positive constants R_1 and R_2 such that

$$\forall\, c \in \ell(\Gamma) : \quad R_1 \|c\|_\ell \le \|f \circledast c\|_F \le R_2 \|c_2\|_\ell. \tag{7.5}$$

Next, one of the concepts that allows us to obtain wavelet decompositions of functions is introduced.

Definition 7.2 The sequence of function spaces $\{V_k\}_{k\in\mathbb{Z}}$ forms a multiresolution analysis of F iff the following conditions hold:

M1. $\forall\, k \in \mathbb{Z} : V_k \subset V_{k+1}$;

M2. $\overline{\bigcup_{k\in\mathbb{Z}} V_k}^F = F$;

M3. $\bigcap_{k\in\mathbb{Z}} V_k = \{0\}$.

M4. $f \in V_k$ iff $f(\delta\,\cdot\,) \in V_{k+1}$, or equivalently, $V_k = U_\delta^{-k} V_0$.

M5. V_0 is τ-invariant, i.e., $f \in V_0 \Longrightarrow \forall\, \gamma \in \Gamma : \tau_\gamma f \in V_0$.

M6. There exists a finite set of functions $\{\phi^a \,|\, a \in \{1,\dots,A\} \wedge A \in \mathbb{N}\} \subset V_0$ such that for all $a \in \{1,\dots,A\}$ ϕ^a has stable shifts and $V_0 = V[\phi^1,\dots,\phi^A]$.

Remarks.

1. The functions $\phi^1, \dots \phi^A$ in M6 are not unique.

2. If P_k denotes the orthogonal projection of $F = L^2(\mathbb{R}^n)$ onto V_k, then condition M1 can be restated as

$$P_k P_{k'} = P_{k'} P_k = P_k, \qquad k' \le k.$$

Moreover, conditions M2 and M3 now read

$$\lim_{k \to \infty} P_k f = f, \quad \text{and} \quad \lim_{k \to -\infty} P_k f = 0,$$

for all $f \in L^2(\mathbb{R}^n)$.

3. Not all the conditions given in Definition 7.2 are independent. For instance, conditions M1, M2, and M3 are really consequences of M4. For a complete description of the relations among the items in Definition 7.2 the reader is referred to [121] and also [100].

4. In view of M1 and M4 the scaling functions $\phi^1, \dots, \phi^A$ satisfy the following *matrix dilation equation*:

$$\phi(x) = \sum_{\gamma \in \Gamma} \mathbf{c}_\gamma \phi(\delta x - \gamma), \tag{7.6}$$

for some matrix-valued sequence $(\mathbf{c})_{\gamma \in \Gamma}$ and with $\phi := (\phi^1, \dots, \phi^A)^t \in V_0^A$. The function ϕ will be called a *vector scaling function*. Note that if only a finite number of $\mathbf{c}$ is non-zero, then ϕ is compactly supported. The collection of matrices $\{\mathbf{c}_\gamma\}_\Gamma$ is sometimes also called a *(matrix) mask*.

5. One mostly deals with the case $F = L^2(\mathbb{R}^n)$. In this setting, ℓ is chosen to be $\ell^2(\Gamma)$, the Banach space of all square-summable matrix-valued sequences $\mathbf{c} : \Gamma \to \mathbb{R}^{A^2}$ with norm

$$\|\mathbf{c}\|_{\ell^2} := \left(\sum_{\gamma \in \Gamma} \|\mathbf{c}_\gamma\|^2 \right)^{1/2},$$

where $\mathbf{c} = (\mathbf{c}_\gamma)_{\gamma \in \Gamma}$ and $\|\cdot\|$ denotes a matrix norm. M6 is then usually expressed in the following manner:

M6*. There exists a function $\phi \in V_0$ such that $\{\tau_\gamma \phi\}_{\gamma \in \Gamma}$ is a Riesz basis for V_0.

The reader may recall that in a separable Hilbert space H, a *Riesz basis* is obtained from an orthonormal basis by means of a bounded invertible operator. Thus, if $H = L^2(\mathbb{R}^n)$, then M6* is equivalent to the following:

There exist positive constants R_1 and R_2, called the *Riesz bounds*, such that

$$R_1 \|\mathbf{c}\|_{\ell^2} \le \| \sum_{\gamma \in \Gamma} \mathbf{c}_\gamma \, \phi(\cdot - \gamma)\|_{L^2(\mathbb{R}^n)} \le R_2 \|\mathbf{c}\|_{\ell^2}. \tag{7.7}$$

Remark. In the case $F = L^2(\mathbb{R}^n)$ one sometimes requires $\{\tau_\gamma\phi\}_{\gamma\in\Gamma}$ to be an orthonormal basis for its closed linear span rather than a Riesz basis. This assumption is based on the fact that one can construct an orthonormal basis of $L^2(\mathbb{R}^n)$ from a Riesz basis $\{\tau_\gamma\phi\}_{\gamma\in\Gamma}$. This is shown in great detail in [90] for $n = 1$ and $A = 1$. The case $n \in \mathbb{N}$ and $A = 1$ can be found in [39] or [121], the general case in [77]. If $\{\tau_\gamma\phi\}_{\gamma\in\Gamma}$ is an orthonormal basis of $L^2(\mathbb{R}^n)$, then the constants R_1 and R_2 in Eq. (7.7) are equal to 1. By taking the Fourier transform of Eq. (7.7), one obtains the following equivalent condition for $\{\tau_\gamma\phi\}_{\gamma\in\Gamma}$ to be orthonormal:

$$\sum_{\gamma\in\Gamma} \hat{\phi}(\xi + 2\pi\gamma)\hat{\phi}^\dagger(\xi + 2\pi\gamma) = I_A, \qquad (7.8)$$

for all $\gamma \in \Gamma$. Here † denotes the Hermitian conjugate and I_A is the $A \times A$ identity matrix.

An example of a multiresolution analysis is given by the *Haar system*.

Example 7.1 Let $\Gamma := \mathbb{Z}$, let $\delta := 2$, and let $V_k := \{f \in L^2(\mathbb{R}) \,|\, f = \text{constant on } [\ell 2^{-k}, (\ell+1)2^{-k}), \ell \in \mathbb{Z}\}$. Clearly, conditions M1—M6 in Definition 7.2 are satisfied if $\phi := \chi_{[0,1)}$. It is straightforward to check that $\phi(\cdot - \ell)$, $\ell \in \mathbb{Z}$, forms an *orthonormal* basis of V_0. The *scalar* dilation equation that ϕ has to satisfy is easily seen to be

$$\phi(x) = \phi(2x) + \phi(2x - 1).$$

Furthermore, the projection P_k is given by

$$P_k f(x) = 2^{-k} \int_{2^k \ell}^{2^k(\ell+1)} f(t)\, dt,$$

with $x \in [\ell 2^{-k}, (\ell + 1)2^{-k})$. It is worthwhile to remark that because of the piecewise continuity of the functions in V_k, the projection $P_k f$ converges very slowly to the function f. To obtain better convergence one has to impose a higher degree of regularity on the scaling function ϕ.

This leads to the following definition.

Definition 7.3 Let $r \in \mathbb{N}$. A multiresolution analysis is called r-regular iff there exist positive constants $C_{m,\alpha}$ such that

$$\|\mathfrak{D}^\alpha \phi(x)\| \leq C_{m,\alpha}(1 + \|x\|)^{-m}, \qquad (7.9)$$

for all $x \in \mathbb{R}^n$ and $m \in \mathbb{N}_0$, and all multi-indices α with $|\alpha| \leq r$.

Remark. Note that $\mathfrak{D}^\alpha \phi(x)$ is only assumed to be in $L^\infty(\mathbb{R}^n)$ and *not* necessarily in $C^0(\mathbb{R}^n)$.

An example of an r-regular multiresolution analysis is given by the *spline space* $S_{r-1}^r(\mathbb{R})$: $V_0 := S_r^{r-1}(\mathbb{R}) = \{f \in L^2(\mathbb{R}) \mid f \in C^{r-1}(\mathbb{R}) \wedge f\mid_{(\ell,\ell+1)}$ is a polynomial of degree at most $r\}$, $\ell \in \mathbb{Z}$. It is not hard to show that the space V_0 generates a multiresolution analysis of $L^2(\mathbb{R})$ if the scaling function ϕ is taken to be $\phi := \chi_{[0,1]}^{*r+1}$, where $*r+1$ denotes $(r+1)$-fold convolution. Clearly, ϕ is r-regular.

Assume that a multiresolution analysis on F is given. Let ϕ be the associated vector scaling function. For $k \in \mathbb{Z}$, let W_k be such that $V_k \oplus W_k = V_{k-1}$ (here $\oplus$ denotes direct sum). If F is a separable Hilbert space, then W_k is the direct orthogonal difference of the spaces V_{k-1} and V_k. In all but one case, F is assumed to be the separable Hilbert space $L^2(\mathbb{R}^n)$, and this will therefore be the choice for F for the remainder of this section. The spaces W_k, $k \in \mathbb{Z}$, are obviously mutually disjoint and orthogonal: $W_k \cap W_{k'} = \emptyset$ and $W_k \perp W_{k'}$ for $k \neq k'$. Moreover,

$$L^2(\mathbb{R}^n) = \bigoplus_{k \in \mathbb{Z}} W_k.$$

The spaces $\{W_k\}_{k \in \mathbb{Z}}$ are called *wavelet spaces*. Let E be the set of all extreme points of the unit cube $[0,1]^n$ in $\mathbb{R}^n$, i.e.,

$$E = \{(v_1, \ldots, v_n) \in \mathbb{R}^n \mid \forall\, i \in \{1, \ldots, n\}: v_i \in \{0, 1\}\}.$$

The proof of following theorem can be found in [83, 137] (for A =1) and in [100] (for general A).

Theorem 7.1 *Suppose* $\{V_k\}_{k \in \mathbb{Z}}$ *is an r-regular multiresolution analysis on* $L^2(\mathbb{R}^n)$ *with vector scaling function* $\phi = (\phi^1, \ldots, \phi^A)^t$. *Then there exist* $B := (2^n - 1)A$ *functions* ψ^b, $b \in (E \setminus \{0\})^A$, *such that each* ψ^b *is r-regular, and the collection* $\{\psi^1, \ldots, \psi^B\}$ *together with its Γ-translates forms an orthonormal basis of* W_0. ∎

Since the wavelet spaces W_k are nested, it is a direct consequence of the preceding theorem that for a fixed $k \in \mathbb{Z}$, $\{(\det \delta)^{nk/2} \psi^b(\delta^k \cdot -\gamma) \mid \gamma \in \Gamma \wedge b \in \{1, \ldots, B\}\}$ is an orthonormal basis of W_k, i.e.,

$$W_k = \overline{\text{span}\{(\det \delta)^{nk/2} \psi^b(\delta^k \cdot -\gamma) \mid \gamma \in \Gamma \wedge b \in \{1, \ldots, B\}\}}^{\,L^2(\mathbb{R}^n)}.$$

Hence, $\{(\det \delta)^{nk/2}\psi^b(\delta^k x - \gamma) \,|\, k \in \mathbb{Z} \wedge \gamma \in \Gamma \wedge b \in \{1,\ldots,B\}\}$ is an orthonormal basis of $L^2(\mathbb{R}^n)$. The functions $\psi^1,\ldots,\psi^B$ are called *wavelets*.

This allows the unique representation of a function $f \in L^2(\mathbb{R}^n)$ in terms of a *wavelet series* or *wavelet decomposition*:

$$f(x) = \sum_{k \in \mathbb{Z}} \sum_{\gamma \in \Gamma} \mathbf{a}_{k\gamma}^t \psi_{k\gamma}(x), \qquad (7.10)$$

where the *vector wavelet* ψ is defined by $\psi := (\psi^1,\ldots,\psi^B)^t$ and

$$\psi_{k\gamma}(x) := (\det \delta)^{nk/2}\psi(\delta^k x - \gamma),$$

for all $k \in \mathbb{Z}$ and $\gamma \in \Gamma$.

Notation. In what follows, the notation $f_{k\gamma}$ will always mean the δ^k-dilate and γ-translate of a function f.

The coefficients $\mathbf{c}_{k,\gamma}$ are given by

$$\mathbf{a}_{k\gamma} = (\langle f, \psi_{k\gamma}^1 \rangle, \ldots, \langle f, \psi_{k\gamma}^B \rangle)^t.$$

Here, and in what follows, $\langle\,,\,\rangle$ denotes the L^2-inner product on $\mathbb{R}^n$. The mapping

$$L^2(\mathbb{R}^n) \ni f \longmapsto (\langle f, \psi_{k\gamma}^1 \rangle, \ldots, \langle f, \psi_{k\gamma}^B \rangle)^t$$

is called the *discrete wavelet transform*. It clearly agrees with the wavelet transform W_ψ introduced in the previous chapter. Moreover, if $f \in L^2(\mathbb{R}^n)$, then $\{\mathbf{c}_{k\gamma}\} \in \ell^2(\Gamma)^{B^2}$. Furthermore, since $W_0 \subset V_1$, there exists a sequence of $B \times A$ matrices $\{\mathbf{d}\}_{\gamma \in \Gamma}$ such that

$$\psi(x) = \sum_{\gamma \in \Gamma} \mathbf{d}_\gamma \phi(\delta x - \gamma). \qquad (7.11)$$

Hence, ψ also satisfies a matrix dilation equation. This dilation equation is used in the proof of Theorem 7.1 to obtain the wavelets.

A family $\{\psi_{k\gamma} \,|\, k \in \mathbb{Z} \wedge \gamma \in \Gamma\}$ is called *orthonormal* iff

$$\langle \psi_{k\gamma}, \psi_{k'\gamma'} \rangle = \int_{\mathbb{R}^n} \psi_{k\gamma}\psi_{k'\gamma'}^\dagger \, d^n x = \delta_{kk',\gamma\gamma'}\, I_B, \qquad (7.12)$$

for all $k, k' \in \mathbb{Z}$ and $\gamma, \gamma' \in \Gamma$. The "two-component" Kronecker delta $\delta_{kk',\gamma\gamma'}$ is defined by

$$\delta_{kk',\gamma\gamma'} := \delta_{kk'}\delta_{\gamma\gamma'},$$

$k, k' \in \mathbb{Z}$, $\gamma, \gamma' \in \Gamma$.

***Example* 7.2** In this example it is shown how the wavelets can be constructed from a multiresolution analysis when $n = 1 = A$, $\Gamma = \mathbb{Z}$, and $\phi := \text{sinc}(\pi x)$. (Recall that the sinc-function is defined by $\text{sinc}(x) := \sin(x)/x$.) Although this is a rather special setup it does nevertheless contain all the ingredients of the general construction and is thus of an exemplary nature.

Let $V_0 := V[\phi] = \{f \in L^2(\mathbb{R}) \mid \text{supp } \hat{f} \subset [-\pi, \pi]\}$. It is easy to verify that $\{\phi_{0,\ell} \mid \phi = \text{sinc}(\pi \cdot) \wedge \ell \in \mathbb{Z}\}$ is an orthonormal basis for V_0. Since $\phi \in V_0 \subset V_1$, there exists an $\ell^2(\mathbb{Z})$ sequence $\{c_\ell\}$ such that

$$\phi = \sum_{\ell \in \mathbb{Z}} c_\ell \phi_{1,\ell}, \qquad (7.13)$$

and $c_\ell := \langle \phi, \phi_1 \rangle$. The dilates of V_0 are then explicitly given by

$$V_k = \{f \in f \in L^2(\mathbb{R}) \mid \text{supp } \hat{f} \subset [-2^{-k}\pi, 2^k\pi]\}, \qquad k \in \mathbb{Z}.$$

The orthonormality of ϕ implies that $\|\{c_\ell\}\|_{\ell^2(\mathbb{Z})} = 2$. Applying the Fourier transform to Eq. (7.13) leads to

$$\hat{\phi}(\xi) = m_0(\xi/2)\hat{\phi}(\xi/2), \qquad (7.14)$$

where

$$m_0(\xi/2) := \frac{1}{2} \sum_{\ell \in \mathbb{Z}} c_\ell e^{-i\ell\xi}. \qquad (7.15)$$

The *symbol* $m_0(\xi)$ is 2π-periodic and of class C^∞. For $\phi = \text{sinc}(\pi \cdot)$, one obtains

$$c_0 = 1, \qquad c_\ell = \begin{cases} c_{2\ell} = 0 & 0 \neq \ell \in \mathbb{Z} \\ c_{2\ell+1} = \frac{2(-1)\ell}{\pi(2\ell+1)} & \ell \in \mathbb{Z} \end{cases}$$

In the present setup, Eq. (7.8) gives

$$\sum_{\ell \in \mathbb{Z}} |\hat{\phi}(\xi + 2\pi\ell)|^2 = 1 \quad \text{a.e.,}$$

and thus, using Equation (7.15),

$$\sum_{\ell \in \mathbb{Z}} |m_0(\xi/2 + \pi\ell)|^2 |\hat{\phi}(\xi + 2\pi\ell)|^2 = 1 \quad \text{a.e.;}$$

breaking the preceding sum into odd and even ℓ, using the periodicity of $m_0(\xi)$ and Eq. (7.14), finally yields

$$|m_0(\xi/2)|^2 + |m_0(\xi/2 + \pi)|^2 = 1 \quad \text{a.e.} \qquad (7.16)$$

The fact that $\{2^{-1/2}\phi(\cdot/2 - \ell)\}$ forms an orthonormal basis of V_{-1} leads to

$$\hat{V}_{-1} = \{m_f(2\xi)\hat{\phi}(2\xi) \,|\, m \text{ is } 2\pi\text{-periodic and locally in } L^2\}$$

$$= \{m_f(2\xi)m_0(\xi)\hat{\phi}(\xi) \,|\, m \text{ is } 2\pi\text{-periodic and locally in } L^2\}.$$

In this characterization of V_{-1} the fact was used that the Fourier transform of any $f \in V_0$ can be expressed as $\hat{f} = m_f(\xi)\hat{\phi}(\xi)$, for a 2π-periodic symbol $m_f(\xi) = m(\xi; f)$ with the property that $\|f\|_{L^2(\mathbb{R})} = \|m_f\|_{L^2([0,2\pi])}$, i.e., the mapping $f \mapsto m_f$ is unitary. To characterize W_{-1}, all 2π-periodic functions $n(\xi)$ that are orthogonal to all the functions $m_f(2\xi)m_0(\xi)$ have to be found: $f \in W_{-1}$ is equivalent to $f \in V_0$ and $f \perp V_{-1}$. Hence,

$$\int_{\mathbb{R}} e^{i\ell'(2\xi)} \hat{f}\, \overline{\hat{\phi}(2\xi)} \, d\xi = 0,$$

for all $\ell' \in \mathbb{Z}$, or equivalently,

$$\int_{[0,2\pi]} e^{i\ell'\xi} \sum_{\ell \in \mathbb{Z}} \hat{f}(2\xi + 2\pi\ell)\, \overline{\hat{\phi}(2\xi + 2\pi\ell)} \, d\xi = 0;$$

and thus

$$\sum_{\ell \in \mathbb{Z}} \hat{f}(2\xi + 2\pi\ell)\, \overline{\hat{\phi}(2\xi + 2\pi\ell)} = 0, \qquad (7.17)$$

where the sum converges uniformly in $L^1([-\pi/2, \pi/2])$. Upon substitution and regrouping into odd and even ℓ, one obtains the following condition on the symbol $n(\xi)$:

$$\int_{[0,\pi]} \left(m_f(2\xi)m_0(\xi)\overline{n(\xi)} + m_f(2\xi)m_0(\xi + \pi)\overline{n(\xi + \pi)} \right) d\xi = 0;$$

hence,

$$m_0(\xi)\overline{n(\xi)} + m_0(\xi + \pi)\overline{n(\xi + \pi)} = 0, \quad \text{a.e. on } [0,\pi]. \qquad (7.18)$$

Since the vector with components $m_0(\xi/2)$ and $m_0(\xi/2 + \pi)$ has length 1, the preceding orthogonality requirement is equivalent to

$$\begin{pmatrix} n(\xi) \\ n(\xi + \pi) \end{pmatrix} = \nu(\xi)e^{-i\xi} \begin{pmatrix} \overline{m_0(\xi + \pi)} \\ -\overline{m_0(\xi)} \end{pmatrix} \quad \text{a.e. on } [0,\pi],$$

for some function $\nu(\xi)$. (For convenience purposes that will become clear later, the factor $e^{-i\xi}$ was added.) It follows directly from Eq. (7.16) that

$\|n\|_{L^2([0,2\pi])} = \|\nu\|_{L^2([0,\pi])}$. It is not hard to see that the preceding matrix equations reduces to

$$n(\xi) = \nu(\xi)e^{-i\xi}\,\overline{m_0(\xi + \pi)} \quad \text{a.e.,}$$

for some π-periodic function ν. Therefore, the space W_{-1} consists of functions of the form $n(\xi)\hat{\phi}(\xi)$, where $n(\xi)$ satisfies the preceding equation. Using the equality of the foregoing norms, an orthonormal basis of $\hat{W}_{-1}$ can be obtained from an orthonormal basis of $L^2([0,\pi])$. To this end, choose $\{2^{-1/2}e^{2i\ell\xi}\}_{\ell\in\mathbb{Z}}$ as an orthonormal basis of $L^2([0,\pi])$ and set

$$\nu(\xi) := 2^{-1/2}e^{2i\ell\xi}.$$

This leads to an orthonormal basis of $\hat{W}_{-1}$ of the form

$$2^{-1/2}e^{-i\xi}\,\overline{m_0(\xi + \pi)}\,e^{2i\ell\xi},$$

for $\ell \in \mathbb{Z}$. An orthonormal basis of W_{-1} is then given by $\{\psi_{-1,2\ell}\,|\,\ell \in \mathbb{Z}\}$, where $\hat{\psi}_{-1} = 2^{-1/2}e^{-i\xi}\,\overline{m_0(\xi + \pi)}$. Dilation yields an orthonormal basis $\{\psi_{0,\ell}\,|\,\ell \in \mathbb{Z}\}$ of W_0 with

$$\hat{\psi}(\xi) = e^{-i\xi/2}m_0(\xi/2 + \pi)\hat{\phi}(\xi/2). \tag{7.19}$$

This form of this equation clearly shows that ψ satisfies a scalar dilation equation of the form $\psi(x) = \sum_{\ell\in\mathbb{Z}} d_\ell\,\phi(2x - \ell)$, for some $\ell^2(\mathbb{Z})$ sequence $\{d_\ell\}$. It should be noted that the choice of $\hat{\psi}$ is not unique; any function of the form $\rho(\xi)\hat{\psi}(\xi)$ with ρ 2π-periodic and $|\rho(\xi)| = 1$, a.e., will also do.

Applying this last formula to the scaling function $\phi = \text{sinc}(\pi \cdot)$ yields

$$\hat{\psi}(\xi) = \begin{cases} e^{-i(\xi/2+\pi)}, & \text{for } \pi \le |\xi| \le 2\pi \\ \\ 0, & \text{otherwise.} \end{cases}$$

The inverse Fourier transform of $\hat{\psi}$ is easily computed and gives

$$\psi(x) = \frac{2\sqrt{2}(\cos(\pi x) - \sin(2\pi x))}{\pi(1 - x)}.$$

The reader may want to construct the wavelets for the multiresolution analysis considered in Example 7.1 and for $\{V_k\}_{k\in\mathbb{Z}}$ where $V_0 := S_0^1$. In the former case, the wavelet is given by

$$
\psi(x) = \begin{cases} 1, & 0x \in [0, 1/2), \\ -1, & x \in [1/2, 1), \\ 0 & \text{otherwise.} \end{cases}
$$

The last case is presented in [41]. It should be noted that $\phi := \chi_{[0,1)} * \chi_{[0,1]})$ is *not* a scaling function; ϕ is not orthogonal to its integer translates. However, there is a way to obtain an orthonormal basis from a Riesz basis. This method will be presented in the case $n = 1 = A$ and $\Gamma = \mathbb{Z}$.

So suppose that $\{\phi_{0,\ell}\}_{\ell \in \mathbb{Z}}$ is a Riesz basis of V_0. Taking the Fourier transform of Eq. (7.7) for $n = 1 = A$ and $\Gamma = \mathbb{Z}$ yields

$$
0 < R_1 \leq \sum_{\ell \in \mathbb{Z}} |\hat{\phi}(\xi + 2\pi\ell)|^2 \leq R_2 < \infty. \tag{7.20}
$$

This is easily proved using *Parsival's Identity*. Recall that Parsival's Identity states that for any two $L^2(\mathbb{R})$ functions f and g,

$$
\langle f, g \rangle = \frac{1}{2\pi} \langle \hat{f}, \hat{g} \rangle.
$$

Note that $\{\phi_{0,\ell}\}_{\ell \in \mathbb{Z}}$ is an orthonormal basis of V_0 iff $R_1 = R_2 = 1$. Therefore, defining

$$
\hat{\tilde{\phi}}(\xi) := \frac{\hat{\phi}(\xi)}{\left(\sum_{\ell \in \mathbb{Z}} |\hat{\phi}(\xi + 2\pi\ell)|^2\right)^{1/2}} \tag{7.21}
$$

leads to an orthonormal basis via the following arguments:

$$
\begin{aligned}
\tilde{V}_0 &= \left\{ f \in L^2 \,|\, f = \sum_{\ell \in \mathbb{Z}} \tilde{c}_\ell \tilde{\phi}_{0,\ell} \wedge \{\tilde{c}_\ell\}_{\ell \in \mathbb{Z}} \in \ell^2(\mathbb{Z}) \right\} \\
&= \left\{ f \in L^2 \,|\, \hat{f} = \tilde{m}\, \hat{\tilde{\phi}} \text{ for some } 2\pi\text{-periodic } \tilde{m} \in L^2([0, 2\pi]) \right\} \\
&= \left\{ f \in L^2 \,|\, \hat{f} = m\, \hat{\phi} \text{ for some } 2\pi\text{-periodic } m \in L^2([0, 2\pi]) \right\} \\
&= \left\{ f \in L^2 \,|\, f = \sum_{\ell \in \mathbb{Z}} c_\ell \phi_{0,\ell} \wedge \{c_\ell\}_{\ell \in \mathbb{Z}} \in \ell^2 \right\} = V_0.
\end{aligned}
$$

In the preceding presentation, the multiresolution analysis and the associated (vector) scaling function provided the starting point for the construction of the (vector) wavelets. As seen earlier, the symbol $m_0(\xi)$ played an important role in the construction of the wavelets. This importance is re-emphasized in the second approach to the construction of wavelets: Instead of beginning with a multiresolution analysis and associated (vector) scaling function, one considers the (matrix) dilation equation

$$\phi(x) = \sum_{\gamma \in \Gamma} c_\gamma \phi(\delta x - \gamma)$$

and finds conditions for the coefficients c_γ that allow the construction of a (vector) scaling function for a multiresolution analysis. This method is briefly presented here for $n = 1$ and $\Gamma = \mathbb{Z}$. Most of the results stated are taken from [40, 41].

Theorem 7.2 *Suppose the dilation equation*

$$\phi(x) = \sum_{\ell \in \mathbb{Z}} c_\ell \phi(2x - \ell)$$

is given. Assume that the sequence $\{c_\ell\}_{\ell \in \mathbb{Z}}$ satisfies

(i) $\exists \, \varepsilon > 0 : \sum_{\ell \in \mathbb{Z}} |c_\ell| \, |\ell|^\varepsilon < \infty;$

(ii) $\forall \, i, j \in \mathbb{Z} : \sum_{\ell \in \mathbb{Z}} c_{\ell-2i} \, c_{\ell-2j} = \delta_{ij};$

(iii) $\sum_{\ell \in \mathbb{Z}} c_\ell = 2.$

Furthermore, assume that the symbol $m_0(\xi) := 1/2 \sum_{\ell \in \mathbb{Z}} c_\ell \, e^{i\ell\xi}$ can be written as

$$m_0(\xi) = \left(\frac{1}{2}(1 + e^{i\xi}) \right)^N \left(\sum_{\ell \in \mathbb{Z}} f_\ell \, e^{i\ell\xi} \right),$$

with

(iv) $\exists \, \varepsilon > 0 : \sum_{\ell \in \mathbb{Z}} |f_\ell| \, |\ell|^\varepsilon < \infty;$

(v) $\sup \{ |\sum_{\ell \in \mathbb{Z}} f_\ell \, e^{i\ell\xi}| < 2^{N-1} \, | \, \xi \in \mathbb{R} \}.$

If

$$d_\ell \quad := \quad (-1)^\ell c_{1-\ell},$$

$$\hat{\phi}(\xi) \quad := \quad \prod_{j=1}^{\infty} m_0(2^{-j}\xi),$$

$$\psi(x) \quad := \quad \sum_{\ell \in \mathbb{Z}} d_\ell \, \phi(2x - \ell),$$

then the family $\{\phi_{k\ell} \,|\, k, \ell \in \mathbb{Z}\}$ defines a multiresolution analysis of L^2 with associated orthonormal wavelet basis $\{\psi_{k\ell} \,|\, k, \ell \in \mathbb{Z}\}$.

Proof. Only an outline of the proof will be presented. Condition (ii) is essentially Eq. (7.16) and thus necessary for orthogonality. Condition (i) guarantees that the infinite product $\prod_{j=1}^{\infty} m_0(2^{-j}\xi)$ converges pointwise for all $\xi \in \mathbb{R}$, and uniformly on compact subsets of $\mathbb{R}$. The dilation equation together with the required orthogonality imply condition (iii). Note that condition (iii) is also equivalent to $m_0(0) = 1$. Finally, conditions (iv) and (v) imply that the iterates T^m. $m \in \mathbb{N}$, of the operator $T : L^2(\mathbb{R}) \to L^2(\mathbb{R})$, defined by

$$Tf(x) := \sum_{\ell \in \mathbb{Z}} c_\ell f(2x - \ell), \tag{7.22}$$

converge pointwise to a continuous function ϕ whose Fourier transform is given by $\prod_{j=1}^{\infty} m_0(2^{-j}\xi)$. (As an initial function one may take $\chi_{[-1/2,1/2)}$.) The choice of coefficients d_ℓ for the wavelet is motivated by Eq. (7.19). For a more detailed proof see, for instance, [40]. ∎

The integer N in the preceding proof serves a twofold purpose: On the one hand, it is used in connection with the construction of *compactly supported* wavelets (all wavelets so far considered with the exception of the Haar system, were infinitely supported); on the other hand, it is a measure for the *regularity* of the scaling function and the wavelet. These remarks are now made more precise.

Since Eq. (7.16) has played a pivotal role throughout the previous presentation, it is of no surprise that the requirement of compact support is expressible in terms of the symbol $m_0(\xi)$: If the scaling function is to be compactly supported, then $m_0(\xi)$ must be a trigonometric polynomial, i.e., the sum over ℓ must be finite. The converse is easily seen to be true also. Hence, all solution of Eq. (7.16) with only a finite number of non-zero terms c_ℓ have to be characterized. The basic idea is to look for solutions of the form $m_0(\xi) = \left(\frac{1}{2}[1 + e^{i\xi}]\right)^N Q(e^{i\xi})$, where Q is a polynomial. Since Eq. (7.16) resembles the well-known identity between the sine and cosine functions, one rewrites it in the form

$$|m_0(\xi)|^2 = |\cos^2 \xi/2|^N |Q(e^{i\xi})|^2.$$

Using this particular form, it is not very difficult to show that there exist solutions $|m_0(\xi)|^2$ of Eq. (7.16). To determine $m_0(\xi)$, the following lemma which is due to Riesz must be applied:

Lemma 7.1 (Riesz) *Let $A(\xi) = \sum_{\ell=0}^{N} a_\ell \cos(\ell\xi)$, $a_\ell \in \mathbb{R}$, be a positive trigonometric polynomial. Then there exists a trigonmetric polynomial $B(\xi)$ of order N with real coefficients, such that $|B(\xi)|^2 = |A(\xi)|$.*

Proof. The simple and elegant proof can be found in [40]. ∎

This then leads to an explicit characterization of all trigonometric polynomials satisfying Eq. (7.16).

Theorem 7.3 *Any trigonometric polynomial solution of*

$$|m_0(\xi)|^2 + |m_0(\xi + \pi)|^2 = 1$$

is of the form

$$m_0(\xi) = \left(\frac{1}{2}[1 + e^{i\xi}]\right)^N Q(e^{i\xi}),$$

where $N \in \mathbb{N}$, and where $Q(e^{i\xi})$ is a polynomial satisfying

$$|Q(e^{i\xi})|^2 = \sum_{\ell=0}^{N-1} \binom{N-1+\ell}{\ell} \sin^{2\ell} \xi/2 + (\sin^{2N} \xi/2) R\left(\frac{1}{2}\cos\xi\right).$$

Here R is an odd polynomial.

Proof. See, for instance, [40]. ∎

Next the regularity of scaling functions and wavelets is investigated. Smoothness in our context is most easily described in terms of the Fourier transform. Recall that, if the Fourier transform of a given function $f \in L^2(\mathbb{R})$ satisfies

$$|\hat{f}(\xi)| \leq c(1 + |\xi|)^{-r-1-\varepsilon}, \qquad \varepsilon > 0,$$

then f is of class C^r. If a scaling function satisfies this inequality then the associated wavelet will also satisfy it. (This follows from $\hat{\psi}(\xi) = e^{-i\xi/2}m_0(\xi/2 + \pi)\hat{\phi}(\xi/2)$.) In order to derive a regularity result we need the next theorem whose straightforward proof can be found in [40].

Theorem 7.4 *Assume that $f, g : \mathbb{R} \to \mathbb{K}$ are two functions, not identically constant, satisfying*

$$\langle f_{k,\ell}, g_{k',\ell'} \rangle = \delta_{kk',\ell\ell'},$$

for all $k, k', \ell, \ell' \in \mathbb{Z}$. Let $r \in \mathbb{N}$ and suppose that $|g(x)| \le c(1 + |x|)^{-\alpha}$, for some $c > 0$ and for $\alpha > r + 1$. Moreover, assume that $f \in C^r(\mathbb{R}, \mathbb{K})$ and that $f^{(m)}$ is bounded for all $m \le r$. Then

$$\int_{\mathbb{R}} x^m g(x) \, dx = 0, \quad for \ m \in \{0, 1, \dots, r\}.$$

∎

Applying this theorem to $f = g = \psi$ yields

Corollary 7.1 *Suppose that $\{\psi_{k\ell} \mid k, \ell \in \mathbb{Z}\}$ is an orthonormal set in $L^2(\mathbb{R})$ such that for all $k, \ell \in \mathbb{Z}$, $|\psi_{k\ell}| \le C(1 + |x|)^{-r-1-\varepsilon}$, $C \in \mathbb{R}$, $\psi \in C_{\mathbb{R}}^m(\mathbb{R})$, and for all $m = 0, 1, \dots, r$, $|\psi^{(m)}| \le C_1$, for some $C_1 \in \mathbb{R}$. Then*

$$\int_{\mathbb{R}} x^m \psi(x) \, dx = 0,$$

for $m = 0, 1, \dots, r$.

Note that by this corollary, $d^m \hat{\psi}/d\xi^m(0) = 0$, for all $m \in \{0, 1, \dots, r\}$. This, together with Eq. (7.19) and the fact that $\hat{\phi}(0) = \int_{\mathbb{R}} \phi \, dx \ne 0$ (for otherwise this would imply that $\phi \equiv 0$), now implies

$$\frac{d^m m_0}{d\xi^m}(0) = 0,$$

for all $m \in \{0, 1, \dots, r\}$. In other words, $m_0(\xi)$ has a zero of order $r + 1$ at $\xi = \pi$:

$$m_0(\xi) = \left(\frac{1 + e^{i\xi}}{2}\right)^{r+1} M(\xi),$$

with $M \in C^r(\mathbb{R}, \mathbb{K})$. Hence, the following theorem holds.

Theorem 7.5 *Suppose $\{\psi_{k\ell} \mid k, \ell \in \mathbb{Z}\}$ is an orthonormal wavelet basis of $L^2(\mathbb{R})$ derived from a scaling function ϕ. If $|\phi|, |\psi| \le C(1 + |x|)^{-r-1-\varepsilon}$, $C \in \mathbb{R}$, and $\psi \in C_{\mathbb{R}}^r(\mathbb{R})$ with $\psi^{(m)}$ bounded for all $m = 0, 1, \dots, r$, then the symbol m_0, as defined in Eq. (7.15), factors as*

$$m_0(\xi) = \left(\frac{1 + e^{i\xi}}{2}\right)^{r+1} M(\xi), \tag{7.23}$$

where $M \in C_{\mathbb{R}}^r(\mathbb{R})$ is 2π-periodic.

Now it is natural to ask what conditions must be imposed on the symbol $m_0(\xi)$ to guarantee that ϕ, and thus ψ, is of class C^α, $0 \leq \alpha < r + 1$. So suppose that ϕ is a finitely supported scaling function satisfying the two-scale dilation equation

$$\phi(x) = \sum_{\ell \in \mathbb{Z}} c_\ell \, \phi(2x - \ell),$$

with $\sum_{\ell \in \mathbb{Z}} c_\ell = 2$. Without loss of generality it may also be assumed that $\hat{\phi}(0) = \int_{\mathbb{R}} \phi(x) \, dx = 1$. Taking the Fourier transform of this equation and using the definition of the symbol m_0 gives

$$\hat{\phi}(\xi) = m_0(\xi/2)\hat{\phi}(\xi/2), \tag{7.24}$$

and thus

$$\hat{\phi}(\xi) = \prod_{j=1}^{\infty} m_0(2^{-j}\xi).$$

Since ϕ is assumed to be finitely supported, m_0 is a trigonometric polynomial that allows the factorization

$$m_0(\xi) = \left(\frac{1 + e^{-i\xi}}{2}\right)^r M(\xi),$$

for some trigonometric polynomial $M(\xi)$. Substituting this last equation into Eq. (7.24) yields

$$\hat{\phi}(\xi) = \left(\frac{1 - e^{-i\xi}}{i\xi}\right)^r \prod_{j=1}^{\infty} M(2^{-j}\xi). \tag{7.25}$$

It is therefore necessary to estimate the growth of the infinite product $\prod_{j=1}^{\infty} M(2^{-j}\xi)$ to determine the regularity of ϕ. The next result states a condition of $M(\xi)$ that ensures that ϕ is of class C^α.

Proposition 7.1 *Suppose that* $\sup\{|M(\xi)| \, | \, \xi \in [0, 2\pi)\} < 2^{r-\alpha-1}$. *Then ϕ is of class C^α.*

Proof. In this proof C will denote a generic real-valued constant whose numerical value may differ from context to context.

Since m_0, and thus $M(\xi)$, is a trigonometric polynomial with $m_0(0) = M(0) = 1$, there exists a constant C such that $|M(\xi)| \leq 1 + C|\xi|$. Hence,

$$\sup_{|\xi| \leq 1} \prod_{j=1}^{\infty} |M(2^{-j}\xi)| \leq \sup_{|\xi| \leq 1} \prod_{j=1}^{\infty} \exp(C2^{-j}|\xi|) \leq \exp(C).$$

Now take ξ with $|\xi| \geq 1$. Then one may find a $j_0 \in \mathbb{N}$ such that $2^{j_0-1} \leq |\xi| < 2^{j_0}$. Thus,

$$\prod_{j=1}^{\infty} M(2^{-j}\xi) = \prod_{j=1}^{j_0} M(2^{-j}\xi) \prod_{j=1}^{\infty} M(2^{-j-j_0}\xi)$$

$$\leq (\sup\{|M(\xi)| \,|\, \xi \in [0, 2\pi)\})^{j_0} \, e^C \leq C \, 2^{j_0(r-\alpha-1-\varepsilon)}$$

$$\leq C \,(1+|\xi|)^{r-\alpha-1-\varepsilon}.$$

Combining this with the equation for $\hat{\phi}$ gives $|\hat{\phi}| \leq C \,(1+|\xi|)^{-\alpha-1-\varepsilon}$. ∎

Remarks.

1. It should be noted that the higher the regularity of a scaling function (and wavelet), the larger the support (see Theorem 7.3 and notice the N is also the number of non-zero coefficients in the dilation equation): If ϕ and ψ are of class C^r, $r \in \mathbb{N}$, then $\operatorname{supp}\phi = [0, 2r-1]$ and $\operatorname{supp}\psi = [1 - r, r]$.

2. The family of functions $\{_r\phi, _r\psi\}$, where r denotes the regularity, is referred to as the *Daubechies scaling functions and wavelets*.

3. It can be shown ([40]) that the Haar wavelet is the only symmetric wavelet in the Daubechies family of wavelets.

4. For $r = 2$, the scaling function $_2\phi$ and the wavelet $_2\psi$ is in $C^\alpha(\mathbb{R})$ with $0 \leq \alpha < 1$. In fact, one can show ([41]) that there exists a set of full measure on which $_2\phi$ is differentiable. At the dyadic rationals in $\operatorname{supp}_2\phi = [0, 3]$, one can prove that $_2\phi$ is left differentiable but has Hölder exponent 0.55 when such a dyadic rational is approached from the right. Hence, the graph of $_2\phi$ exhibits fractal-like features. In fact, I. Daubechies has shown that $_2\phi$ is an *affine fractal function* generated by two affine mappings λ_1 and λ_2 (cf. [42]).

5. As seen earlier, a convenient way of constructing (vector) wavelets is via a multiresolution analysis. However, it should be emphasized that a (vector) wavelet ψ is really nothing but a special basis of $L^2(\mathbb{R}^n)$; namely, a function $\psi \in L^2(\mathbb{R}^n)$ such that the two-parameter family $\{(\det \delta)^{nk} \, \tau \circ U_\delta^{-k}(\psi) \,|\, k \in \mathbb{Z} \wedge \gamma \in \Gamma\}$ forms a Riesz basis for $L^2(\mathbb{R}^n)$.

Since there are no known explicit formulae for the Daubechies scaling functions (and wavelets), with the exception of the Haar wavelet, of course, the question arises of how to obtain a graphical representation of these functions. One approach that yields an efficient and quick algorithm is based on Eq. (7.22) and the fact that the iterates of $T^m \chi_{[-1/2,1/2)}$ converge pointwise to ϕ. This method is now briefly described for ${}_3\phi$.

Recall that $\operatorname{supp}{}_3\phi = [0,5]$ and define

$$V_k([0,5]) := \{{}_3\phi_{k\ell} \in V_k \mid \operatorname{supp}{}_3\phi_{k\ell} \subseteq [0,5]\}.$$

Then the linear space $V_k([0,5])$ has dimension $5(2^k - 1) + 1$; in other words, only if $\ell \in \{0, 1, \ldots, 5(2^k - 1) + 1\}$ is the support of ${}_3\phi_{k\ell}$ entirely contained in $V_k([0,5])$. Also note that ${}_3\phi_{k,0}(0) = 0 = {}_3\phi_{k,5(2^k-1)+1}$, for all $k \in \mathbb{Z}$. Furthermore, recall that

$$\forall x \in \mathbb{R}: \quad \phi(x) = \lim_{n\to\infty} T^n \chi_{[-1/2,1/2)}(x).$$

Note that T maps V_0 into V_1, and its restriction to functions in $V_0([0,5])$ likewise maps into $V_1([0,5])$. Moreover, the nth approximation, $T^n \chi_{[-1/2,1/2)}$, of ${}_3\phi$ is a step function supported on intervals of length 2^{-n}.

Now let $n \in \mathbb{Z}$, set $J_n := [-2^{-(n+1)}, 2^{-(n+1)})$, and let $\mathfrak{X}_n([0,5]) := \{\chi_{J_n}(\cdot - 2^{-n}\ell) \mid \ell \in \{0, 1, \ldots, 5(2^n - 1) + 1\}\}$. Then the mappings $j_n : \mathfrak{X}_n([0,5]) \to \mathbb{Z}/2^n$,

$$\chi_{J_n} \xmapsto{\ j_n\ } 2^{-n}\ell$$

are isomorphisms. Let $e_i^{(0)} \in \{0,1\}^{\mathbb{Z}}$ be that 6-vector satisfying $e_i^{(0)}(j) = \delta_{ij}$ and, in general, let $e_i^{(n)}$ be that $5(2^n - 1) + 1$-vector in $\{0,1\}^{\mathbb{Z}/2^n}$ satisfying $e_i^{(n)}(j/2^n) = \delta_{ij}$. Then the action of T onto, for example, χ_{J_0} corresponds to the action of a 6×11 matrix $A^{(1)}$ onto the 6-vector $e_i^{(0)} \in \{0,1\}^{\mathbb{Z}}$ yielding the 11-vector $e_j^{(1)}$ in $\{0,1\}^{\mathbb{Z}/2}$:

$$A^{(1)}e_i^{(0)} = \sum_{j=0}^{5} e_j^{(1)}.$$

It is not hard to find the matrix representation of T; it is the 6×11 matrix $A^{(1)}$:

$$
\begin{pmatrix}
c_0 & 0 & 0 & 0 & 0 & 0 \\
c_1 & c_0 & 0 & 0 & 0 & 0 \\
c_2 & c_1 & c_0 & 0 & 0 & 0 \\
c_3 & c_2 & c_1 & c_0 & 0 & 0 \\
c_4 & c_3 & c_2 & c_1 & c_0 & 0 \\
c_5 & c_4 & c_3 & c_2 & c_1 & c_0 \\
0 & c_5 & c_4 & c_3 & c_2 & c_1 \\
0 & 0 & c_5 & c_4 & c_3 & c_2 \\
0 & 0 & 0 & c_5 & c_4 & c_3 \\
0 & 0 & 0 & 0 & c_5 & c_4 \\
0 & 0 & 0 & 0 & 0 & c_5
\end{pmatrix} .
$$

The just-introduced isomorphisms j_n allow the following interpretation: Choose $x \in [0,5)$. Then x belongs to exactly one interval of the form $[-2^{-n-1} + 2^{-n}\ell, 2^{-n-1} + 2^{-n}\ell)$, for some $n \in \mathbb{N}$ and some $\ell \in \{0, 1, \ldots, 5(2^n - 1) + 1\}$. The function value of the nth approximation of $_3\phi(x)$ is then equal to the ℓth component of the $5(2^n - 1) + 1$-vector $e_i^{(n)}$. Using this interpretation, one only has to iterate the banded matrix $A^{(1)}$ to quickly compute $_3\phi$ on $[0,5]$: The input vector is $e_i^{(0)}$, and the jth component of the output vector is then given by

$$
(A^{(1)}e_i^{(0)})_j = \sum_{\max\{0, [\![(j-1)/2]\!]\}}^{[\![(j-1)/2]\!]} c_{j-2i}e_{i+1}^{(0)}.
$$

The same procedure does also apply to the computation of the derivate of $_3\phi$. The derivative ϕ' is the fixed point of the linear operator T' defined by

$$
T'f = 2 \sum_{\ell=0}^{5} c_\ell f(2 \cdot -\ell)
$$

and therefore the pointwise limit of the iterates $(T')^n s_0$ of the step function $s_0 := d/dx(\chi_{[0,1)} * \chi_{[0,1)}) = \chi_{[-1,0)} - \chi_{[0,1)}$ under T'. As before, there exist isomorphisms between characteristic functions supported on intervals of length 2^{-n} and now *shifted* sets $-2^{-n-1} + \mathbb{Z}/2^n$, since the function s_0 is not a characteristic but rather a step function. The vectors $e_i^{(n)}$ as introduced earlier now break up into two parts: one which corresponds to $\chi_{[-1,0)}$, $e_i^{(n)+}$, and one, $e_i^{(n)-}$, which corresponds to $\chi_{[0,1)}$. The shift in the sets $-2^{-k-1} + \mathbb{Z}/2^k$ introduces an additional component into these vectors. However, as before, the approximation to the derivative $_3\phi'$ at a given $x \in [0,5)$ is the corresponding component of $A^{(1)}e_i^{(n)+} - A^{(1)}e_i^{(n)-}$.

It should be clear that quite similar procedures are to be used when the corresponding wavelet and its derivative need to be computed.

Next, the *reconstruction* and *decomposition algorithm* for wavelets is introduced. This will be done in the more general setting of an arbitrary lattice Γ with dilation operator δ and $V_0 = V[\phi^1, \ldots, \phi^A] \subset L^2(\mathbb{R}^n)$.

Let $f_k \in V_k$, $k \in \mathbb{Z}$. Then, since $V_k = V_{k-1} \oplus W_{k-1}$, one may decompose f_k into an averaged or "blurred" component $f_{k-1} \in V_{k-1}$ and a "fine structure" or error component $g_{k-1} \in W_{k-1}$:

$$
\begin{aligned}
f_k &= \sum_{\gamma \in \Gamma} \mathbf{a}_{k,\gamma}^t \, \phi(\delta \cdot - \gamma) = f_{k-1} + g_{k-1} \\
&= \sum_{\gamma \in \Gamma} \mathbf{a}_{k-1,\gamma}^t \, \phi(\delta \cdot - \gamma) + \sum_{\gamma \in \Gamma} \mathbf{b}_{k-1,\gamma}^t \, \psi(\delta \cdot - \gamma),
\end{aligned}
$$

with $\mathbf{a}_{k,\gamma}$, $\mathbf{a}_{k-1,\gamma}$, $\mathbf{b}_{k-1,\gamma} \in \ell^2(\Gamma)^B$. If the vector scaling functions $\phi_{k\gamma}$ and the vector wavelets $\psi_{k\gamma}$ are fully orthogonal, i.e., $\langle \phi_{k\gamma}, \phi_{k'\gamma'} \rangle = \delta_{kk', \gamma\gamma'} I_A$, $\langle \psi_{k\gamma}, \psi_{k'\gamma'} \rangle = \delta_{kk', \gamma\gamma'} I_B$, and $\langle \phi_{k\gamma}, \psi_{k'\gamma'} \rangle = O_{A \times B}$ (here $O_{A \times B}$ denotes the $A \times B$ zero matrix), for all $k, k' \in \mathbb{Z}$ and $\gamma, \gamma' \in \Gamma$, then the coefficient matrices $\mathbf{a}_{k-1,\gamma}$ and $\mathbf{b}_{k-1,\gamma}$ can be obtained easily from the coefficient matrices $\mathbf{a}_{k\gamma}$ via the *Mallat transform*

$$
M : \ell^2(\Gamma)^{B^2} \to \ell^2(\Gamma)^{A^2} \times \ell^2(\Gamma)^{B^2},
$$

$$
M(\{\mathbf{a}_k\}) := (\{\mathbf{a}_{k-1}\}, \{\mathbf{b}_{k-1}\}), \tag{7.26}
$$

with

$$
\mathbf{a}_{k-1,\gamma} = \sum_{\gamma' \in \Gamma} \mathbf{c}_{\gamma'-\delta\gamma} \, \mathbf{a}_{k\gamma'} \quad \text{and} \quad \mathbf{b}_{k-1} = \sum_{\gamma' \in \Gamma} \mathbf{d}_{\gamma'-\delta\gamma} \, \mathbf{a}_{k\gamma'}. \tag{7.27}
$$

Proceeding with the decomposition of f_{k-1} into $f_{k-2} \in V_{k-2}$ and $g_{k-2} \in W_{k-2}$, one obtains after j steps

$$
f_k = f_{k-j} + \sum_{i=1}^{j} g_{k-i} \in V_{k-j} \oplus \bigoplus_{i=1}^{j} W_{k-i}. \tag{7.28}
$$

This equation describes the *decomposition algorithm* for f_k. In terms of the sequences of coefficient matrices, this decomposition can also be expressed as follows:

$$
\begin{array}{ccccccc}
\{\mathbf{a}_k\} & \to & \{\mathbf{a}_{k-1}\} & \to & \{\mathbf{a}_{k-2}\} & \to \cdots \to & \{\mathbf{a}_{k-j}\} \\
& \searrow & & \searrow & & \searrow \quad \searrow & \\
& & \{\mathbf{b}_{k-1}\} & & \{\mathbf{b}_{k-2}\} & \cdots & \{\mathbf{b}_{k-j}\}
\end{array}
$$

To reverse the preceding process, one may reconstruct the original function f_k by taking — at decomposition level j — the "blurred" approximation f_{k-j} and its fine structure correction g_{k-j} to obtain $f_{k-j+1} = f_{k-j} + g_{k-j} \in V_{k-j+1}$, etc., until one arrives at f_k. The coefficient matrices for this *reconstruction algorithm* are given by the *inverse Mallat transform*

$$M^{-1} : \ell^2(\Gamma)^{A^2} \times \ell^2(\Gamma)^{B^2} \to \ell^2(\Gamma)^{A^2},$$

$$M^{-1}(\{\mathbf{a}_{k-1}\}, \{\mathbf{b}_{k-1}\}) = \{\mathbf{a_k}\}, \tag{7.29}$$

with

$$\mathbf{a}_{k,\gamma} = \sum_{\gamma' \in \Gamma} \mathbf{a}^t_{k-1,\gamma'}\, \mathbf{c}_{\gamma-\delta\gamma'} + \mathbf{d}^t_{k-1,\gamma'}\, \mathbf{d}_{\gamma-\delta\gamma'}. \tag{7.30}$$

This reconstruction can also be graphically represented by a diagram that is *dual* to the decomposition diagram:

$$\{\mathbf{a}_k\} \;\leftarrow\; \{\mathbf{a}_{k-1}\} \;\leftarrow\; \{\mathbf{a}_{k-2}\} \;\leftarrow\; \cdots \;\leftarrow\; \{\mathbf{a}_{k-j}\}$$
$$\nwarrow \qquad\qquad \nwarrow \qquad\qquad \nwarrow \qquad\qquad \nwarrow$$
$$\{\mathbf{b}_{k-1}\} \qquad\quad \{\mathbf{b}_{k-2}\} \qquad\quad \cdots \qquad\quad \{\mathbf{b}_{k-j}\}$$

The decomposition and reconstruction algorithm can be used to approximate a given function $f \in L^2(\mathbb{R}^n)$. Without loss of generality one may assume that $f \in V_0$. Then one decomposes f into a finite series $f = f_{k-j} + g_{k-1} + g_{k-2} + \ldots + g_{k-j}$ consisting of a "blurred" or coarse scale approximation f_{k-j} and a finite number of fine-scale corrections $g_{k-1}, \ldots, g_{k-j}$. These summand functions are known from the expansion of f in terms of the vector scaling functions ϕ, and the coefficient matrices are obtained by applying the Mallat transform successively. For the reconstruction, however, one does not use all the coefficient matrices $\mathbf{b}$, but only those that are above a predetermined threshold. In this way one decompresses the graph of f to obtain a satisfactory approximation $\tilde{f}$. This procedure has been successfully applied to image compression and signal analysis, and the interested reader is referred to literature on this subject. Some references may be found in the literature listed in the bibliography.

Finally, a few approximation properties of the Daubechies scaling functions and wavelets are presented. These properties play an important role in the application of wavelets to the theory of numerical solutions of differential equations.

For the remainder of this subsection, I denotes either an open interval (a, b) of $\mathbb{R}$ or the real line $\mathbb{R}$ itself. Let $\mathcal{D}(I)$ denote the linear space of all infinitely differentiable functions with compact support contained in I. Let

$$H_0^q(I) := \overline{\mathcal{D}(I)}^{W_q^2(I)}, \tag{7.31}$$

where $W_q^2(I)$ is the Sobolev space of functions defined on I. The restriction of functions in V_k and W_k to I is denoted by $V_k(I)$ and $W_k(I)$, respectively. To state the next result, the concept of best approximation must be introduced. To this end, let $f : I \to \mathbb{R}$ be any function and let $\mathcal{P}^m$ denote the set of all real polynomials of degree less than or equal to m. Define

$$E_m(f; I) := \min_{p \in \mathcal{P}^m} \max_{x \in I} |f(x) - p(x)|. \tag{7.32}$$

The following result is well-known in approximation theory and usually referred to as an *estimate of Whitney-Jackson type*. These theorems estimate the approximation entirely in terms of the approximant and not the approximating elements.

Proposition 7.2 *Let* $n \in \mathbb{N}$ *and* $f \in \mathcal{D}(\mathbb{R})$. *If* $C := \|f^n\|_\infty$, *then for any compact interval* I,

$$E_{n-1}(f; I) \le \frac{2C|I|^n}{4^n n!}. \tag{7.33}$$

∎

The preceding proposition, Corollary 7.1, and Schwarz's inequality imply the next result. (The details are left to the reader.)

Proposition 7.3 *Let* $f \in \mathcal{D}(\mathbb{R})$, *let* $C := \|f^n\|_\infty$, *and let* $_r\psi$ *be an* r-*regular Daubechies wavelet. Then*

$$\forall\, k, \ell \in \mathbb{Z} : \; |\langle f, \psi_{k\ell} \rangle| \le C_1\, 2^{-k(r+1/2)}, \tag{7.34}$$

where $C_1 := 2C(2r-1)^{r+1/2}/(4^r r!)$. ∎

Now consider the Sobolev space $H_0^q(\mathbb{R})$. Since the Daubechies' scaling functions and wavelets form an r-regular family, one can find an $r \ge q$ such that $_r\phi$ and $_r\psi \in H_0^q$. Let J be the smallest interval such that supp $f + 2r \subseteq J$, and let C_1 be as in Proposition 7.2.

Proposition 7.4 *Let* $f \in \mathcal{D}(\mathbb{R})$ *and let* $r \in \mathbb{N}_0$ *be such that* $_r\phi, _r\psi \in H_0^q$. *Let* $k \in \mathbb{N}$. *Then*

$$\|f - P_k f\|_{W_q^2(\mathbb{R})} \le \sum_{\kappa \ge k} \sum_{\ell \in \mathbb{Z}} |\langle f, \psi_{\kappa\ell} \rangle| \, \|\psi_{\kappa\ell}\|_{W_q^2(\mathbb{R})} \le C_2 2^{-k(r-q)}, \tag{7.35}$$

where $P_k : L^2(\mathbb{R}) \twoheadrightarrow \underline{\Rightarrow}_{\underline{2}}$ denotes the orthogonal projection of f onto V_k, and $C_2 := C_1 \sqrt{|J|} \, (2r - 1)(1 - 2^{-(r-q)})^{-1} \, \|\psi\|_{W_q^2(\mathbb{R})}$.

Proof. HINT: Use Proposition 7.3 and the fact that $\|\psi_{k\ell}\|_{W_q^2(\mathbb{R})} \leq 2^{kq} \|\psi\|_{W_q^2(\mathbb{R})}$.   ∎

Using the classical result that $\{f_{|I} \mid f \in \mathcal{D}(\mathbb{R})\}$ is dense in H_0^q (see, for instance, [2]) and the just-stated proposition, one arrives at the following approximation theorem for Daubechies scaling functions and wavelets. This result lays the foundation for the application of wavelets to Galerkin methods ([81]).

Theorem 7.6 *Let $q \in \mathbb{N}_0$, let $r \in \mathbb{N}_0$ be such that $_r\phi, _r\psi \in H_0^q$, and let $f \in H_0^q$. Then*

$$\forall \varepsilon > 0 \; \exists k \in \mathbb{Z} : \; \|f - \varphi\|_{W_q^2(\mathbb{R})} < \varepsilon, \tag{7.36}$$

where $\varphi \in V_k(I)$.   ∎

7.2   Fractal Function Wavelets

In this section it is shown how fractal functions can be used to generate a multiresolution analysis on $L^2(\mathbb{R})$ and $C_b(\mathbb{R})$, respectively. The type of fractal functions that allows this construction was already introduced in Example 5.2. The space V_0 will consist of *piecewise fractal functions* parametrized by free parameters $s_1, \ldots, s_N$ and can thus be thought of as a *parametrized spline space*. It will be seen that for particular values of the parameters $s_1, \ldots, s_N$, the space V_0 reduces to a known spline space. The spaces V_0 result from a linear isomorphism between the space of real-valued bounded functions on compact subsets of $\mathbb{R}$ and a certain function space Λ whose elements are sequences $\boldsymbol{\lambda} := \{\lambda_{\ell,i} \mid i \in \{0, 1, \ldots, N\} \wedge \ell \in \mathbb{Z}\}$ of functions bounded on $[0, 1]$. The lift of the dilation operator U_δ to the function space Λ gives then rise to dilation- and shift-invariant linear spaces of sequences $\boldsymbol{\lambda}$. They are then used to obtain the scaling functions and wavelets. However, because of the construction, the spaces V_0 will be generated by more than one (fractal) scaling function. Necessary and sufficient conditions on the sequences $\boldsymbol{\lambda}$ are given to ensure that the resulting vector scaling functions and vector wavelets are of class C^r, $r \in \mathbb{N}_0$. The vector wavelets derived from these vector scaling functions are, however, not fully orthogonal; they are so-called *pre-wavelets*: Only for $k \neq k'$,

$$\langle \boldsymbol{\psi}_{k\gamma}, \boldsymbol{\psi}_{k'\gamma'} \rangle = \delta_{kk',\gamma\gamma'} I_B.$$

The free parameters $s_1, \ldots, s_N$ allow the construction of continuous, compactly supported, and orthogonal wavelets. These fractal vector scaling functions and wavelets and their properties are then compared to Daubechies scaling functions and wavelets.

7.2.1 The general construction

Let $1 \neq N \in \mathbb{N}$, $I := [0,1)$ and let $B_{\mathbb{R}}(I) := B(I, \mathbb{R})$ denote the Banach space of all bounded $\mathbb{R}$-valued functions on I endowed with the sup-norm. Let $B^N := \prod_{j=0}^{N-1} B_{\mathbb{R}}(I)$ denote the N-fold direct product of $B_{\mathbb{R}}(I)$. Elements of B^N will be denoted by $\lambda := (\lambda_0, \ldots, \lambda_{N-1})$, where each $\lambda_j \in B_{\mathbb{R}}(I)$, $j = 0, 1, \ldots, N-1$. For such a λ, let $b : [0,1] \to [0,1]$ be defined by $b(x) = \sum_i u_i^{-1}(x)\chi_{Ii}(x)$ and $v(x, \cdot) : \mathbb{R} \to \mathbb{R}$ by $v(x,y) = \sum_i (\lambda_i(x) + s_i y)\chi_{I_i}(x)$, where $I_i := u_i(I)$, with $u_i : I \to I$ given by $u_i(x) := (1/N)(x+i)$, $i = 0, 1, \ldots, N-1$. The unique fixed point $f_\lambda := f_{\Phi_\lambda}$ of the associated Read-Bajraktarević operator Φ_λ is an element of $B_{\mathbb{R}}(I)$. In the event that each component λ_i of λ is continuous and

$$v_{i+1}(0, f_\lambda(0)) = v_i(1^-, f_\lambda(1^-)), \qquad i = 0, 1, \ldots, N-1,$$

$f_\lambda \in C_{\mathbb{R}}(I)$ (also Chapter 5). The graph G_λ of f_λ satisfies

$$G_\lambda = \bigcup_{j=0}^{N-1} w_i(G_\lambda),$$

where $w_i : I \times \mathbb{R} \to I \times \mathbb{R}$ is given by $w_i(x,y) := (u_i(x), v_i(x,y))$, for $i = 0, 1, \ldots, N-1$. Note that each image $w_i(G_\lambda)$ satisfies a similar equation, namely,

$$w_i(G_\lambda) = \bigcup_{j=0}^{N-1} (w_i \circ w_j \circ w_i^{-1})(w_i(G_\lambda)). \tag{7.37}$$

Now define the *rescaling functions* $\varrho_i : I_i \times \mathbb{R} \to I \times \mathbb{R}$ by

$$\varrho_i := u_i^{-1} \times \mathrm{id}_{\mathbb{R}}, \qquad i = 0, 1, \ldots, N-1.$$

The application of ϱ_i to Eq. (7.37) gives

$$\varrho_i \circ w_i(G_\lambda) = \bigcup_{j=0}^{N-1} w_{i,j}(\varrho_i \circ w_i(G_\lambda)), \tag{7.38}$$

with $w_{i,j} := (\varrho_i \circ w_i) \circ w_j \circ (\varrho_i \circ w_i)^{-1}$. It is rather straightforward to show that the maps $w_{i,j} : I \times \mathbb{R} \to I \times \mathbb{R}$ are equal to $w_{i,j} = (u_j, v_{i,j})$, where

$$v_{i,j} \quad = \quad \lambda_i \circ u_j + s_i \lambda_j - s_j \lambda_i + s_j \mathrm{id}_{\mathbb{R}} \tag{7.39}$$

$$=: \quad \lambda_{i,j} + s_j \mathrm{id}_{\mathbb{R}}. \tag{7.40}$$

Now, if $\lambda(i) := (\lambda_{i,0}, \lambda_{i,1}, \ldots, \lambda_{i,N-1}) \in B^N$, then the preceding calculations and the fixed-point property of f_λ imply

$$\text{graph } f_{\lambda(i)} = \varrho_i \circ w_i(G_\lambda) = \text{graph } f_\lambda \circ u_i. \tag{7.41}$$

In other words, the horizontally scaled image of f_λ, $f_\lambda \circ u_i$, is a fractal function of its own generated by maps $\lambda(i) \in B^N$. The immediate goal is to find a representation of the dilation operator U_δ, where $\delta := N \mathrm{id}_{\mathbb{R}} : \Gamma \to \Gamma$ in terms of the functions λ_i and the associated sequences λ. (Here and in what follows, we assume without loss of generality that $\Gamma = \mathbb{Z}$.) To this end, let $\overline{\Delta} : B^N \to \prod_{j=0}^{N-1} B^N$ be given by

$$\overline{\Delta}(\lambda) := (\lambda(0), \lambda(1), \ldots, \lambda(N-1)). \tag{7.42}$$

Now let $\prod B^N := \prod_{\mathbb{Z}} B^N$, and let $\boldsymbol{\lambda} \in \prod B^N$. Define $\Delta : \prod B^N \to \prod B^N$ by

$$(\Delta \boldsymbol{\lambda})_{Nn+j} := \overline{\Delta}(\lambda_n)_j, \tag{7.43}$$

for all $j \in \{0, 1, \ldots, N-1\}$ and $n \in \mathbb{Z}$. Using Eqs. (7.39) and (7.42), this last equation can also be expressed in the following form:

$$(\Delta \boldsymbol{\lambda})_{Nn+j,i} = \lambda_{n,j} \circ u_i + s_j \lambda_{n,i} - s_i \lambda_{n,j}. \tag{7.44}$$

Shortly it will be proven that the operator Δ is the representation of the dilation operator U_δ in terms of the bi-infinite seqences $\boldsymbol{\lambda}$. However, before this can be done, the following result, which gives the basic correspondance between the elements in B^N and functions in $B_{\mathbb{R}}(I)$, is needed.

Theorem 7.7 *The mapping* $\lambda \overset{\theta}{\longmapsto} f_\lambda$ *is a linear isomorphism from* B^N *to* $B_{\mathbb{R}}(I)$.

Proof. From the definitions of the maps u_i, v_i, and the Read-Bajraktarević operator, it should be clear that $\alpha f_\lambda + f_{\lambda'}$ is a fixed point of the Read-Bajraktarević operator $\Phi_{\alpha f_\lambda + f_{\lambda'}}$, for all $\alpha \in \mathbb{R}$ and $\lambda, \lambda' \in B^N$. The uniqueness of the fixed point of $\Phi_{\alpha f_\lambda + f_{\lambda'}}$ implies that $\alpha f_\lambda + f_{\lambda'} = f_{\alpha \lambda + \lambda'}$. Hence, linearity is proven.

Injectivity follows directly from the observation that $f_\lambda \equiv 0$ iff $\lambda = 0$. To show surjectivity, let $\lambda_i(f) := f \circ u_i - s_i f$, for $i = 0, 1, \ldots N - 1$. Then $\lambda(f) \in B^N$ whenever $f \in B_{\mathbb{R}}(I)$. Moreover, $f_{\lambda(f)} = f$. Hence, θ is an isomorphism. ∎

Next, one has to identify elements in $\prod B^N$ with $\mathbb{R}$-valued functions on $\mathbb{R}$. To this end, let $(B_c(\mathbb{R}), \|\cdot\|_\infty)$ denote the Banach space of $\mathbb{R}$-valued functions that are bounded on any compact subset of $\mathbb{R}$. An element $\boldsymbol{f} := \{f_n\}_{n \in \mathbb{Z}} \in \prod B^N$ is identified with an $f \in B_c(\mathbb{R})$ via the linear isomorphism

$$\boldsymbol{f} \overset{\tau}{\longmapsto} \sum_{n \in \mathbb{Z}} f_n(\cdot - n)\chi_{[n,n+1)}. \tag{7.45}$$

Note that the linear isomorphism θ defined in Theorem 7.7 canonically induces a linear isomorphism $\prod \theta : \prod B^N \to \prod B^N$. To summarize, a bi-infinite sequence of functions $\{\lambda(n)\}_{n \in \mathbb{Z}}$ where each λ_n is itself a finite sequence of functions $\{\lambda_{n,0}, \lambda_{n,1}, \ldots, \lambda_{n,N-1}\}$ gives rise to a piecewise fractal function $f_\lambda : \mathbb{R} \to \mathbb{R}$, with the property that the restriction of f_λ to any interval $[n, n+1)$, $n \in \mathbb{Z}$, is itself a fractal function generated by the Read-Bajraktarević operator $\Phi_{\lambda(n)}$. The basic idea is to use obtain properties for the fractal function f_λ by imposing conditions on the bi-infinite sequence functions $\boldsymbol{\lambda}$. Figure 7.1 shows an example of such a piecewise fractal function. The following theorem now gives the representation of the dilation operator U_δ on $\prod B^N$.

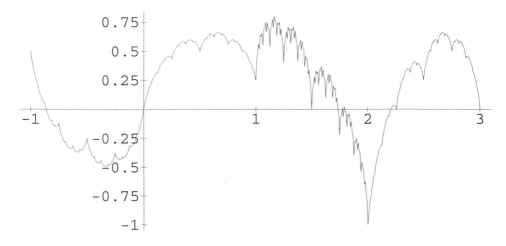

Figure 7.1: A piecewise fractal function.

Theorem 7.8 *Let $\prod B_{\mathbb{R}}(I) := \prod_{\mathbb{Z}} B_{\mathbb{R}}(I)$, and let $\Delta : \prod B^N \to \prod B^N$ be as defined in Eq. (7.42). Then the following diagram commutes:*

$$
\begin{array}{ccc}
\prod B^N & \xmapsto{\Delta} & \prod B^N \\
\downarrow \prod \theta & & \downarrow \prod \theta \\
\prod B_{\mathbb{R}}(I) & & \prod B_{\mathbb{R}}(I). \\
\downarrow \tau & & \downarrow \tau \\
B_c(\mathbb{R}) & \xmapsto{U_\delta} & B_c(\mathbb{R})
\end{array}
\qquad (7.46)
$$

Proof. Let $\boldsymbol{f} = \{f_n\}_{n \in \mathbb{Z}} \in \prod B^N$. Then

$$
\begin{aligned}
(U_\delta \circ \tau)(\boldsymbol{f})(x) &= \sum_{n \in \mathbb{Z}} f_n\left(\frac{x - nN}{N}\right) \chi_{[n,n+1)}\left(\frac{x}{N}\right) \\
&= \sum_{i \in \mathbb{Z}} \sum_{j=0}^{N-1} f_i\left(\frac{x - iN}{N}\right) \chi_{[iN+j,iN+j+1)}(x).
\end{aligned}
$$

Hence,

$$
(\tau^{-1} \circ U_\delta \circ \tau(\boldsymbol{f}))_{iN+j} = f_i\left(\frac{\cdot + j}{N}\right) = f_i \circ u_j, \qquad (7.47)
$$

for all $x \in I$. Equations (7.41) and (7.43) then give

$$
\Delta = \left(\tau \circ \prod \theta\right)^{-1} \circ U_\delta \circ \left(\tau \circ \prod \theta\right).
$$

∎

The shift or translation operator $\tau_\gamma : B_c(\mathbb{R}) \to B_c(\mathbb{R})$ can also easily be lifted to $\prod B^N$. It is not hard to see that τ_γ corresponds to the right shift operator $\prod \tau_\gamma : \prod B^N \to \prod B^N$ given by

$$
\{\lambda_n\}_{n \in \mathbb{Z}} \longmapsto \{\lambda_{n-1}\}_{n \in \mathbb{Z}}, \qquad (7.48)
$$

that is,

$$
\prod \tau_\gamma = \left(\tau \circ \prod \theta\right)^{-1} \circ \tau_\gamma \circ \left(\tau \circ \prod \theta\right). \qquad (7.49)
$$

Now it natural to ask what happens to certain subspaces of $B_c(\mathbb{R})$ under the lift $(\tau \circ \prod \theta)^{-1}$. This question is answered for a particular subspace, namely, $C_{\mathbb{R}}^r(I)$. (This subspace is the most natural to consider, and its characterization is also needed for later developments.) Let $f \in C_{\mathbb{R}}^r(I)$, $r \in \mathbb{N}$, and let $\lambda = (\tau \circ \prod \theta)^{-1} f$. As $f|_I$ is the unique fixed point of the Read-Bajraktarević operator $\Phi_{(\lambda)_0}$, one has

$$f \circ u_i = \lambda_{0,i} + s_i f, \qquad (7.50)$$

for $x \in I$ and $i = 0, 1, \ldots, N - 1$.

Now let $\tilde{C}_{\mathbb{R}}^r(I) := \{g \,|\, g = f|_I, \text{ where } f \in C_{\mathbb{R}}^r(\bar{I})\}$. Since $f \in \tilde{C}_{\mathbb{R}}^r(I)$, it is immediate that $(\lambda)_0 \in \prod_{j=0}^{N-1} \tilde{C}_{\mathbb{R}}^r(I)$. Taking the mth derivative, $m = 0, 1, \ldots, r$, of Eq. (7.50) with respect to x gives

$$f^{(m)} \circ u_i = N^m \lambda_{0,i}^{(m)} + s_i N^m f^{(m)}, \qquad (7.51)$$

$i = 0, 1, \ldots, N - 1$. Using the continuity of $f^{(m)}$ at 0 and $u_{i-1}(1^-) = u_i(0) = i/N$ yields

$$L_0^m \lambda := \frac{1 - s_{N-1} N^m}{1 - s_0 N^m} \lambda_{0,0}^{(m)}(0) - \lambda_{-1,N-1}^{(m)}(1^-) = 0, \qquad (7.52)$$

respectively,

$$L_i^m \lambda := \lambda_{0,i}^{(m)}(0) - \lambda_{0,i-1}^{(m)}(1) + \frac{s_i N^m}{1 - s_0 N^m} \lambda_{0,0}^{(m)}(0)$$

$$- \frac{s_{i-1} N^m}{1 - s_{N1} N^m} \lambda_{0,N1}^{(m)}(1) = 0. \qquad (7.53)$$

Here the linear operators $L_i^m : \prod B^N \to \mathbb{R}$, $i = 0, 1, \ldots, N - 1$ and $m = 0, 1, \ldots, r$, were introduced. Since for all $n \in \mathbb{Z}$, $\tau_\gamma^n f \in C_{\mathbb{R}}^r(\mathbb{R})$, it follows that

$$\left(\prod \tau\right)^n \lambda \in \bigcap_{m=0}^{r} \bigcap_{i=0}^{N-1} \ker L_i^m. \qquad (7.54)$$

The reader is asked to compare these results to the ones obtained in Chapter 5. Now let $\prod C^r := \{\lambda \in \prod_{\mathbb{Z}} \prod_{j=0}^{N-1} \tilde{C}_{\mathbb{R}}^r(I) \,|\, \forall\, n \in \mathbb{Z} : (\prod \tau)^n \lambda \in \bigcap_{m=0}^{r} \bigcap_{i=0}^{N-1} \ker L_i^m\}$. It then follows from the foregoing that

$$(\tau \circ \prod \theta)^{-1} C_{\mathbb{R}}^r(\mathbb{R}) \subseteq \prod C^r.$$

The reverse set containment is implied by the next result.

Theorem 7.9 *Suppose that* $N^r \max_i \{|s_i|\} < 1$. *Then*

$$(\tau \circ \prod \theta)^{-1} C_{\mathbb{R}}^r(\mathbb{R}) = \prod C^r.$$

Proof. The proof follows from Theorem 5.8 and the following arguments. Let $\lambda \in \prod C^1$ and let $f_\lambda := (\tau \circ \prod \theta)^{-1} \lambda$. Since λ_n, $n \in \mathbb{Z}$, satisfies the hypothesis of Theorem 5.8, $f_\lambda|_{[n,n+1)} \in C^1_{\mathbb{R}}([n, n+1))$. Observe that

$$f'_\lambda(n^+) = \frac{N\lambda'_{n,0}(0)}{1 - s_0 N},$$

and

$$f'_\lambda(n^-) = \frac{N\lambda'_{n-1,N-1}(1^-)}{1 - s_{N-1}N}.$$

Hence, by Eq. (7.52), $f_\lambda \in C^1_{\mathbb{R}}(\mathbb{R})$. This, together with the arguments given in the proof of Theorem 5.8, yields the result for $r = 1$. The general case now follows by induction on r. (Note that $N\lambda' \in \prod C^m$, with s_i replaced by Ns_i, whenever $\lambda \in \prod C^{m+1}$.) ∎

Corollary 7.2 $\triangle \prod C^r \subseteq \prod C^r$. ∎

Example 7.3 Let $\mathcal{P}^d(I)$ denote the linear space of all polynomials of degree d or less supported on I, and let $\prod \mathcal{P}^d := \prod_\mathbb{Z} \prod_{j=0}^{N-1} \mathcal{P}^d(I)$. Note that if $\lambda_i, \lambda_j \in \mathcal{P}^d(I)$, then clearly $\lambda_{i,j} = \lambda_i \circ u_j + s_i\lambda_j - s_j\lambda_i \in \mathcal{P}^d(I)$. Hence, Eq. (7.43) implies that $\triangle \prod \mathcal{P}^d \subseteq \prod \mathcal{P}^d$. Setting $\prod \mathcal{P}^d_r := \prod \mathcal{P}^d \cap \prod C^r$, that is, $\lambda \in \prod \mathcal{P}^d_r$ iff $\lambda \in \prod(\mathcal{P}^d)$ and satisfies Eq. (7.53), Theorem 7.9 immediately gives $\triangle \prod \mathcal{P}^d_r \subseteq \prod \mathcal{P}^d_r$.

For most of the remainder of this chapter the case $d = 1$ is of particular interest. In this situation the functions λ_i are affine, and the continuous fractal functions generated by these λ_i belong to the class $\mathcal{A}$ of affine fractal functions supported on I. Since $f_\lambda \in C_{\mathbb{R}}(I)$, the space $\theta\left(\prod_{j=0}^{N-1} \mathcal{P}^1\right) \cap C_{\mathbb{R}}(I) =: \mathfrak{C}$ is easily seen to be $(N + 1)$-dimensional. Hence, the elements g of $\mathfrak{C}$ are completely determined by $g(i/N)$, $i = 0, 1, \ldots, N - 1$. This allows the following geometric interpretation: Given a $y := (y_0, y_1, \ldots, y_{N-1})^t \in \mathbb{R}^{N+1}$, there exists a unique continuous fractal function f_y having the interpolation property with respect to $\{(i/N, y_i) \mid i = 0, 1, \ldots, N - 1\}$ (cf. Example 5.4). The affine functions λ_i are then determined by Eqs. (5.17) and (5.19). The linear form of these equations implies the following — easy-to-prove — result.

Proposition 7.5 *The mapping* $\mathbb{R}^{N+1} \ni y \longmapsto \lambda \in \mathfrak{C}$ *is a linear isomorphism.*

Later it will be seen that this proposition adds a geometric component to the multiresolution analysis generated by fractal functions.

The following example illustrates how $\lambda \in \prod_{j=0}^{N-1} \mathcal{P}^d \cap C_{\mathbb{R}}(I)$ can be obtained.

Example 7.4 Let $N := 2$, $d := 2$, and $r := 1$. Furthermore, let $s_i := s$, $i = 0, 1$ with $|s| < 1/2$. The linear space $\mathcal{P}^2(I)$ is identified with $\mathbb{R}^3$ via the correspondence $a + bx + cx^2 \mapsto (a, b, c)$. Therefore, $\lambda = (\lambda_0, \lambda_1) \in \prod_{j=0}^{1} \mathcal{P}^2(I)$ can be identified with $\mathbb{R}^6$ via

$$\lambda_i(x) = a_i + b_i x + c_i x^2 \mapsto v := (a_0, b_0, c_0, a_1, b_1, c_1),$$

and $\lambda \in \prod \mathcal{P}^2$ with $\prod_{\mathbb{Z}} \mathbb{R}^6$:

$$\lambda \mapsto v_n := \{(a_{n,0}, b_{n,0}, c_{n,0}, a_{n,1}, b_{n,1}, c_{n,1})\}_{n \in \mathbb{Z}}.$$

Note that a function $f \in C_{\mathbb{R}}(I)$ iff $\lambda \in \prod_{j=0}^{1} \mathcal{P}^2$ satisfies Eq. (7.52) for $m = 0$ and $m = 1$. These equations, however, are equivalent to the "more geometric" conditions

$$\langle \mathfrak{n}_m, v \rangle_E = 0, \qquad m = 0, 1, \tag{7.55}$$

where

$$\mathfrak{n}_0 := (-1, -1, -1, 1, 0, 0,)^t + s(2, 1, 1, -2, -1, -1,)^t,$$

$$\mathfrak{n}_1 := (0, -1, -2, 0, 1, 0)^t + s(0, 4, 4, 0, -4, -4,)^t,$$

and $\langle \cdot, \cdot \rangle_E$ denotes Euclidean inner product in $\mathbb{R}^6$. Hence, the space $S_1 := \prod_{j=0}^{1} \mathcal{P}^d \cap C_{\mathbb{R}}(I)$ is a four-dimensional vectorspace, and each element $f \in S_1$ is uniquely determined by $f(0)$, $f'(0)$, $f(1)$, and $f'(1)$. Suppose the base elements $\{f_1, f_2, f_3, f_4\}$ of S_1 are chosen according to

$$f_1(0) = f_2'(0) = f_3'(0) = f_4'(0) = 1,$$

and

$$f_1'(0) = f_1(1) = f_1'(1) = f_2(0) = \ldots = f_4(1) = 0.$$

Then the corresponding functions $\lambda(k)$, $k = 1, 2, 3, 4$, are given by

$$\begin{aligned}
\lambda(1) &= \theta^{-1} f_1 = (1 - s, 0, s - 1/2, 1/2 - s, 2s - 1, 1/2 - s), \\
\lambda(2) &= \theta^{-1} f_2 = (0, 1/2 - s, s - 3/8, 1/8, -1/4, 1/8), \\
\lambda(3) &= \theta^{-1} f_3 = (0, 0, 1/2 - s, 1/2, 1 - 2s, s - 1/2), \\
\lambda(4) &= \theta^{-1} f_4 = (0, 0, s - 3/8, s - 1/4, -s, 1/4),
\end{aligned}$$

Now the time has come to use the preceding results to define a nested sequence of function spaces necessary for a multiresolution analysis. For this purpose, let

$$V_0 := (\tau \circ \prod \theta) \prod \mathcal{P}_r^d, \tag{7.56}$$

or

$$
\begin{aligned}
V_0 :=\ & \{f : \mathbb{R} \to \mathbb{R} \mid \forall j \in \mathbb{Z}\ \exists g \in \mathcal{F}^r([j, j+1], \mathbb{R}) \mid \\
& f|_{(j,j+1)} = g|_{(j,j+1)}\},
\end{aligned}
\tag{7.57}
$$

and define V_k, $k \in \mathbb{Z}$, by requiring

$$f \in V_k \iff U_\delta^{-k} f \in V_0, \qquad k \in \mathbb{Z}. \tag{7.58}$$

Proposition 7.6 *The $\{V_k\}_{k \in \mathbb{Z}}$ defined earlier form a nested sequence of linear spaces.*

Proof. It is clear that the V_k, $k \in \mathbb{Z}$, are linear spaces. The nestedness follows from the observation that

$$U_\delta V_k \subseteq V_k \iff \Delta \prod \mathcal{P}_r^d \subseteq \prod \mathcal{P}_r^d,$$

and the validity of the second statement was shown in Example 7.3. ∎

Example 7.5 This is a continuation of Example 7.4 with the notation and terminology used there. Set, as above, $V_0 := (\tau \circ \prod \theta) \prod (\mathcal{P}^2)^1$. Then $f \in V_0$ is uniquely determined by specifying the values of f and f' at the integers. Note the the restriction of elements in V_0 to any interval of the form $[n, n+1)$, $n \in \mathbb{Z}$, defines a two-dimensional linear space. Thus, if $\phi^1, \phi^2 \in V_0$ are such that $\text{supp}\,\phi^i \subseteq [-1, 1]$, $i = 1, 2$, $\phi^1(0) = \phi^{2\prime}(1) = 1$ and $\phi^{1\prime}(0) = \phi^2(0) = 0$, then any $f \in V_0$ can be expressed in the form

$$f = \sum_{\ell \in \mathbb{Z}} f(\ell)\phi^1(\cdot - \ell) + f'(\ell)\phi^2(\cdot - \ell).$$

The functions ϕ^1 and ϕ^2 can be constructed in the following way: Let $\boldsymbol{\lambda}^1, \boldsymbol{\lambda}^2 \in \prod C^1$ be defined by

$$
\lambda_i^1 = \begin{cases} \lambda(3), & i = -1, \\ \lambda(1), & i = 0, \\ 0, & \text{otherwise,} \end{cases}
$$

and

$$\lambda_i^1 = \begin{cases} \lambda(4), & i = -1, \\ \lambda(2), & i = 0, \\ 0, & \text{otherwise.} \end{cases}$$

Fig. 7.2 shows the graphs of ϕ^1 and ϕ^2 for $s = 3/10$. This example will be continued later; in particular, it will be shown that ϕ^1 and ϕ^2 generate a multiresolution analysis on $L^2(\mathbb{R})$.

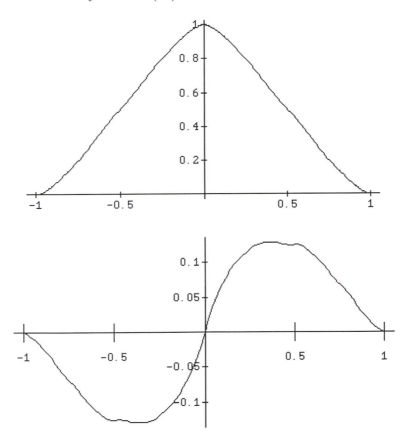

Figure 7.2: The graphs of ϕ^1 and ϕ^2.

In order to establish that the nested linear function spaces V_k as defined in Eq. (7.56), and respectively (7.57), form a multiresolution analysis of $L^2(\mathbb{R})$, a few general results concerning the conditions for a multiresolution analysis have to be proven.

To this end, suppose that $\boldsymbol{\phi} := (\phi^1, \ldots, \phi^A)^t$ is a finite vector function whose components are bounded and compactly supported functions in $L^2(\mathbb{R})$. Let

$$V_0 := \overline{\text{span}\left\{\phi^a(\cdot - \ell) \mid a \in \{1, \ldots, A\} \wedge \ell \in \mathbb{Z}\right\}}^{L^2(\mathbb{R})},$$

and

$$V_k := U_\delta^{-k} V_0, \qquad k \in \mathbb{Z}.$$

The next proposition gives sufficient conditions on $\boldsymbol{\phi}$ guaranteeing that the *density condition* M2 and the *separation condition* M3 hold.

Proposition 7.7 *Let* $\{\phi^a\}_{a \in A}$ *be a finite set of bounded and compactly supported functions in* $L^2(\mathbb{R})$.

1. *(Separation)* $\bigcap_{k \in \mathbb{Z}} V_k = \{0\}$.

2. *(Density) Assume that there exists an* $\mathbf{a} = \{\mathbf{a}_a\}_{a \in A}$ *such that*

$$\sum_{a=1}^{A} \sum_{\ell \in \mathbb{Z}} \mathbf{a}_a \phi^a(x - \ell) = 1, \quad a.e. \ x \in \mathbb{R}.$$

Then $\bigcup_{k \in \mathbb{Z}} V_k$ *is dense in* $L^2(\mathbb{R})$.

Proof.

1. Let $I_n := [n, n+1]$, $n \in \mathbb{Z}$, and let $U := \{f \circ \chi_{I_0} \mid f \in V_0\}$. The boundedness and compact support of the functions ϕ^a implies that U is a finite-dimensional linear space over $\mathbb{R}$. Hence, the norms $\|\cdot\|_2$ and $\|\cdot\|_\infty$ are equivalent on U. Therefore, there exists a positive constant c such that $\|f\|_\infty \le c\|f\|_2$, for all $f \in U$. Using the translation invariance of the space V_0, one obtains $\|f \circ \chi_{I_n}\|_\infty \le c\|f \circ \chi_{I_n}\|_2$, for any $f \in V_0$. Hence,

$$\|f\|_\infty = \sup_{n \in \mathbb{Z}}\{\|f \circ \chi_{I_n}\|_\infty\} \le c \sum_{n \in \mathbb{Z}} \|f \circ \chi_{I_n}\|_2 = c\|f\|_2,$$

    for all $f \in V_k$. Following [167], note that

$$\|f\|_\infty \le c N^{-k/2} \|f\|_2,$$

    for all $f \in V_k$. Hence, if $f \in \bigcap_{k \in \mathbb{Z}} V_k$, then $\|f\|_\infty = 0$.

2. Because of the translation and dilation invariance of $\bigcup_{k\in\mathbb{Z}} V_k$, it suffices to show that $\chi_{[0,1]} \in \overline{\bigcup_{k\in\mathbb{Z}} V_k}^{L^2(\mathbb{R})}$. Without loss of generality, suppose then that supp $\phi^a \in [0, M]$, $a \in A$. Choose a $j \in \mathbb{N}$ such that $N^j > M$, and let $S_{N^j} := \sum_{\ell=0}^{N^j} \sum_{a=1}^{A} \mathbf{a}_a \phi^a(\cdot - \ell)$. Then supp $S_{N^j} \subseteq [0, N^j + M]$, $S_{N^j} \equiv 1$ on $[M, N^j]$, and

$$\|S_{N^j}\|_\infty \leq MA \max_{a\in A}\{\|\mathbf{a}_a \phi^a\|_\infty\}.$$

This now implies that $\|S_{N^j}(N^j \cdot) - \chi_{[0,1]}\|_\infty \to 0$. As $S_{N^j}(N^j \cdot) \in \bigcup_{k\in\mathbb{Z}} V_k$, the result follows. ∎

Next, necessary and sufficient conditions are given for the set of translates of $\{\phi^a\}$ to be a Riesz basis of V_0. To this end, let

$$E_\phi(\xi) := \sum_{\ell\in\mathbb{Z}} \hat{\phi}(\xi + 2\pi\ell)\hat{\phi}^\dagger(\xi + 2\pi\ell). \tag{7.59}$$

Note that E_ϕ is an $A \times A$ matrix. Using the *Poisson Summation Formula*, it is easy to verify that

$$
\begin{aligned}
E_\phi(\xi) &= \sum_{\ell\in\mathbb{Z}} \left(\int_{\mathbb{R}} \phi(y - \ell)\phi^\dagger(y)dy\right) e^{i\xi\ell} \\
&= \sum_{\ell=-M+1}^{M-1} \left(\int_{\mathbb{R}} \phi(y - \ell)\phi^\dagger(y)dy\right) e^{i\xi\ell},
\end{aligned}
\tag{7.60}
$$

for some $M > 0$. Observe that in the last equality the fact was used that ϕ has compact support. (Recall that if a function $f \in L^1(\mathbb{R}^n)$ is such that

1. $\sum_{\ell\in\mathbb{Z}} f(x + 2\pi\ell)$ converges everywhere (in L^1) to a continuous function, and

2. $\sum_{\ell\in\mathbb{Z}} \hat{f}(\ell)e^{ix\ell}$ converges everywhere,

then

$$\sum_{\ell\in\mathbb{Z}} f(x + 2\pi\ell) = \sum_{\ell\in\mathbb{Z}} \hat{f}(\ell)e^{ix\ell}. \tag{7.61}$$

Equation (7.60) is called Poisson's Summation Formula.)

Theorem 7.10 *The collection* $\mathcal{B}_\phi := \{\phi^a(\cdot - \ell) \,|\, a \in \{1,\ldots,A\} \wedge \ell \in \mathbb{Z}\}$ *forms a Riesz basis for* V_0 *iff* E_ϕ *is non-singular for* $\xi \in [0, 2\pi]$.

Proof. The proof consists of finding the Riesz bounds R_1 and R_2 (Eq. (7.7)). Note that by Parsival's inequality and the shift property of the Fourier transform,

$$\int_{\mathbb{R}} \left| \sum_{\ell \in \mathbb{Z}} \mathbf{c}_\ell^\dagger \phi(x - \ell) \right|^2 dx \; = \; \frac{1}{2\pi} \int_{\mathbb{R}} \left| \sum_{\ell \in \mathbb{Z}} e^{i\xi \ell} \mathbf{c}_\ell^\dagger \hat{\phi}(\xi) \right|^2 d\xi$$

$$= \; \frac{1}{2\pi} \int_{\mathbb{R}} \left| \langle \, \hat{\mathbf{c}}^\dagger(\xi), \hat{\phi}(\xi) \, \rangle \right|^2 d\xi,$$

where $\hat{\mathbf{c}}(\xi) := \sum_{\ell \in \mathbb{Z}} e^{i\xi \ell} \mathbf{c}_\ell^\dagger$. Since the Fourier transform of the symbol $\mathbf{c}(\xi)$ is 2π-periodic, the last of the preceding equations can be rewritten as

$$\frac{1}{2\pi} \sum_{\ell \in \mathbb{Z}} \int_{[0,2\pi]} \left| \hat{\mathbf{c}}^\dagger(\xi) \hat{\phi}(\xi + 2\pi\ell) \right|^2 d\xi$$

$$= \frac{1}{2\pi} \sum_{\ell \in \mathbb{Z}} \int_{[0,2\pi]} \left(\hat{\mathbf{c}}^\dagger(\xi) \hat{\phi}(\xi + 2\pi\ell) \hat{\phi}^\dagger(\xi + 2\pi\ell) \hat{\mathbf{c}}(\xi) \right) d\xi$$

$$= \frac{1}{2\pi} \int_{[0,2\pi]} \hat{\mathbf{c}}^\dagger \mathbf{E}_\phi(\xi) \hat{\mathbf{c}}(\xi) d\xi.$$

Using Parsival's identity again gives

$$\| \{\mathbf{c}_\ell\}_{\ell \in \mathbb{Z}} \|^2_{L^2(\mathbb{R})} \; = \; \frac{1}{2\pi} \int_{[0,2\pi]} \hat{\mathbf{c}}^\dagger(\xi) \hat{\mathbf{c}}(\xi) d\xi =: \frac{1}{2\pi} \| \hat{\mathbf{c}} \|^2_{L^2(\mathbb{R})}. \tag{7.62}$$

Now let $L^2([0, 2\pi]; \mathbb{C}^A) := \{\mathbf{f} : [0, 2\pi] \to \mathbb{C}^A \, | \, \| \mathbf{f} \|_{L^2(\mathbb{R})} < \infty\}$. Then Eq. (7.7) is equivalent to

$$R_1 \| \hat{\mathbf{c}} \|^2_{L^2(\mathbb{R})} \leq \int_{[0,2\pi]} \hat{\mathbf{c}}^\dagger \mathbf{E}_\phi(\xi) \hat{\mathbf{c}}(\xi) d\xi \leq R_2 \| \hat{\mathbf{c}} \|^2_{L^2(\mathbb{R})}, \tag{7.63}$$

for all $\hat{\mathbf{c}} \in L^2([0, 2\pi]; \mathbb{C}^A)$.

From Eq. (7.59) it is clear that $\mathbf{E}_\phi(\xi)$ is self-adjoint and positive, and thus has real non-negative eigenvalues λ_a, $a = 1, \ldots, A$. Moreover, $\mathbf{E}_\phi$ is easily seen to be continuous in ξ, implying that the eigenvalues λ_a are also continuous functions in ξ. Let $m(\xi) := \min_{a \in \{1,\ldots,A\}} \{\lambda_a\}$ and let $M(\xi) := \max_{a \in \{1,\ldots,A\}} \{\lambda_a\}$. Then

$$m(\xi) \hat{\mathbf{c}}^\dagger(\xi) \hat{\mathbf{c}}(\xi) \leq \hat{\mathbf{c}}^\dagger(\xi) \mathbf{E}_\phi(\xi) \hat{\mathbf{c}}(\xi) \leq M(\xi) \hat{\mathbf{c}}^\dagger(\xi) \hat{\mathbf{c}}(\xi),$$

and so Eq. (7.63) holds with

$$R_1 := \min\{m(\xi) \, | \, \xi \in [0, 2\pi]\} \quad \text{and} \quad R_2 := \max\{m(\xi) \, | \, \xi \in [0, 2\pi]\}.$$

Observe that since Eq. (7.63) holds for all $\hat{c} \in L^2([0, 2\pi]; \mathbb{C}^A)$, these are the best possible bounds. Now, as $M(\xi)$ is always finite, $\mathcal{B}_\phi$ is a Riesz basis iff $\min\{m(\xi) \mid \xi \in [0, 2\pi]\} > 0$, which is true iff the symbol $\boldsymbol{E}_\phi(\xi)$ is non-singular for $\xi \in [0, 2\pi]$. ∎

Using a "pseudo-counting" argument, it is not hard to see that there are $B = A(N - 1)$ wavelets that span $W_0 = V_1 \ominus V_0$.

Corollary 7.3 *The collection* $\mathcal{B}_\psi := \{\psi^b(\cdot - \ell) \mid b \in \{1, \ldots, B\} \wedge \ell \in \mathbb{Z}\}$ *forms a Riesz basis for W_0 iff*

1. $W_0 \subseteq \overline{\text{span} \, \mathcal{B}_\psi}^{L^2(\mathbb{R})}$;

2. $\boldsymbol{E}_\psi := \sum_{\ell \in \mathbb{Z}} \hat{\psi}(\xi + 2\pi\ell) \hat{\psi}^\dagger(\xi + 2\pi\ell)$ *is non-singular for $\xi \in [0, 2\pi]$.* ∎

Since ψ is a linear combination of ϕ, there must exist a relation between the symbols $\boldsymbol{E}_\phi$ and $\boldsymbol{E}_\psi$. The next theorem derives this relation.

Theorem 7.11 *Let $\zeta := e^{-i\xi/N}$ and let $D(\zeta) := (1/N)\sum_{\ell \in \mathbb{Z}} \boldsymbol{d}_\ell \zeta^\ell$. Then,* $\boldsymbol{E}_\psi(\xi) = \sum_{j=0}^{N-1} D(e^{(-2\pi ij\zeta)/N}) \boldsymbol{E}_\phi(\frac{\xi + 2\pi j}{N}) D(e^{(-2\pi ij\zeta)/N})^\dagger$ *and, if $\boldsymbol{E}_\phi(\xi)$ is non-singular, then the symbol $\boldsymbol{E}_\psi(\xi)$ is also non-singular for all ζ, provided that $\bigcap_{j=0}^{N-1} \text{null} \, D(e^{(-2\pi ij\zeta)/N})^\dagger = \{0\}$, where* null *denotes the null space of $D(\zeta)$.*

Proof.

1. Equation (7.11) and the definition of $\boldsymbol{E}_\psi(\xi)$ give

$$
\begin{aligned}
\boldsymbol{E}_\psi(\xi) &= \sum_{\ell \in \mathbb{Z}} \left(\frac{1}{N} \sum_{\ell' \in \mathbb{Z}} \boldsymbol{d}_{\ell'} e^{-i(\xi + 2\pi\ell)\ell'/N} \hat{\phi}\left(\frac{\xi + 2\pi\ell}{N}\right) \right) \\
&\quad \times \left(\frac{1}{N} \sum_{\ell' \in \mathbb{Z}} \boldsymbol{d}_{\ell'} e^{-i(\xi + 2\pi\ell)\ell'/N} \hat{\phi}\left(\frac{\xi + 2\pi\ell}{N}\right)^\dagger \right) \\
&= \sum_{j=0}^{N-1} D(e^{(-2\pi ij\zeta)/N} \sum_{\ell'' \in \mathbb{Z}} \hat{\phi}\left(\frac{\xi + 2\pi j}{N} + 2\pi\ell''\right) \\
&\quad \times \hat{\phi}^\dagger\left(\frac{\xi + 2\pi j}{N} + 2\pi\ell''\right) D(e^{(-2\pi ij\zeta)/N})^\dagger,
\end{aligned}
$$

where $\ell = N\ell'' + j$.

2. Let $E(\xi, \zeta; j, N) := D(e^{(-2\pi ij\zeta)/N})\boldsymbol{E}_\phi(\frac{\xi+2\pi j}{N})D(e^{(-2\pi ij\zeta)/N})^\dagger$. Since $D(\zeta)\boldsymbol{E}_\phi(\xi)D(\zeta)^\dagger$ is a positive matrix for all ξ, it follows from 1 that $\boldsymbol{E}_\psi$ is non-singular if $\bigcap_{j=0}^{N-1}$ null $E(\xi, \zeta; j, N)$ is trivial. By the hypotheses, however, null $E(\xi, \zeta; j, N) =$ null $D(e^{(-2\pi ij\zeta)/N})^\dagger$. ∎

Example 7.6 This is the continuation of the two preceding examples. Here it is shown that the scaling functions ϕ^1 and ϕ^2 defined in Example 7.5 generate a multiresolution analysis of $L^2(\mathbb{R})$. As before, one defines

$$V_0 := \overline{\text{span}\{\phi^a(\cdot - \ell) \mid a = 1, 2 \wedge \ell \in \mathbb{Z}\}}^{L^2(\mathbb{R})}.$$

Since $\lambda(1) + \lambda(3) = (1-s, 0, 0, 1-s, 0, 0)$ one has $\theta(\lambda(1) + \lambda(3)) \equiv 1$, implying that ϕ^1 forms a partition of unity:

$$\sum_{\ell \in \mathbb{Z}} \phi^1(x - \ell) \equiv 1.$$

As both scaling functions are bounded, compactly supported, and satisfy condition 2 in Proposition 7.7 with $\mathbf{a} = (1, 0)$, the density and separation properties hold. It remains to be shown that the integer translates of the vector scaling function ϕ form a Riesz basis of V_0. This can be done via Theorem 7.10 by proving that the symbol $\boldsymbol{E}_\phi$ is non-singular for all $\xi \in [0, 2\pi]$. Note that for this example, the number M in Eq. (7.59) is equal to 2, since ϕ^1 and ϕ^2 are both supported on $[-1, 1]$.

Setting $e_\ell := \int_\mathbb{R} \phi(y - \ell)\phi^\dagger(y)dy$, $\ell = -1, 0, 1$, one can show — after considerable algebra and using Proposition 5.9 — that these 2×2 matrices are given by

$$e_0 = \begin{pmatrix} \frac{92-69s-78s^2+56s^3}{60(2-s)(1-s^2)} & 0 \\ 0 & \frac{5-4s}{240(1-s^2)} \end{pmatrix},$$

$$e_{-1}^t = e_1 = \begin{pmatrix} \frac{28+9s-42s^2+4s^3}{120(2-s)(1-s^2)} & \frac{13+9s-27s^2+4s^3}{240(2-s)(1-s^2)} \\ \frac{-13-9s+27s^2-4s^3}{240(2-s)(1-s^2)} & \frac{-3-4s+8s^2}{480(1-s^2)} \end{pmatrix}.$$

Hence,

$$\boldsymbol{E}_\phi(\xi) = e_{-1}e^{-i\xi} + e_0 + e_1 e^{i\xi}, \qquad \xi \in [0, 2\pi],$$

and, therefore,

$$\det \boldsymbol{E}_\phi(\xi) = \frac{r_0 + r_1\zeta + r_2\zeta^2 + r_1\zeta^3 + r_0\zeta^4}{57,000(s-2)^2(s^2-1)\zeta^2},$$

where $\zeta = e^{i\xi}$ and

$$
\begin{aligned}
r_0 &:= -1 - 40s - 147s^2 + 200s^3 - 16s^4, \\
r_1 &:= 544 + 640s - 3912s^2 + 3520s^3 - 896s^4, \\
r_2 &:= -3,006 + 8,400s - 7,722s^2 + 2,160s^3 - 96s^4.
\end{aligned}
$$

Note that the denominator in the preceding expression for $\det \boldsymbol{E}_\phi$ does not vanish for $|s| < 1$, and that the roots of the quartic equation in the numerator are of the form $\rho \pm \sqrt{\rho^2 - 1}$, with

$$
\rho = -(r_1/4r_0) \pm \left(\sqrt{2 + (r_1/2r_2)^2 - (r_2/r_0)} \right) / 2,
$$

assuming $r_0 \neq 0$. It follows directly from the form of the zeros of $\det \boldsymbol{E}_\phi$ that they lie on the unit circle in $\mathbb{C}$ iff $\rho \in \mathbb{R}$ and $\rho^2 \leq 1$. This, however, is equivalent to

$$
R(s) := (2r_0 + r_2)^2 - 4r_1^2 \leq 0.
$$

Now, the polynomial $R(s)$ allows the factorization

$$
R(s) = 30,720(2 - s)^3(2s - 1)^2(32 - 39s - 18s^2 + 26s^3).
$$

Methods from elementary calculus can be used to establish that the cubic polynomial in the preceding factorization is strictly positive for $|s| < 1$. Hence, $R(s) > 0$, for $s \neq 1/2$.

Therefore, ϕ generates a multiresolution analysis of $L^2(\mathbb{R})$ for any $s \in (-1, 1) \setminus \{1/2\}$. If, in addition, $|s| < 1/2$, then ϕ^1 and ϕ^2 are also of class $C^1(\mathbb{R})$. In this case, the coefficient matrices $\mathbf{c}_\ell$ can be calculated by using Eq. (7.50) and the characterization of a function $f \in V_0$ given in Example 7.5. A straightforward computation gives

$$
\mathbf{c}_{-1} = \begin{pmatrix} 1/2 & 1 - 2s \\ -1/8 & s - 1/4 \end{pmatrix}, \quad \mathbf{c}_0 = \begin{pmatrix} 1 & 0 \\ 0 & 1/2 \end{pmatrix},
$$

and

$$
\mathbf{c}_1 = \begin{pmatrix} 1/2 & 2s - 1 \\ 1/8 & 1/4 - s \end{pmatrix}.
$$

Next, the vector wavelet associated with the vector scaling function in the preceding example is constructed. It will be seen that it is possible to construct wavelets supported on $[-1, 2]$. This corresponds to choosing $\mathbf{d}_\ell = 0$ for $\ell \notin \{-1, 0, 1, 2, 3\}$. Let $\psi \in V_1$ be one of the two components of $\boldsymbol{\psi}$. (Since

$N = A = 2$, one also has $B = 2$.) Then there exist coefficients $d_\ell^a \in \mathbb{R}$ such that

$$\psi = \sum_{\ell=-1}^{3} \sum_{a=0}^{1} d_\ell^a \phi^a (2x - \ell). \qquad (7.64)$$

A necessary condition for $\psi \in W_0 \subseteq V_{-1}$ is

$$\langle \phi^a (\cdot - \ell), \psi \rangle = 0,$$

for $a = 0, 1$ and $\ell = -1, 0, 1, 2$. (That only these values for ℓ have to be considered follows from the lengths of the supports of ϕ^a and ψ.) Using the two-scale dilation equation for ϕ, this last condition can be rewritten as

$$\sum_{\ell-1} \sum_{\ell'=-1}^{3} \sum_{a',a''=0}^{1} c_{\ell'}^{a,a''} \langle \phi^{a''} (2 \cdot -2\ell - \ell'), \phi^{a'} (2 \cdot -\ell) \rangle d_\ell^{a'} = 0, \qquad (7.65)$$

for $a = 0, 1$ and $k = -1, 0, 1, 2, 3$. (Here $\mathbf{c}_\ell = (c_\ell^{a,a'})$.) Observe that the preceding equations form a linear system of eight equations in the 10 unknowns d_ℓ^a. A tedious calculation shows that there exist two solutions $\mathbf{d}^0 = (d_\ell^{0,a'})$ and $\mathbf{d}^1 = (d_\ell^{1,a'})$ that form a basis of the null space of this linear system. Now let $\mathbf{d}_\ell = (d_\ell^{a,a'})_{a,a'=0,1}$, and define the vector wavelet $\boldsymbol{\psi} = (\psi_0, \psi_1)^t$ by

$$\boldsymbol{\psi}(x/N) := \sum_{\ell=-1}^{3} \mathbf{d}_\ell \, \boldsymbol{\phi}(x - \ell). \qquad (7.66)$$

Clearly, $\psi^0, \psi^1 \in W_0$. Moreover, it can be shown ([77]) that $\mathcal{B}_\psi$ forms a Riesz basis for W_0. But the vector wavelet $\boldsymbol{\psi}$ is *not* fully orthogonal; ψ^0 and ψ^1 are orthogonal to their dilates but not their translates, i.e.,

$$\langle \psi_{k\ell}^a, \psi_{k'\ell'}^{a'} \rangle = 0, \qquad k \neq k',$$

where $a, a' = 0, 1$, $k, k', \ell, \ell/ \in \mathbb{Z}$.

However, it will be shown shortly that there exists a way of obtaining fully orthogonal vector wavelets. This construction is given in Section 7.3.

7.2.2 A multiresolution analysis of $C_b(\mathbb{R})$

Here a multiresolution analysis of $C_b(\mathbb{R})$ is presented. To this end, let $1 \neq N \in \mathbb{N}$ and let $s_i = s$, $i = 1, \ldots, N$, $|s| < 1$. Let

$$\begin{aligned} V_0 \; := \; & \{ f : \mathbb{R} \to \mathbb{R} \, | \, \forall j \in \mathbb{Z} \, \exists g \in \mathcal{F}([j, j+1], \mathbb{R}) \, | \\ & f|_{(j,j+1)} = g|_{(j,j+1)} \}. \end{aligned} \qquad (7.67)$$

Define

$$V_0 := \mathcal{V}_0 \cap C_b(\mathbb{R}), \qquad (7.68)$$

and

$$f \in V_k \iff f(N^k \cdot) \in V_0, \qquad k \in \mathbb{Z}. \qquad (7.69)$$

Let $X := \{0, 1, \ldots, N\}$ and let $\mathfrak{e}_n \in \mathbb{Z}^X$, $n = 0, 1, \ldots, N$, be such that $\mathfrak{e}_n(j) = \delta_{nj}$, for $j = 0, 1, \ldots, N$. Define an interpolating set Δ_n by

$$\Delta_n := \{(j/N, \mathfrak{e}_n(j)) \mid j = 0, 1, \ldots, N\}, \qquad n \in \{0, 1, \ldots, N\},$$

and let $f^n \in \mathcal{F}|\Delta_n([0, 1], \mathbb{R})$ be the affine fractal function interpolating Δ_n. The set $\{f^0, f^1, \ldots, f^N\}$ will be used to define a basis for V_0. It is not difficult to see that the functions

$$\phi^0 := f^0 + f^N_{0,-1}, \qquad \phi^i := f^i, \quad i = 1, \ldots, N-1, \qquad (7.70)$$

form a basis for V_0. (Note that $C_b(\mathbb{R})$ is separable.) Thus,

$$V_k = \operatorname{span} \{\phi^a_{k\ell} \mid n = 0, 1, \ldots, N \wedge k, \ell \in \mathbb{Z}\} \cap C_b(\mathbb{R}).$$

Theorem 7.12 *The set of functions $\{\phi^0, \phi^1, \ldots, \phi^N\}$ generates a multiresolution analysis of $C_b(\mathbb{R})$.*

Remark. Condition M6* is equivalent to

$$\forall k \in \mathbb{Z} \, \exists R_1, R_2 \in \mathbb{R}^+ : \; R_1 \|\mathbf{c}\|_{\ell^\infty(Z)} \le \|\phi \circledast \mathbf{c}\| \le \gtrless \prec \|\mathbf{c}\|_{\ell^\infty(\bar{\lambda})},$$

where $\phi := (\phi^0, \phi^1, \ldots, \phi^N)^t$ and $Z := \{0, 1, \ldots, N-1\} \times \mathbb{Z}$.

Proof. That the V_k, $k \in \mathbb{Z}$, are linear spaces follows directly from Proposition 7.5, the nestedness from Eqs. (7.36) and (7.40).

By choosing collinear interpolation points on each interval of the form $[\ell N^k, (\ell+1)N^k]$, $k, \ell \in \mathbb{Z}$, a bounded piecewise linear function on $\mathbb{R}$ is generated. Thus, the space V_k contains all bounded functions that are linear on $[\ell N^k, (\ell+1)N^k]$, $k, \ell \in \mathbb{Z}$. Since $N^k \to 0$ as $k \to -\infty$, $\bigcup_{k \in \mathbb{Z}} V_k$ is dense in $C_b(\mathbb{R})$. This gives the density property.

The separation property follows from the following argument: Clearly, all constant functions are elements of V_k, $k \in \mathbb{Z}$. Let $f \in \bigcap_{k \in \mathbb{Z}} V_k$. For $k \in \mathbb{N}_0$, let Φ_k be the Read-Bajraktarević operator generating that fractal function on $I_k := [0, N^k]$ which interpolates the points $(iN^{k-1}, f(iN^{k-1}))$, $i = 0, 1, \ldots, N$. Let L_k denote the linear interpolant through $(0, f(0))$ and $(N^k, f(N^k))$. Since

$f \in V_{k-1} \cap V_k$, it is straightforward to show that $\Phi_k(L_k)|_{[0,N^{k-1}]} = L_{k-1}|_{[0,N^{k-1}]}$, and thus for any $0 \leq m \leq k$,

$$\Phi_m \circ \cdots \circ \Phi_k(L_k)|_{[0,N^{k-1}]} = L_m|_{[0,N^{m-1}]}.$$

Note that $f|_{[0,N^j]}$ is a fixed point of Φ_j, for all $j \in \mathbb{N}_0$. Hence,

$$\|(f - L_m)|_{[0,N^{m-1}]}\|_\infty$$

$$= \|\Phi_m \circ \cdots \circ \Phi_k(f)|_{[0,N^{m-1}]} - \Phi_m \circ \cdots \circ \Phi_k(L_m)|_{[0,N^{m-1}]}\|_\infty$$

$$\leq s^{k-m+1} \|(f - L_k)|_{[0,N^k]}\|_\infty \leq 2s^{k-m+1} \|f\|_\infty.$$

Now, letting $k \to \infty$ gives $f|_{[0,N^{m-1}]} = L_m|_{[0,N^{m-1}]}$, and thus $f|_{[0,\infty)} = f(0)$. Similarly, one shows $f|_{(-\infty,0]} = f(0)$. This implies the result.

Finally, to show that the set $\{\phi_{k\ell}^a \mid a \in \{0,1,\ldots,N\} \wedge \ell \in \mathbb{Z}\}$ is a Riesz basis for V_k, $k \in \mathbb{Z}$, one only needs to set

$$R_1 := \min\left\{ \left\| \sum_{n=0}^{N} c_n f^n \right\|_\infty \;\middle|\; \max_{0 \leq n \leq N} |c_n| = 1 \right\} > 0,$$

and

$$R_2 := \max\left\{ \left\| \sum_{n=0}^{N} c_n f^n \right\|_\infty \;\middle|\; \max_{0 \leq n \leq N} |c_n| = 1 \right\}.$$

∎

The wavelet spaces W_k are defined as the orthogonal complement of V_k in V_{k+1}. A possible choice for W_0 is given by

$$W_0 := \operatorname{span}\{\phi_{1,\ell}^a \mid a = 1,\ldots,N-1 \wedge \ell \in \mathbb{Z}\}. \tag{7.71}$$

Hence, one can choose the functions $\phi_{1,\ell}^a$ that are supported on $[0,1]$ and their integer-translates to generate the wavelet space W_0. Therefore,

$$\psi^b := \phi_{1,\ell}^a, \tag{7.72}$$

where $b := \ell(N-1) + a$ and $\ell = 0, 1, \ldots, N-1$. Hence,

$$W_k = \operatorname{span}\{\psi_{k\ell}^b \mid b = 1,\ldots,N(N-1) \wedge \ell \in \mathbb{Z}\}, \qquad k \in \mathbb{Z}. \tag{7.73}$$

It is worthwhile mentioning that any function $f \in V_0$ has a very simple expansion in terms of the scaling functions ϕ^a:

$$f = \sum_{a=0}^{N} \sum_{\ell=-\infty}^{+\infty} f\left(\ell + \frac{a}{N}\right) \phi_{0,\ell}^a. \tag{7.74}$$

7.3   Orthogonal Fractal Function Wavelets

In this subsection it is shown that a proper choice of the scaling factors s_i yields fully orthogonal fractal function wavelets. This construction is presented for $N := 2$ and $d := 1$, and the underlying fractal functions are obtained as in Example 5.4. In this case, the space $\mathfrak{C} = \theta\left(\prod_{j=0}^{1}\mathcal{P}^1\right) \cap C_{\mathbb{R}}(I)$ is three-dimensional. The first objective is to find an orthonormal basis for $\mathfrak{C}$. To this end, let y_1, y_2, and y_3 be vectors in $\mathbb{R}^3$, and let $e_i := e_{y_i}$, $i = 1, 2, 3$, denote the fractal function interpolating the set $\{(j/2, y_{i,j}) \mid j = 0, 1, 2\}$, with $y_i = (y_{i,0}, y_{i,1}, y_{i,2})^t$. Furthermore, it is assumed that $e_1(0) = e_1(1) = 0$, $e_2(0) = 0$, and $e_3(1) = 0$. These functions may, without loss of generality, also be rescaled so that $e_1(1/2) = 1$, $e_2(1) = 1$, and $e_3(0) = 1$. Thus, one may take $y_1 = (0, 1, 0)^t$, $y_2 = (0, p, 1)^t$, and $y_3 = (1, q, 0)^t$, for some $p, q \in \mathbb{R}$. The idea is to determine p, q, and the scaling factors s_i so that the functions e_1, e_2, and e_3 are mutually orthogonal. Using, for instance, Eqs. (5.17) and (5.18), one obtains the following formulae for the functions λ_i, $i = 0, 1$: For e_1,

$$\lambda_0 = \mathrm{id}_{\mathbb{R}}, \qquad \lambda_1 = -\mathrm{id}_{\mathbb{R}} + 1;$$

for e_2,

$$\lambda_0 = (p - s_0)\mathrm{id}_{\mathbb{R}}, \qquad \lambda_1 = (1 - s_1 - q)\mathrm{id}_{\mathbb{R}} + (1 - s_0);$$

and for e_3,

$$\lambda_0 = (q + s_0 - 1)\mathrm{id}_{\mathbb{R}} + (1 - s_0), \qquad \lambda_1 = (s_1 - q)\mathrm{id}_{\mathbb{R}} + (q - s_1).$$

In other words,

$$B^2 \ni \lambda(1) = (\mathrm{id}_{\mathbb{R}}, -\mathrm{id}_{\mathbb{R}} + 1),$$

$$\lambda(2) = ((p - s_0)\mathrm{id}_{\mathbb{R}}, (1 - s_1 - q)\mathrm{id}_{\mathbb{R}} + (1 - s_0)),$$

and

$$\lambda(3) = ((q + s_0 - 1)\mathrm{id}_{\mathbb{R}} + (1 - s_0), (s_1 - q)\mathrm{id}_{\mathbb{R}} + (q - s_1)).$$

Requiring that

$$\int_{[0,1]} e_i e_j \, dx = 0, \quad \text{for } i \neq j,$$

yields — using Proposition 5.9 and some rather tedious algebra — the following equations for the unknowns p and q:

$$p = \frac{-(4 - 6s_0 - 2s_0 s_1 - 4s_0^2 - 4s_1^2 + 3s_0^3 + 3s_0^2 s_1)}{16 + 4s_0 s_1 - 4s_0^2 - 4s_1^2}, \tag{7.75}$$

and

$$q = \frac{-(4 - 6s_1 - 2s_0s_1 - 4s_0^2 - 4s_1^2 + 3s_1^3 + 3s_0s_1^2)}{16 + 4s_0s_1 - 4s_0^2 - 4s_1^2}, \tag{7.76}$$

as well as the following algebraic equation to be satisfied by s_0 and s_1:

$$\begin{aligned}
\mathfrak{p}(s_0, s_1) \quad := \quad & 2s_1^4 + 6s_1^3 - 7s_0s_1^3 + 18s_0s_1^2 - 29s_1^2 \\
& -7s_0^3s_1 + 18s_0^2s_1 - 14s_0s_1 + 12s_1 \\
& +2s_0^2 + 6s_0^3 - 28s_0^2 + 12s_0 + 8 = 0. \tag{7.77}
\end{aligned}$$

The preceding arguments prove the next result.

Proposition 7.8 *The affine fractal functions e_1, e_2, and e_3 constitute an orthogonal basis for $\mathfrak{C}$ only for those scaling factors s_i, $i = 0, 1$, which satisfy $|s_i| < 1$ and $\mathfrak{p}(s_0, s_1) = 0$.* ∎

Corollary 7.4 *The only scaling factors s_i, $i = 0, 1$, such that the basis $\{\tilde{e}_{\tilde{y}_1}, \tilde{e}_{\tilde{y}_2}, \tilde{e}_{\tilde{y}_3}\}$ with $\tilde{y}_1 := (0, 1, a)^t$, $\tilde{y}_2 := (0, b, 1)^t$, and $\tilde{y}_3 := (1, c, 0)^t$, $a, b, c \in \mathbb{R}$, constitutes an orthogonal basis of $\mathfrak{C}$ are those which satisfy $|s_i| < 1$ and $\mathfrak{p}(s_0, s_1) = 0$. The same is true for bases of the form $\{\tilde{e}_{\tilde{y}_1}, \tilde{e}_{\tilde{y}_2}, \tilde{e}_{\tilde{y}_3}\}$ with $\tilde{\tilde{y}}_1 = (a, 1, 0)^t$, $\tilde{\tilde{y}}_2 = (0, b, 1)^t$, and $\tilde{\tilde{y}}_3 = (1, c, 0)^t$.*

Proof. Let s_0 and s_1 be scaling factors that satisfy the hypotheses of the corollary, and let $\{\tilde{e}_{\tilde{y}_1}, \tilde{e}_{\tilde{y}_2}, \tilde{e}_{\tilde{y}_3}\}$ with $\tilde{y}_1 := (0, 1, a)^t$, $\tilde{y}_2 := (0, b, 1)^t$, and $\tilde{y}_3 := (1, c, 0)^t$, $a, b, c \in \mathbb{R}$, be the corresponding orthogonal basis. Suppose that $a \neq 0$, for otherwise the result follows from Proposition 7.8. Let $\{\tilde{e}_{\tilde{y}_1}^*, \tilde{e}_{\tilde{y}_2}^*, \tilde{e}_{\tilde{y}_3}^*\}$ be the orthonormal basis constructed from $\{\tilde{e}_{\tilde{y}_1}, \tilde{e}_{\tilde{y}_2}, \tilde{e}_{\tilde{y}_3}\}$. If $\tilde{e}_{\tilde{y}_1}^*(1) = a'$ and $\tilde{e}_{\tilde{y}_2}^*(1) = b'$ set $w_1 := b'\tilde{e}_{\tilde{y}_1}^* - a'\tilde{e}_{\tilde{y}_2}^*$ and $w_2 := a'\tilde{e}_{\tilde{y}_1}^* + b'\tilde{e}_{\tilde{y}_2}^*$. Note that $a'^2 + b'^2 = 1$. Then $\{w_1, w_2, \tilde{e}_{\tilde{y}_3}^*\}$ is an orthonormal basis of $\mathfrak{C}$ with $w_1(0) = w_1(1) = 0$ and $w_2(0) = 0$. The result now follows from Proposition 7.8.

An analogous argument can be applied to obtain the second statement. ∎

Definition 7.4 A triple $(e_{y_1}, e_{y_2}, e_{y_3})$ of mutually orthogonal affine fractal functions, each of which possesses the interpolation property with respect to $\{(j/2, y_{i,j}) \,|\, j = 0, 1, 2\}$, $i = 1, 2, 3$, will be called a *fundamental basis* for $\mathfrak{C}$.

Now suppose that the scaling factors s_i, $i = 0, 1$, are such that $|s_i| < 1$ and $\mathfrak{p}(s_0, s_1) = 0$. Set

$$\phi^0(x) := \begin{cases} e_{y_1}(x), & x \in [0, 1], \\ 0, & \text{otherwise,} \end{cases} \tag{7.78}$$

$$\phi^1(x) := \begin{cases} e_{y_2}(x), & x \in [0,1], \\ e_{y_3}(x-1), & x \in [1,2], \\ 0, & \text{otherwise}, \end{cases} \qquad (7.79)$$

and

$$\phi^{i*} := \frac{\phi^i}{\|\phi^i\|_{L^2(\mathbb{R})}}, \qquad i = 0, 1.$$

Theorem 7.13 *Suppose that the scaling factors s_i, $i = 0, 1$, admit the existence of an orthogonal basis of affine fractal functions. Then the functions ϕ^{i*} generate a multiresolution analysis of $L^2(\mathbb{R})$.*

Proof. Let $\phi_{k\ell}^{i*} := 2^{-k/2}\phi^{i*}(2^k \cdot - \ell)$, $i = 0, 1$, $k, \ell \in \mathbb{Z}$, and define V_k as usual:

$$V_k = \overline{\text{span}\{\phi_{k\ell}^{i*} \mid i = 0, 1 \wedge k, \ell \in \mathbb{Z}\}}^{L^2(\mathbb{R})} \cap L^2(\mathbb{R}).$$

It is a direct consequence of $\{e_{y_1}, e_{y_2}, e_{y_3}\}$ being a fundamental basis for $V_0([0,1))$ that $\{e_{y_1}(2\cdot), e_{y_2}(2\cdot), e_{y_3}(2\cdot)\}$ and $\{e_{y_1}(2\cdot -1), e_{y_2}(2\cdot -1), e_{y_3}(2\cdot -1)\}$ are fundamental bases for $V_{-1}([0, \frac{1}{2}))$ and $V_1([\frac{1}{2}, 1))$, respectively. Therefore,

$$\phi(x) = \begin{pmatrix} \phi^{0*}(x) \\ \phi^{1*}(x) \end{pmatrix} = \sum_{\ell=0}^{3} \mathbf{c}_\ell \phi(2x - \ell). \qquad (7.80)$$

Consequently, $V_0 \subseteq V_1$, and thus $V_k \subseteq V_{k+1}$, $k \in \mathbb{Z}$.

Since $\chi_{[0,1]} = (1 - p - q)e_{y_1} + e_{y_2} = e_{y_3}$, it is immediate that

$$\sum_{\ell \in \mathbb{Z}} (1 - p - q)\phi^{0*}(x - \ell) + \phi^{1*}(x - \ell) = 1, \qquad x \in \mathbb{R}.$$

The result now follows from Proposition 7.7. ∎

The coefficient matrices $\mathbf{c}_\ell$, $\ell = 0, 1, 2, 3$, may be computed directly from Eq. (7.80) either by using the inner product

$$\langle \phi, \phi \rangle = \int_{\mathbb{R}} \phi\phi^\dagger \, dx = \int_{\mathbb{R}} \begin{pmatrix} \langle \phi^{0*}, \phi^{0*} \rangle & \langle \phi^{0*}, \phi^{1*} \rangle \\ \langle \phi^{1*}, \phi^{0*} \rangle & \langle \phi^{1*}, \phi^{1*} \rangle \end{pmatrix} dx,$$

or by evaluating the equation at $x = j/4$, $j = 1, \ldots, 8$:

$$\mathbf{c}_0 = \begin{pmatrix} s_0 - p + 1/2 & 1 \\ 1/2(p - s_0) + p(s_0 - p) & p \end{pmatrix},$$

$$\mathbf{c}_1 = \begin{pmatrix} s_1 - q + 1/2 & 0 \\ 1/2(1 - p - s_1) + p(s_1 - q) & 1 \end{pmatrix},$$

$$\mathbf{c}_2 = \begin{pmatrix} 0 & 0 \\ 1/2(1 - p - s_0) + q(s_0 - p) & q \end{pmatrix},$$

$$\mathbf{c}_3 = \begin{pmatrix} 0 & 0 \\ 1/2(q - s_1) + q(s_1 - q) & 0 \end{pmatrix}.$$

Next it will be shown that, even if longer supports are considered, compactly supported continuous and orthogonal scaling functions can only be constructed from those scaling factors which allow the existence of a fundamental basis.

Proposition 7.9 *Suppose that for a given finite-dimensional subspace V of $C_{\mathbb{R}}^r([0,1])$ there is no orthonormal basis with more than one function vanishing either at $x = 0$ or $x = 1$. Then any pair of functions $\phi^1, \phi^2 \in C_{\mathbb{R}}^r([0,1])$ with compact support that is a linear combination of the basis elements of V and their integer-translates and is constructed so that the set $F := \{\phi^a(x - \ell) \mid a = 1, 2 \wedge \ell \in \mathbb{Z}\}$ spans V and $\langle \phi^a, \phi^{a'}(\cdot - \ell) \rangle = \delta_{aa',\ell 0}$, $a.a' \in \{1,2\}$, $\ell \in \mathbb{Z}$, must have the property that the leftmost non-zero components of ϕ^1 and ϕ^2 are linearly dependent, as are their rightmost non-zero components.*

Proof. Suppose that $\operatorname{supp} \phi^a = [0, M_a]$, $a = 1, 2$. By continuity, $\phi^a(0) = 0$ and $\phi^a(x + M_a - 1)|_{x=1} = 0$. Furthermore, $\langle \phi^a|_{[0,1]}, \phi^{a'}(\cdot + M_{a'} - 1)|_{[0,1]} \rangle = 0$ for $a, a' = 1, 2$. Since F spans V, the result now follows.   ∎

It should be noted that — by a rotation — we may always assume that $M_1 \neq M_2$.

Proposition 7.10 *Suppose that the hypotheses of the preceding proposition are satisfied. Then there exists no such pair of functions ϕ^1 and ϕ^2.*

Proof. Suppose that $\dim V = k$, that $M_1 > M_2$ and that ϕ^1 and ϕ^2 do exist. Let $y := \phi^1/\|\phi^1\|_{L^2(\mathbb{R})}$ and $z := \phi^2/\|\phi^2\|_{L^2(\mathbb{R})}$. Then y and z are orthonormal and there exists a basis in which y and z have the following representation:

$$\begin{aligned} y &= (y_1, y_2, \ldots, y_{M_1}, 0, \ldots, 0)^t \in V \\ &= ((y_{1,1}, 0^*), (y_{2,1}, y_2^*), \ldots, (0^*, y_{M_1,k}), (0, 0^*), \ldots, (0, 0^*))^t, \end{aligned}$$

and

$$\begin{aligned} z &= (z_1, z_2, \ldots, z_{M_2}, 0, \ldots, 0)^t \in V \\ &= ((z_{1,1}, 0^*), (z_{2,1}, y_2^*), \ldots, (0^*, z_{M_2,k}))^t, \end{aligned}$$

where $2 \leq M_1 < M_2$, and y_i^*, $i = 2, 3, \ldots, M_1$, z_j^*, $j = 2, 3, \ldots, M_2$, and 0^* each have $k - 1$ components, and $z_{1,1} > 0$.

Now consider the following rotation $R : V \to V$ defined on pairs of vectors of the preceding form:

$$R(y, z) := (\tilde{y}, \tilde{z})^t = \frac{1}{\sqrt{y_{1,1}^2 + z_{1,1}^2}} (z_{1,1}y - y_{1,1}z, \; y_{1,1}y + z_{1,1}z)^t.$$

Note that $\tilde{y}$ and $\tilde{z}$ can be written as

$$\tilde{y} = ((0, 0^*), (\tilde{y}_{2,1}, \tilde{y}_2^*), \dots, (\tilde{y}_{M_1,1}, \tilde{y}_{M_1}^*))^t$$

$$\tilde{z} = ((\sqrt{y_{1,1}^2 + z_{1,1}^2}, 0^*), (\tilde{z}_{2,1}, \tilde{z}_2^*), \dots, (\tilde{z}_{M_2,1}, \tilde{z}_{M_2}^*))^t.$$

Define a right shift map $\sigma_+ : \operatorname{range} R \to V$ by

$$\sigma_+(\tilde{y}, \tilde{z}) := ((\tilde{y}_2, \tilde{y}_3, \dots, \tilde{y}_{M_1}, 0), (\tilde{z}_1, \tilde{z}_2, \dots, \tilde{z}_{M_2}))^t = (\hat{y}, \hat{z})^t.$$

Clearly, both R and σ_+ are continuous and $\|\hat{y}\| = \|\hat{z}\| = 1$. (Here $\| \cdot \|$ denotes a norm on the finite-dimensional vectorspace V.) Moreover, $\hat{y}$ and $\hat{z}$ are orthogonal. It should also be obvious that the functions corresponding to the vectors $\hat{y}$ and $\hat{z}$ are in $C_{\mathrm{IR}}^r(\mathrm{IR})$. Proposition 7.9 implies that $\hat{y}_1$ and $\hat{z}_1$ are linearly dependent, and thus $\hat{y}_1^* = 0$. Hence σ_+ maps into range R. This allows the iteration of the map $\sigma_+ \circ R$ on the pairs (y, z). In this way a sequence $\{(y^{(j)}, z^{(j)})\}_{j \in \mathrm{IN}}$ of pairs of vectors is produced and, since this sequence is contained in a compact set, it has a limit point $(\bar{y}, \bar{z})$ in range R. The continuity of the inner product now implies that $\|\bar{y}\| = \|\bar{z}\| = 1$, that $\bar{y}$ and $\bar{z}$ are orthogonal, and that the functions corresponding to $\bar{y}$ and $\bar{z}$ are elements of $C_{\mathrm{IR}}^r(\mathrm{IR})$.

Observe that the sequence $\{z_{1,1}^{(j)}\}$ is monotone increasing in j and, $z_{1,1}^{(j)}$ being a component of a unit vector, bounded above by one. Let $\bar{z}_{1,1} := \lim_{j \to 0} z_{1,1}^{(j)}$. Hence, the sequence $\{(z_{1,1}^{(j)})^2\}_{j \in \mathrm{IN}}$ converges to $(\bar{z}_{1,1})^2$. Therefore, the sequence $\{(y_{1,1}^{(j)})^2\}_{j \in \mathrm{IN}}$ must converge to zero. Using the following induction argument, it can be shown that $\{(y_{i,1}^{(j)})^2\}_{j \in \mathrm{IN}}$ converges to zero for all $i = 2, 3, \dots, M_1$. So, suppose that $\{(y_{i',1}^{(j)})^2\}$ converges to zero for some i'. Then, for all $j \in \mathrm{IN}$,

$$|y_{i',1}^{(j+1)}| = \left| \frac{z_{1,1}^{(j)} y_{i'+1,1}^{(j)} - y_{1,1}^{(j)} z_{i'+1,1}^{(j)}}{\sqrt{(y_{1,1}^{(j)})^2 + (z_{1,1}^{(j)})^2}} \right|.$$

Note that the denominator in the preceding equation is bounded above by $\sqrt{2} < 2$. Thus,

$$|y_{i',1}^{(j+1)}| \geq (1/2) \left(|z_{1,1}^{(j)} y_{i'+1,1}^{(j)}| - |y_{1,1}^{(j)} z_{i'+1,1}^{(j)}| \right),$$

or,

$$|z_{1,1}^{(j)} y_{i'+1,1}^{(j)}| \leq 2|y_{i',1}^{(j+1)}| + |y_{1,1}^{(j)} z_{i'+1,1}^{(j)}|$$

$$\leq 2|y_{i',1}^{(j+1)}| + |y_{1,1}^{(j)}|.$$

Since $z_{1,1}^{(j)} \geq z_{1,1} > 0$ and since the left-hand side of the last expression converges to zero, the claim is proven.

Because of this last result, $\hat{y}_{i,1} = 0$, for $i = 1, \ldots, M_1$. By Proposition 7.9, the leftmost non-zero tuple in $\hat{y}$, say $\hat{y}_p$, is linearly dependent on $\hat{z}_1 = (\hat{z}_{1,1}, 0^*)$. Hence, $\hat{y}_p$ hs to be of the form $\hat{y}_p = (\hat{y}_{p,1}, 0^*)$, implying that $\hat{y}_p = 0$. This contradiction proves the proposition.   ∎

Combining the results in Propositions 7.8, 7.9 and 7.10 yields the next theorem.

Theorem 7.14 *If s_0 and s_1 do not satisfy the algebraic equation $\mathfrak{p}(s_0, s_1) = 0$, then there exist no continuous, compactly supported scaling functions ϕ^1 and ϕ^2 formed from affine fractal functions generated with these scaling factors.*

Proof. By Proposition 7.10, only scaling factors s_0 and s_1 that allow an orthogonal basis of $\mathfrak{C}$ in which at least two basis elements vanish at either zero or one have to be considered. But, in Propositions 7.8 and 7.9, it was shown that this can occur only when $\mathfrak{p}(s_0, s_1) = 0$.   ∎

The preceding theorem shows that, in order to construct continuous and compactly supported scaling functions, the zero-set of the polynomial $\mathfrak{p}(s_0, s_1)$ has to be investigated. The following proposition gives a characterization of this zero-set.

Proposition 7.11 *The zero-set of the polynomial*

$$\begin{aligned}\mathfrak{p}(s_0, s_1) = \; & 2s_1^4 + 6s_1^3 - 7s_0s_1^3 + 18s_0s_1^2 - 29s_1^2 - 7s_0^3s_1 + 18s_0^2s_1 \\ & -14s_0s_1 + 12s_1 + 2s_0^3 + 6s_0^3 - 28s_0^2 + 12s_0 + 8 = 0\end{aligned}$$

has two connected components C_1 and C_2. The component C_1 is a closed convex curve and C_2 consists of a pair of asymptotically linear curves that intersect transversally at exactly one point.

Proof. A change of variables, $T : \mathbb{R}^2 \to \mathbb{R}^2$, $(s_0, s_1) \mapsto (x - y, x + y)$, transforms the polynomial $\mathfrak{p}(s_0, s_1)$ into a new polynomial $\mathfrak{q}(x, y)$:

$$\mathfrak{q}(x, y) = 18y^4 + (24x^2 - 42)y^2 - 10x^4 + 48x^3 - 70x^2 + 24x + 8.$$

The zero-set of $\mathfrak{p}(s_0, s_1)$ is the pre-image of the zero-set of $\mathfrak{q}(x, y)$ under T. Since T is just a rotation by $\pi/4$ followed by a dilation by $\sqrt{2}$, the essential properties of the zero-set of $\mathfrak{p}(s_0, s_1)$ are preserved under T. Hence, it suffices to find the zero-set of $\mathfrak{q}(x, y)$. Also, note that the polynomial $\mathfrak{q}(x, y)$ is symmetric with respect to y, for all x. To find the zero-set of $\mathfrak{q}$, the equation $\mathfrak{q}(x, y) = 0$ is solved for y in terms of x. This gives

$$y = \pm\sqrt{7/6 - (2/3)x^2 + (1/6)\sqrt{3}\sqrt{12x^4 - 32x^3 + 28x^2 - 16x + 11}},$$

or

$$y = \pm\sqrt{7/6 - (2/3)x^2 - (1/6)\sqrt{3}\sqrt{12x^4 - 32x^3 + 28x^2 - 16x + 11}}.$$

The first solution yields a pair of asymptotically linear curves that intersect transversally at $(x, y) = (2, 0)$ or $(s_0, s_1) = (2, 0)$. This component C_1 consists of pairs (s_0, s_1) that are outside the range of s_i-values necessary to construct fractal functions. The second solution, which is real-valued only for $x \in [-1/5, 1]$, gives two halves of a symmetric closed curve γ.

Denote by γ^+ the positive branch of the curve γ. If it can be established that γ^+ is the graph of a convex function, then — by symmetry — γ is a convex curve. To show that γ^+ is indeed the graph of a convex function f, it suffices to prove that $f \cdot f$ is a convex function. This can be done by showing that $f''|_{[-1/5,1]}$ is non-positive, i.e., that

$$\frac{3}{4} + \frac{\sqrt{3}\,(48x^3 - 96x^2 + 56x - 16)^2}{24\mathfrak{p}_1^{3/2}} - \frac{\sqrt{3}\,(144x^2 - 192x + 56)}{12\mathfrak{p}_1^{1/2}} \leq 0,$$

where $\mathfrak{p}_1(x) := 12x^4 - 32x^3 + 28x^2 - 16x + 11$. Since $\mathfrak{p}_1 > 0$, for all $x \in \mathbb{R}$, the preceding inequality is equivalent to

$$-4\mathfrak{p}_1^{3/2} - \sqrt{3}\,(-144x^6 + 576x^5 - 888x^4 + 832x^3 - 780x^2 + 528x - 122) \leq 0.$$

To prove that this inequality holds for $x \in [-1/5, 1]$, the expression $\sqrt{\mathfrak{p}_1}$ is first approximated by the linear polynomial $\mathfrak{p}_2 := \sqrt{3}\,(7/4 - (3/4)x)$. (Note that $\mathfrak{p}_1 - \mathfrak{p}_2^2 = (1/16)(x - 1)(192x^3 - 320x^2 + 101x - 29) \geq 0$ for $x \in [-1/5, 1]$, and thus $\mathfrak{p}_2 \leq \sqrt{\mathfrak{p}_1}$ on $[-1/5, 1]$.) Replacing $\mathfrak{p}_1^{3/2}$ in the preceding inequality by $\mathfrak{p}_1\mathfrak{p}_2$ and expanding the resulting expression gives

$$-4\mathfrak{p}_1^{3/2} - \sqrt{3}\,(-144x^6 + 576x^5 - 888x^4 + 832x^3 - 780x^2 + 528x - 122)$$
$$\leq \tfrac{-4\sqrt{3}}{3125}(5x - 4)^2(4{,}500x^4 - 11{,}925x^3$$

$+11,415x^2 - 9,729x + 9,128) \leq 0.$

The last inequality is justified since the second and third factor is non-negative for all $x \in \mathbb{R}$.   ■

Now let $(\bar{s}_0, \bar{s}_1)$ be that pair which solves $\mathfrak{p}(s_0, s_1) = 0$ and also $d\mathfrak{p}/ds_0 = 0$. Then, a careful examination of the polynomial $\mathfrak{p}(s_0, s_1)$ using Proposition 7.11 shows that there is at least one but at most two values of $s_1 \in (-1, 1)$ with $\mathfrak{p}(s_0, s_1) = 0$. If $s_1 = s_0$, then $x = 0$ in the polynomial $\mathfrak{q}(x, y)$, and thus $0 = \mathfrak{q}(0, y) = -2(5y + 1)(-2 + y)^2$. Hence, $y = s_0 = s_1 = -1/5$, implying $p = q = 3/10$. Consequently, for $s_0 = s_1 = -1/5$ there exists a fundamental basis $(e_{y_1}, e_{y_2}, e_{y_3})$ of $\mathfrak{C}$ with $y_1 = (0, 1, 0)^t$, $y_2 = (0, -3/10, 1)^t$, and $y_3 = (1, -3/10, 0)^t$. The associated scaling functions ϕ^1 and ϕ^2 are depicted in Fig. 7.3. Note that the graph of ϕ^1 is symmetric about the line $x = 1/2$,

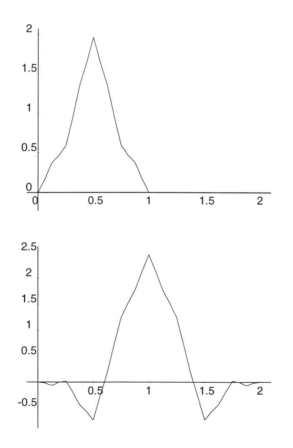

Figure 7.3: The scaling functions for $s_0 = s_1 = -1/5$.

while the graph of ϕ^2 is symmetric about the line $x = 1$. Normalizing the vector scaling function $\boldsymbol{\phi} = (\phi^1, \phi^2)^t$ gives the following matrix coefficients:

$$c_0 = \begin{pmatrix} 3\sqrt{2}/10 & 4/5 \\ -1/20 & -3\sqrt{2}/20 \end{pmatrix}, \quad c_1 = \begin{pmatrix} 3\sqrt{2}/10 & 0 \\ 9/20 & \sqrt{2}/2 \end{pmatrix},$$

$$c_2 = \begin{pmatrix} 0 & 0 \\ 9/20 & -3\sqrt{2}/20 \end{pmatrix}, \quad c_3 = \begin{pmatrix} 0 & 0 \\ -1/20 & 0 \end{pmatrix}.$$

Before the wavelets are constructed, the regularity and approximation order of the scaling functions ϕ^1 and ϕ^2 as defined in Eqs. (7.78) and (7.79) is briefly discussed. But first a definition is needed.

Definition 7.5 Let $I \subseteq \mathbb{R}$ be an interval on the real line and let $f \in C_{\mathbb{R}}(I)$. The Hölder exponent α_x of f at $x \in I$ is defined as

$$\alpha_x := \liminf_{\varepsilon \to 0} \left\{ \frac{\log |f(x) - f(y)|}{\log |x - y|} \,\Big|\, y \in B_\varepsilon(x) \right\}, \tag{7.81}$$

where $B_\varepsilon(x)$ denotes the ball of radius $\varepsilon > 0$ centered at $x \in I$. The Hölder exponent α of f on I is then defined by

$$\alpha := \inf\{\alpha_x \,|\, x \in I\}. \tag{7.82}$$

Definition 7.6 For a given closed subspace S of $L^2(\mathbb{R})$ define $E(f; S) := \min\{\|f - s\|_{L^2(\mathbb{R})} \,|\, s \in S\}$. Let $h > 0$ and denote by $U_h : L^2(\mathbb{R}) \to L^2(\mathbb{R})$ the dilation operator $U_h f := f(\cdot/h)$. The subspace S is said to provide approximation order k iff for all $f \in W_2^k(\mathbb{R})$

$$E(f : S^h) \leq C\, h^k \|f\|_{W_2^k(\mathbb{R})}, \tag{7.83}$$

where $S^h := \{U_h s \,|\, s \in S\}$ and $C > 0$.

Proposition 7.12 Let s_i, $|s_i| < 1$, $i = 0, 1$, be a given pair of scaling factors, and let $f \in \mathfrak{C}$. If $|s_i| < 1/2$, $i = 0, 1$, then f is of class $\mathrm{Lip}^1(\mathbb{R})$. If $\max\{|s_i| \,|\, i = 0, 1\} > 1/2$ and if graph f is not a straight line, then f has Hölder exponent $\alpha = -\log \max\{|s_i| \,|\, i = 0, 1\}/\log 2$.

Remark. The second statement in the preceding proposition has already been proven in [18]. Since this proof is not very involved, it is repeated here.

Proof. Let $\bar{s} := \max\{|s_i| \,|\, i = 0, 1\}$, let $\mathbf{i}(n) = (i_1 \, i_2 \, \ldots \, i_n) \in \{0, 1\}^{\mathbb{N}}$ be a finite code of length $n \in \mathbb{N}$, and let $a(\mathbf{i}(n)) := i_1/2 + i_2/2^2 + \ldots + i_n/2^n$. Let $f \in \mathfrak{C}$. Then, since f is the unique fixed point of a Read-Bajraktarević operator Φ,

$$
\begin{aligned}
f(x) \;=\; & \sum_{k=1}^{n} \prod_{j=1}^{k-1} s_{i_j} \left(2^{k-1} x + b_{i_k} - \sum_{m=1}^{k} \frac{i_m}{2^{m-1}} \right) \\
& + \prod_{j=1}^{n} s_{i_j} \, f \left(2^n x - \sum_{m=1}^{n} 2^{n-m} i_m \right),
\end{aligned}
\tag{7.84}
$$

for all $x \in I_{\mathbf{a}(n)} := [a(\mathbf{i}(n)), a(\mathbf{i}(n)) + 2^{-n}]$. Here b_i, denotes $\lambda_i - \mathrm{lin}\,\lambda_i$, $i = 0, 1$, with lin representing the linear part of the affine function λ_i. Now suppose that $x, y \in I_{\mathbf{a}(n)}$ with $2^{-(n+1)} \le |x - y| \le 2^{-n}$. Then it follows directly from the preceding formula that

$$
|f(x) - f(y)| \le \sum_{k=1}^{n} (2\bar{s})^{k-1} |x - y| + \bar{s}^n \, C,
$$

where $C := 2 \, \|f\|_\infty$.

First assume that $\bar{s} < 1/2$. Then

$$
|f(x) - f(y)| \le \sum_{k=1}^{n} (2\bar{s})^{k-1} |x - y| + (2\bar{s})^n \, (2C) |x - y|,
$$

where the fact that $1 \le 2^{n+1}|x - y|$ was used to obtain the last term in the precedig inequality. Hence,

$$
|f(x) - f(y)| \le \frac{1 + 2C}{1 - 2\bar{s}} \, |x - y|.
$$

Now suppose that x and y satisfy $2^{-(n+1)} \le |x - y| \le 2^{-n}$, but $x \in I_x$ and $y \in I_y$ are not inside the same dyadic interval. Let $x_0 := I_x \cap I_y$. Then, $|f(x) - f(y)| \le |f(x) - f(x_0)| + |f(x_0) - f(y)|$, and thus

$$
|f(x) - f(y)| \le \frac{2(1 + 2C)}{1 - 2\bar{s}} \, |x - y|.
$$

Since this latter inequality holds for all $x, y \in [0, 1]$, f is of class $\mathrm{Lip}^1(\mathbb{R})$.

If $\bar{s} > 1/2$, assume without loss of generality that $f(0) = f(1) = 0$. (This can always be achieved by adding two linear functions L_1 and L_2 to f so that $\tilde{f} := f + L_1 + L_2 \in \mathfrak{C}$ and $\tilde{f}(0) = \tilde{f}(1) = 0$.) Now, if $x, y \in I_{\mathbf{a}(n)}$ with $2^{-(n+1)} \leq |x - y| \leq 2^{-n}$, then — as in the previous case — it is easily seen that

$$|f(x) - f(y)| \leq \bar{s}^n \left(C + \sum_{k=1}^{n} (2\bar{s})^{k-n-1} \right) = \bar{s}^n \left(C + \frac{2\bar{s}}{1 - \frac{1}{2\bar{s}}} \right).$$

Choosing $\alpha := \log \bar{s} / \log 1/2$ and noting that $2^{-(n+1)} \leq |x - y| \leq 2^{-n}$, one finds that $|x - y|^\alpha \geq 2^{-\alpha(n+1)} \geq 2^{-\alpha}\bar{s}^n$. Thus,

$$|f(x) - f(y)| \leq 2^\alpha \left(C + \frac{2\bar{s}}{1 - \frac{1}{2\bar{s}}} \right) |x - y|^\alpha,$$

provided x and y lie in the same dyadic interval of length 2^{-n}. If this is not the case, then as before, one finds

$$|f(x) - f(y)| \leq 2^{\alpha+1} \left(C + \frac{2\bar{s}}{1 - \frac{1}{2\bar{s}}} \right) |x - y|^\alpha,$$

for all $x, y \in [0, 1]$. It remains to be shown that α is the largest possible Hölder exponent. Suppose then without loss of generality that $\bar{s} = |s_0|$. Since f vanishes at $x = 0$ and $x = 1$, and $f(1/2) \neq 0$, there exist $x_0, y_0 \in [0, 1]$ so that $f(x_0) \neq f(y_0)$ and $\mathbf{m} := (x_0 - y_0)/(f(x_0) - f(y_0)) > 0$. (Here use was made of the fact that graph f is not a straight line.) Setting $x_n := x_0 2^{-n}$ and $y_n := y_0 2^{-n}$, $n \in \mathbb{N}$, and using Eq. (7.78) yields

$$f(x_n) - f(y_n) = \sum_{k=1}^{n-1} (2s_0)^{k-1} \left(\frac{x_0 - y_0}{2^n} \right) + s_0^n (f(x_0) - f(y_0)).$$

Therefore,

$$|f(x_n) - f(y_n)| = s_0^n |f(x_0) - f(y_0)| \left| 1 + \mathbf{m} \left(\frac{1}{2s_0} \right)^2 \frac{1 - \left(\frac{1}{2s_0} \right)^{n-2}}{1 - \frac{1}{2s_0}} \right|$$

$$\geq s_0^n C \geq C_1 |x - y|^\alpha.$$

∎

The next theorem states one of the main results in this section.

Theorem 7.15 *Suppose that the scaling factors s_i, $i = 0, 1$, give rise to a fundamental basis. Then the space V_0 allows approximation order two. If $\max_{i=0,1}\{|s_i|\} < 1/2$ then the scaling functions ϕ^{a*}, $a = 1, 2$, are of class $\mathrm{Lip}^1(\mathbb{R})$. On the other hand, if $\max_{i=0,1}\{|s_i|\} > 1/2$, then ϕ^{a*}, $a = 1, 2$, are of class $\mathrm{Lip}^\alpha(\mathbb{R})$, where the exponent $\alpha = -\log_2 \max_{i=0,1}\{|s_i|\}$.*

Proof. Note that, in order to show that V_0 has approximation order two, it suffices to prove that the *hat function* $h : [0, 2] \to [0, 1]$, $h = \chi_{[0,1]} * \chi_{[0,1]}$, is in V_0 ([166]). First it is noted that for $y := (0, 1/2, 1)^t$ and any scaling factors s_i with $|s_i| < 1$, $i = 0, 1$, the affine fractal function f_{e_y} interpolating $\{(j/2, y_j) \,|\, j = 0, 1, 2\}$ is the identity on $[0, 1]$. Consequently,

$$h = (1/2 - p)\phi^{0*} + \phi^{1*} + (1/2 - q)\phi^{0*}(\cdot - 1).$$

If $|s_i| < 1/2$, $i = 0, 1$, then Proposition 7.12 gives the first statement. The second conclusion holds if it can be shown that the graphs of ϕ^{a*}, $a = 1, 2$, are not straight lines. This is clearly true for ϕ^{1*}. To show that it is also true for ϕ^{2*} it suffices to prove that $p \neq 1/2 \neq q$ since only in this case does the graph of ϕ^{2*} degenerate to a line. However, if p was equal to $1/2$, then

$$\int_{[0,1]} e_1 e_2 \, dx = \frac{s_0 - 2}{(s_0 + s_1 - 2)(s_0 + s_1 - 4)} \neq 0.$$

Hence, graph e_2 is not a straight line. A similar argument shows that graph e_3 is not a straight line. ∎

Before commencing with the construction of the wavelets, the Fourier transforms of the previously constructed continuous, compactly supported and orthogonal scaling functions are computed. To this end, set $N := 2$ in Eq. (5.44). Using the expressions for the functions λ_i, $i = 0, 1$, one easily shows that

$$\hat{e}_y(\xi) = \frac{1}{2} \int_{[0,1]} \lambda_0 e^{-i\xi x} \, dx + \frac{1}{2} \int_{[0,1]} \lambda_1 e^{-i\xi x} \, dx$$

$$+ \frac{1}{2}\left(s_0 + s_1 e^{i\xi/2}\right) \hat{e}_y(\xi/2).$$

Hence,

$$\hat{\phi}^{0*}(\xi) = 8 e^{-i\xi/2} \frac{\sin^2 \xi/2}{\xi^2} + \frac{1}{2}\left(s_0 + s_1 e^{-i\xi/2}\right) \hat{\phi}^{0*}(\xi/2). \tag{7.85}$$

Setting $C(\xi) := (1/2)(s_0 + s_1 e^{-i\xi/2})$ and $h_1(\xi) := 8\xi^{-2} e^{-i\xi/2} \sin^2 \xi/2$, and iterating the equation, yields

$$\hat{\phi}^{0*}(\xi) = \sum_{n \in \mathbb{N}_0} \left(\prod_{j=1}^{n} C(2^{-j+1}\xi) \right) h_1(2^{-n}\xi). \tag{7.86}$$

Since $|C(2^{-j}\xi)| \le (1/2)(|s_0| + |s_1|) < 1$, for all $j \in \mathbb{N}$, and since $|\sin x/x| \le 1$, for all $x \in \mathbb{R}$, the series converges uniformly for all $\xi \in \mathbb{R}$.

To compute the Fourier transform of ϕ^{1*}, one writes

$$
\begin{aligned}
\hat{\phi}^{1*}(\xi) &= \int_{[0,2]} e^{-i\xi x} \phi^{1*}(x) \, dx \\
&= \int_{[0,1]} e^{-i\xi x} \phi^{1*}(x) \, dx + \int_{[0,1]} e^{-i\xi(x+1)} \phi^{1*}(x+1) \, dx \\
&= \int_{[0,1]} e^{-i\xi x} e_2(x) \, dx + e^{-i\xi} \int_{[0,1]} e^{-i\xi x} e_3(x) \, dx.
\end{aligned}
$$

Again using Eq. (5.44), one quickly computes the Fourier transforms of e_2 and e_3 as

$$
\begin{aligned}
\hat{e}_2(\xi) = \; &\frac{1}{2} \int_{[0,1]} e^{-i\xi x/2}(p - s_0)x \, dx \\
&+ \frac{1}{2} \int_{[0,1]} e^{-i\xi(x+1)/2}((1 - s_1 - p)x + p) \, dx \\
&+ \frac{1}{2} \left(s_0 + s_1 e^{-i\xi/2} \right) \hat{e}_2(\xi/2)
\end{aligned}
$$

and

$$
\begin{aligned}
\hat{e}_3(\xi) = \; &\frac{1}{2} \int_{[0,1]} e^{-i\xi x/2}((q + s_0 - 1)x + 1 - s_0 \, dx \\
&+ \frac{1}{2} \int_{[0,1]} e^{-i\xi(x+1)/2}((s_1 - q)x + q - s_1) \, dx \\
&+ \frac{1}{2} \left(s_0 + s_1 e^{-i\xi/2} \right) \hat{e}_3(\xi/2).
\end{aligned}
$$

Integration gives

$$\hat{e}_2(\xi) = h_2(\xi) + C(\xi)\hat{e}_2(\xi/2)$$

and

$$\hat{e}_3(\xi) = h_3(\xi) + C(\xi)\hat{e}_3(\xi/2),$$

where

$$h_2(\xi) := \left(\frac{(1 - s_1)e^{-i\xi/2} - s_0}{-i\xi} \right) \left(e^{-i\xi/2} - 4e^{-i\xi/4}\frac{\sin \xi/4}{\xi} \right)$$

and

$$h_3(\xi) := \left(\frac{(s_0 - 1 - s_1)e^{-i\xi/2}}{-i\xi} \right) \left(1 - 4e^{-i\xi/4}\frac{\sin \xi/4}{\xi} \right).$$

Consequently,

$$\hat{\phi}^{1\,*}(\xi) = \sum_{n\in\mathbb{N}_0} \left(\prod_{j=1}^{n} C(2^{-j}\xi) \right) \left(h_2(2^{-n}\xi) + e^{-i\xi}h_3(2^{-n}\xi) \right), \qquad (7.87)$$

and this infinite series converges uniformly for all $\xi \in \mathbb{R}$.

Now the construction of the vector wavelet ψ is presented. This is done in a slightly more general setting than before: It is assumed that $\phi = (\phi^1, \ldots, \phi^A)^t$ and that $\psi = (\psi^1, \ldots, \psi^A)^t$. Furthermore, since $\operatorname{supp}\phi = [0, 2]^A$, it is also assumed that $\operatorname{supp}\psi = [0, 2]^A$. The reason for this approach is that it provides another way of constructing vector scaling functions and vector wavelets, namely, by solving *matrix dilation equations*.

Note that under the preceding assumptions, the coefficient matrices $\{\mathbf{c}_\ell\}_{\ell\in\mathbb{Z}}$ and $\{\mathbf{d}_\ell\}_{\ell\in\mathbb{Z}}$ for ϕ and ψ, respectively, are $A \times A$ matrices. The orthogonality conditions

$$\langle \phi, \phi_{0,\ell} \rangle = \delta_{0,\ell}I_A, \qquad \ell \in \mathbb{Z},$$

and

$$\langle \psi, \psi_{0,\ell} \rangle = \delta_{0,\ell}I_A, \qquad \ell \in \mathbb{Z},$$

can be re-expressed in terms of these coefficient matrices as

$$\sum_{\ell=0}^{3} \mathbf{c}_\ell \mathbf{c}_{\ell-2\ell'}^\dagger = \delta_{0,\ell'} I_A \qquad (7.88)$$

and

$$\sum_{\ell=0}^{3} \mathbf{d}_\ell \mathbf{d}_{\ell-2\ell'}^\dagger = \delta_{0,\ell'} I_A. \qquad (7.89)$$

(Note that, as $\psi \in W_0 \subseteq V_{-1}$, the component functions $\psi^a|_{[j/2,(j+1)/2]}$ are affine fractal functions interpolating on $\mathbb{Z}/4$.)

The condition that W_0 is the orthogonal complement of V_0 in V_1 means

$$\sum_{\ell=0}^{3} \mathbf{c}_\ell \mathbf{d}_{\ell-2\ell'}^\dagger = 0, \qquad \ell' \in \mathbb{Z}, \tag{7.90}$$

and the fact that $V_0 \oplus W_0 = V_1$ means

$$\sum_{\ell=0}^{3} \mathbf{c}_{m-2\ell}^\dagger \mathbf{c}_{n-2\ell'} + \mathbf{d}_{m-2\ell}^\dagger \mathbf{d}_{n-2\ell'} = \delta_{mn} I_A, \qquad m, n \in \mathbb{Z}. \tag{7.91}$$

Setting $\mathbf{h}_1 := (\mathbf{c}_0, \mathbf{c}_1)$, $\mathbf{h}_2 := (\mathbf{c}_2, \mathbf{c}_3)$, $\mathbf{g}_1 := (\mathbf{d}_0, \mathbf{d}_1)$, and $\mathbf{g}_2 := (\mathbf{d}_2, \mathbf{d}_3)$, these equations can be rewritten as

$$\mathbf{h}_1 \mathbf{h}_2^\dagger = 0, \tag{7.92}$$

$$\mathbf{h}_1 \mathbf{h}_1^\dagger + \mathbf{h}_2 \mathbf{h}_2^\dagger = I_{2A}, \tag{7.93}$$

and likewise

$$\mathbf{g}_1 \mathbf{g}_2^\dagger = 0, \tag{7.94}$$

$$\mathbf{g}_1 \mathbf{g}_1^\dagger + \mathbf{g}_2 \mathbf{g}_2^\dagger = I_{2A}, \tag{7.95}$$

$$\mathbf{h}_2 \mathbf{g}_1^\dagger = 0, \tag{7.96}$$

$$\mathbf{h}_1 \mathbf{g}_2^\dagger = 0, \tag{7.97}$$

$$\mathbf{h}_1 \mathbf{g}_1^\dagger + \mathbf{h}_2 \mathbf{g}_2^\dagger = 0, \tag{7.98}$$

and, finally,

$$\mathbf{h}_1 \mathbf{h}_1^\dagger + \mathbf{h}_2 \mathbf{h}_2^\dagger + \mathbf{g}_1 \mathbf{g}_1^\dagger + \mathbf{g}_2 \mathbf{g}_2^\dagger = I_{2A}. \tag{7.99}$$

The general solution to Eq. (7.92) is given by

$$\mathbf{h}_2^\dagger = P_1 \boldsymbol{\eta},$$

where $P_1 : \mathbb{R}^{2A} \to \mathbb{R}^{2A}$ denotes an orthogonal projection onto the null space of $\mathbf{h}_1$, and $\boldsymbol{\eta}$ is an arbitrary $2A \times A$ matrix. Likewise, it follows from Eq. (7.97) that

$$\mathbf{g}_2^\dagger = P_1 \boldsymbol{\xi}$$

is the general solution, with $\boldsymbol{\xi}$ representing an arbitrary $2A \times A$ matrix. Defining

$$\mathbf{h}_1^\dagger := (I_{2A} - P_1) \boldsymbol{\eta},$$

it is easy to verify that Eqs. (7.92) and (7.93) are satisfied, provided the matrix η is chosen in such a way that

$$\eta^\dagger \eta = I_{2A}.$$

If η is chosen as

$$\eta = \mathbf{h}_1^\dagger + \mathbf{h}_2^\dagger,$$

then the above requirement is clearly fulfilled. Eqs. (7.94), (7.95), and (7.96) now suggest the following choice for $\mathbf{g}_1^\dagger$:

$$\mathbf{g}_1^\dagger = (I_{2A} - P_1)\boldsymbol{\xi},$$

with $\boldsymbol{\xi}^\dagger \boldsymbol{\xi} = I_{2A}$. Moreover, Eq. (7.98) implies that

$$\eta^\dagger \boldsymbol{\xi} = 0.$$

Suppose that $\{e_1, \ldots, e_{2A}\}$ is an orthonormal basis for $\mathbb{R}^{2A}$. Let $\eta = (\eta_1\, \eta_2\, \cdots\, \eta_A)$, where η_j, $j = 1, \ldots, A$, denotes the jth column in the matrix η. Define

$$\boldsymbol{\xi} := (\eta_1\, \eta_2\, \cdots\, \eta_A\, e_{A+1}\, \cdots\, e_{2A}).$$

Then it is straightforward to verify that with this choice of η and $\boldsymbol{\xi}$, Eq. (7.99) is also satisfied. These arguments have now proven the following theorem.

Theorem 7.16 *Let $\mathbf{c}_0$, $\mathbf{c}_1$, $\mathbf{c}_2$, and $\mathbf{c}_3$ be given $A \times A$ matrices satisfying $\sum_{\ell=0}^{3} \mathbf{c}_\ell \mathbf{c}_{\ell-2\ell'}^\dagger = \delta_{0,\ell'} I_A$, for all $\ell' \in \mathbb{Z}$. Then there exist $B \times A$ matrices $\mathbf{d}_0$, $\mathbf{d}_1$, $\mathbf{d}_2$, and $\mathbf{d}_3$ such that Eqs. (7.89), (7.90), and (7.91) are satisfied.*

Theorem 7.16 allows the explicit construction of the wavelet vector ψ for $A = 2$. To this end, let

$$\psi^a := \begin{cases} e_{y_1^a}, & x \in [0, 1/2], \\ e_{y_2^a}, & x \in [1/2, 1], \\ e_{y_3^a}, & x \in [1, 3/2], \\ e_{y_4^a}, & x \in [3/2, 2], \end{cases} \qquad a = 0, 1. \qquad (7.100)$$

Here $\mathbb{R}^3 \ni y_j^a = \psi^a((2(j-1) + k)/4)$, $j = 1, 2, 3, 4$, $k = 0, 1, 2$, $a = 0, 1$. In order to ensure continuity, the vectors y_j^a have to be of the following form: $y_1^a = (0, b_1^a, b_2^a)^t$, $y_2^a = (b_2^a, b_3^a, b_4^a)^t$, $y_3^a = (b_4^a, b_5^a, b_6^a)^t$, and $y_4^a = (b_6^a, b_7^a, 0)^t$. The unknowns b_n^a, $n = 1, \ldots, 7$, are determined so that ψ is orthogonal to ϕ, its non-zero integer-translates, and so that $\langle \psi^0, \psi^1 \rangle = 0$. In the case when $s_0 = s_1 = -1/5$, the inner product formula for affine fractal functions gives

$$\int_{[0,2]} \psi^a(x)\phi^0(x)\, dx = \frac{5}{32}b_1^a + \frac{41}{96}b_2^a + \frac{5}{32}b_3^a + \frac{3}{64}b_4^a, \tag{7.101}$$

$$\int_{[0,2]} \psi^a(x)\phi^0(x+1)\, dx = \frac{3}{64}b_4^a + \frac{5}{32}b_5^a + \frac{41}{96}b_6^a + \frac{5}{32}b_7^a, \tag{7.102}$$

$$\int_{[0,2]} \psi^a(x)\phi^1(x)\, dx = -\frac{1}{48}b_1^a - \frac{1}{20}b_2^a + \frac{3}{16}b_3^a + \frac{107}{240}b_4^a$$

$$+ \frac{3}{16}b_5^a - \frac{1}{20}b_6^a - \frac{1}{48}b_4^a, \tag{7.103}$$

$$\int_{[0,2]} \psi^a(x)\phi^1(x+1)\, dx = \frac{3}{16}b_1^a - \frac{1}{20}b_2^a - \frac{1}{48}b_6^a - \frac{1}{1602}b_4^a, \tag{7.104}$$

$$\int_{[0,2]} \psi^a(x)\phi^1(x-1)\, dx = -\frac{1}{160}b_4^a - \frac{1}{48}b_5^a - \frac{1}{20}b_6^a + \frac{3}{16}b_7^a, \tag{7.105}$$

and

$$\int_{[0,2]} \psi^0(x)\psi^1(x)\, dx = \frac{25}{96}(b_1^0 b_1^1 + b_3^0 b_3^1 + b_5^0 b_5^1 + b_7^0 b_7^1)$$

$$+ \frac{73}{192}(b_2^0 b_2^1 + b_4^0 b_4^1 + 2b_6^0 b_6^1)$$

$$+ \frac{3}{128}(b_4^1 b_2^0 + b_4^0 b_2^1 + b_6^0 b_4^1 + b_6^1 b_4^0)$$

$$+ \frac{5}{64}(b_2^0 b_1^1 + b_1^0 b_2^1 + b_3^0 b_2^1$$

$$+ \; b_4^0 b_3^1 + b_3^0 b_4^1 + b_5^0 b_4^1 + b_4^0 b_5^1$$

$$+ \; b_6^0 b_5^1 + b_5^0 b_6^1 + b_7^0 b_6^1 + b_6^0 b_7^1). \tag{7.106}$$

for $a = 0, 1$. Setting Eqs. (7.101) — (7.106) equal to zero determines all but three of the 14 unknowns b_j^a, $a = 0, 1$, $j = 1, \ldots, 7$. Two of these three are then used to normalize the integrals of ψ^0 and ψ^1. One degree of freedom remains because of the existence of a one-parameter family of rotations taking the vector wavelet ψ into other vector wavelets. It is worthwhile remarking that once Eqs. (7.102), (7.104) and (7.105) are satisfied then $\langle \psi, \psi_{0,\ell} \rangle = 0$, for all $\ell \in \mathbb{Z} \setminus \{0\}$. A solution to the preceding equations yielding a normalized vector wavelet ψ with smallest support is given by

$$b_1^0 = -\frac{3\sqrt{2}}{200}, \quad b_2^0 = \frac{9\sqrt{2}}{20}, \quad b_3^0 = -\frac{273\sqrt{2}}{200}, \quad b_4^0 = \frac{\sqrt{2}}{2},$$

$$b_5^0 = \frac{5\sqrt{2}}{200}, \quad b_6^0 = -\frac{3\sqrt{2}}{20}, \quad b_7^0 = \frac{\sqrt{2}}{200},$$

and

$$b_1^1 = 0, \quad b_2^1 = 0, \quad b_3^1 = \frac{3}{10}, \quad b_4^1 = -1,$$

$$b_5^1 = \frac{48}{25}, \quad b_6^1 = -\frac{3}{5}, \quad b_7^1 = \frac{1}{50}.$$

This then yields the following wavelet matrix coefficients:

$$\mathbf{d}_0 = \begin{pmatrix} \sqrt{2}/20 & 3\sqrt{6}/20 \\ 0 & 0 \end{pmatrix}, \quad \mathbf{d}_1 = \begin{pmatrix} -9\sqrt{3}/20 & \sqrt{6}/6 \\ 0 & -\sqrt{3}/3 \end{pmatrix},$$

$$\mathbf{d}_2 = \begin{pmatrix} 3\sqrt{3}/20 & -\sqrt{6}/20 \\ 3\sqrt{6}/10 & -\sqrt{3}/5 \end{pmatrix}, \quad \mathbf{d}_3 = \begin{pmatrix} -\sqrt{3}/60 & 0 \\ -\sqrt{6}/30 & 0 \end{pmatrix}.$$

The graphs of the wavelets ψ^0 and ψ^1 are depicted in Fig. 7.4. Next it is shown that neither component of the vector wavelet ψ constructed from the vector scaling function ϕ defined by Eqs. (7.77) and (7.78) can have support on $[0, 1]$. More precisely, the minimum length of the support of any of its components is at least $3/2$, and at least one component must have support larger than or equal to two.

Proposition 7.13 *Suppose that the scaling factors s_0 and s_1 admit a fundamental basis of $\mathfrak{C}$. Then there does not exist a vector wavelet whose components are supported on $[0, 1]$.*

Proof. Let $U_1 := V_1([0, 1])$ and let $U_0 := \{f \in U_1 \mid f(0) = f(1) = 0\}$. It is easy to see that $\dim U_0 = 5$ and $\dim U_1 = 3$. To simplify notation, the following convention is adopted:

$$\varphi^1 := \phi^1|_{[0,1]}, \qquad \varphi^2 := \phi^1(\cdot + 1)|_{[0,1]},$$

$$\varphi^3 := \psi^1|_{[0,1]}, \qquad \varphi^4 := \psi^1(\cdot + 1)|_{[0,1]}.$$

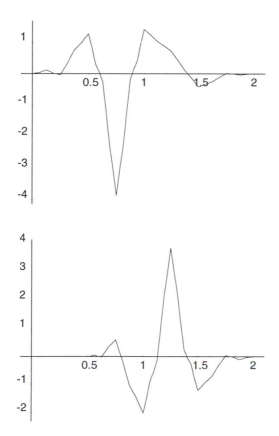

Figure 7.4: The orthogonal wavelets ψ^0 and ψ^1.

Now suppose ψ is a wavelet supported on $[0,1]$. Since the scaling functions and the wavelets are mutually orthogonal, span $\{\varphi^1, \varphi^3\}$, span$\{\varphi^2, \varphi^4\}$, span$\{\phi^0\}$, and span$\{\psi\}$ are mutually orthogonal spaces. If, in addition to φ^2 and φ^4 being linearly independent, φ^1 and φ^3 are also linearly independent, then span $\{\phi^0, \varphi^1, \varphi^2, \varphi^3, \varphi^4\}$ is five-dimensional and hence equals U_1. But then $\psi \in$ span $\{\phi^0, \varphi^1, \varphi^2, \varphi^3, f\varphi^4\}^\perp = \{0\}$, which is not permissible. Therefore, either φ^1 and φ^3 or φ^2 and φ^4 are linearly dependent. Assume that the former is true. Then two cases need to be considered: $\varphi^3 = 0$ or $\varphi^3 = \varphi^1$. (The latter can always be achieved by scaling ψ^1.) If $\varphi^3 = 0$, then $\varphi^4 \in U_0$, and thus $\{\phi^0, \varphi^4, \psi\}$ forms an orthogonal basis of U_0. Since φ^1 is orthogonal to these three vectors, $\varphi^1 \in U_0^\perp$. In the other case, $\varphi^5 := \varphi^2 - \varphi^4 \in U_0$. As $\phi^1 \neq \psi$, $\varphi^5 \neq 0$. Hence, $\{\phi^0, \varphi^5, \psi\}$ forms an orthogonal basis of U_0 and, again, $\varphi^1 \in U_0^\perp$. The conclusion of the theorem follows if we can show that it is not possible for φ^1 to be in $U_0^\perp$. To this end, let $\varphi^6 := \varphi^1(2 \cdot -1) = \phi^1(2 \cdot -1)|_{[0,1]}$. Note that $\varphi^6 \neq \varphi^1$ but $\varphi^6(0) = \varphi^1(0)$ and $\varphi^6(1) = \varphi^1(1)$. Since span $\{\phi^0(2 \cdot), \phi^0(2 \cdot -1), \phi^1(2 \cdot)\}$, it follows that $\varphi^6 \in U_0^\perp$. Now, $\varphi^1 - \varphi^6 \in U_0$ and thus $\varphi^1 = \varphi^6 + (\varphi^1 - \varphi^6) \notin U_0^\perp$. This contradiction shows that φ^1 and φ^3 cannot be linearly dependent.

A similar argument shows that φ^2 and φ^4 cannot be linearly dependent.

■

The preceding proposition implies the next theorem.

Theorem 7.17 *Let s_0 and s_1 be scaling factors that admit the existence of a fundamental basis of $\mathfrak{C}$, and let $\boldsymbol{\psi} = (\psi^0, \psi^1)^t$ be a vector wavelet. Then the support of one of the components of $\boldsymbol{\psi}$ must have length greater than or equal to 3/2, while the support of the other one must be of length greater than or equal to two.*

Proof. The preceding proposition shows that the support of each component ψ^a must have length at least $3/2$. It is also easy to realize that ψ^0 and ψ^1 cannot both be supported on $[0, 3/2]$. If this was the case, $\psi^0|_{[0,1/2]}$ and $\psi^1|_{[0,1/2]}$, or $\psi^0|_{[1,3/2]}$ and $\psi^1|_{[1,3/2]}$ woud have to be linearly dependent. But then a wavelet supported on $[0, 1]$ could be found by simply rotating the components. This contradiction proves the claim.

It remains to establish that there is a vector wavelet whose components are supported on $[0, 3/2]$ and $[0, 2]$, respectively. So suppose that supp $\psi^0 = [0, 3/2]$ and supp $\psi^1 = [0, 2]$. In this case, the wavelet matrix coefficients $\mathbf{d}_\ell$, $\ell = 0, 1, 2, 3$, take the form

$$\mathbf{d}_0 = \begin{pmatrix} d_1 & d_2 \\ 0 & 0 \end{pmatrix}, \qquad \mathbf{d}_1 = \begin{pmatrix} d_3 & d_4 \\ d_5 & d_6 \end{pmatrix},$$

$$\mathbf{d}_2 = \begin{pmatrix} d_7 & 0 \\ d_8 & d_8 \end{pmatrix}, \qquad \mathbf{d}_3 = \begin{pmatrix} 0 & 0 \\ d_{10} & d_{11} \end{pmatrix}.$$

Defining $\mathbf{a} := c_0 \mathbf{d}_2^\dagger + c_1 \mathbf{d}_3^\dagger$, $\mathbf{b} := c_2 \mathbf{d}_0^\dagger + c_3 \mathbf{d}_1^\dagger$, $\mathbf{c} := c_0 \mathbf{d}_0^\dagger + c_1 \mathbf{d}_1^\dagger + c_2 \mathbf{d}_2^\dagger + c_3 \mathbf{d}_3^\dagger$, and using the explicit form of the scaling function matrix coefficients $\mathbf{c}_\ell$, $\ell = 0, 1, 2, 3$, gives

$$A_{11} = 1/2(2s_0 + 1 - 2p)d_7, \quad A_{21} = 1/2(1 - 2p)(p - s_0)d_7,$$

$$B_{22} = (1 - q)(q - s_1)d_5, \quad C_{12} = 1/2(2s_1 + 1 - 2q)d_5.$$

Note that if one sets $d_7 = 0$, then the preceding equations yield $p = 1/2$ and $s_0 = 0$. These values for p and s_0 now imply that $1/2 = (1 + s_1^2)/(4 - s_1^2)$.

This last equation has, however, no solution for $|s_1| < 1$. Likewise, one can show that if d_5 were equal to zero, then s_0 would be outside the admissible domain. Therefore, $d_5 = d_7 = 0$. Now, in order for $\langle \psi^0, \psi^1 \rangle = 0$, $d_4 d_6 = 0$. Hence, one of them must have a support of length one, which is impossible by Proposition 7.13. ∎

Next the question of the symmetry of wavelets is briefly addressed. It was shown in [40] that the Haar wavelet is the only symmetric wavelet in the Daubechies family of continuous, compactly supported and orthogonal wavelets. One of the advantages of the fractal-geometric construction of wavelets over other constructions is that there do exist symmetric wavelets. The next theorem makes this statement more precise.

Theorem 7.18 *Let ϕ be a vector scaling function whose components are affine fractal functions with $s_0 = s_1$. Then there exists a set of interpolation points such that the components of the vector wavelet are symmetric, respectively, anti-symmetric affine fractal functions.*

Proof. Since $s_0 = s_1$, both scaling factors must be equal to $-1/5$. Furthermore, it is easy to see that, if $\psi := (\psi_s, \psi_a)^t$, ψ_s is symmetric to the line $x = 1$, while ψ_a is anti-symmetric with respect to $x = 1$. Bearing this in mind, one sets

$$y_1^s = (0, b_1^s, b_2^s)^t, \quad y_2^s = (b_2^s, b_3^s, b_4^s)^t, \quad y_3^s = (b_4^s, b_3^s, b_2^s)^t,$$

$$y_4^s = (b_2^s, b_1^s, 0)^t,$$

$$y_1^a = (0, b_1^a, b_2^a)^t, \quad y_2^a = (b_2^a, b_3^a, 0)^t, \quad y_3^a = (0, -b_3^a, -b_2^a)^t,$$

and

$$y_4^a = (-b_2^a, -b_1^a, 0)^t.$$

Then Eq. (7.104) is automatically satisfied and Eqs. (7.100) —(7.103) give

$$b_1^s = 1, \quad b_2^s = -30, \quad b_3^s = 111, \quad b_4^s = -100,$$

and

$$b_1^a = 1, \quad b_2^a = -30, \quad b_3^a = 81.$$

If the resulting wavelets ψ^s and ψ^a are normalized, then the wavelet matrix coefficients $\{\mathbf{d}_\ell\}$ are given by

$$\mathbf{d}_0 = \begin{pmatrix} -1/20 & -3\sqrt{2}/20 \\ -\sqrt{2}/20 & -3/10 \end{pmatrix}, \quad \mathbf{d}_1 = \begin{pmatrix} 9/20 & -\sqrt{2}/2 \\ 9\sqrt{2}/20 & 0 \end{pmatrix},$$

$$\mathbf{d}_2 = \begin{pmatrix} 9/20 & -3\sqrt{2}/20 \\ -9\sqrt{2}/20 & 3/10 \end{pmatrix}, \quad \mathbf{d}_3 = \begin{pmatrix} -1/20 & 0 \\ \sqrt{2}/20 & 0 \end{pmatrix}.$$

∎

The graphs of ψ_s and ψ_a are shown in Fig. 7.5.

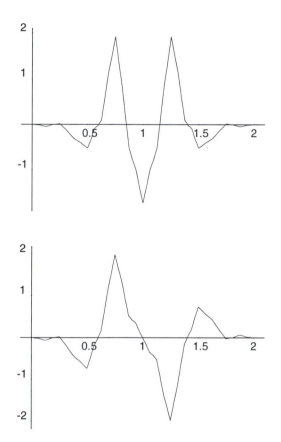

Figure 7.5: The symmetric and antisymmetric wavelets ψ_s and ψ_a.

Recently, wavelets have been used as approximating and test functions for Galerkin-type finite element methods to obtain weak solutions to boundary value problems. (An incomplete list of references is [81, 115, 132].) In this particular setting the wavelet expansion of the approximating weak solution has to be restricted to a compact interval. Hence, there arises the need to consider the restriction of scaling functions and wavelets to compact intervals. Next, it is shown that unlike other wavelets, the vector wavelets constructed via the fractal-geometric approach are very wellsuited for compact intervals.

For this purpose, let

$$\bar{V}_k := V_k \cap L^2([0,1]), k \in \mathbb{Z}, \quad \text{and} \quad \bar{\phi}^a_{k\ell} := \phi^a_{k\ell}|_{[0,1]}, \quad a = 0, 1.$$

Then it is straightforward to show that $\{\bar{V}_k\}_{k \in \mathbb{Z}}$ is a multiresolution analysis on $L^2([0,1])$ and that $\{\bar{\phi}^a_{k\ell} \mid a = 0, 1 \wedge \ell \in \mathbb{Z}\}$ is an orthonormal basis for $\bar{V}_k$. Now define

$$\phi^a_k := \phi^a|_{[k-1,k]} \quad \text{and} \quad \psi^a_k := \psi^a|_{[k-1,k]}, \quad a = 0, 1; \quad k = 1, 2.$$

Let $R := \begin{pmatrix} a & b \\ -b & a \end{pmatrix}$, $|a|^2 + |b|^2 = 1$, be a rotation in $\mathbb{R}^2$ so that

$$\psi^+ := R \circ \psi \tag{7.107}$$

satisfies

$$\langle \psi^{+,0}_k, \phi^1_k \rangle = 0, \qquad k = 1, 2. \tag{7.108}$$

To see that such a rotation exists, note that the preceding equation is equivalent to requiring

$$a\langle \psi^0_1, \phi^1_1 \rangle + b\langle \psi^1_1, \phi^1_1 \rangle = 0$$

and

$$a\langle \psi^0_2, \phi^1_2 \rangle + b\langle \psi^1_2, \phi^1_2 \rangle = 0.$$

These last two equations are not linearly independent, since

$$\langle \psi^0, \phi^1 \rangle = \langle \psi^0_1, \phi^1_1 \rangle + \langle \psi^0_2, \phi^1_2 \rangle = 0$$

and

$$\langle \psi^1, \phi^1 \rangle = \langle \psi^1_1, \psi^1_1 \rangle + \langle \psi^1_2, \phi^1_2 \rangle = 0.$$

Thus, there exist numbers a and b so that Eq. (7.108) is satisfied. Define

$$\bar{\psi}^0_{k\ell} := \begin{cases} 0, & \psi^{+,0}_{k\ell} \cap [0,1]^c \neq 0, \\ \psi^{+,0}_{k\ell}, & \text{otherwise,} \end{cases} \quad \text{and} \quad \bar{\psi}^1_{k\ell} := \psi^1_{k\ell}|_{[0,1]}. \tag{7.109}$$

If $\bar{W}_k := \bar{V}_{k+1} \ominus \bar{V}_k$, then it follows immediately that the collection $\{\bar{\psi}_{k\ell} := (\bar{\psi}^0_{k\ell}, \bar{\psi}^1_{k\ell})^t\}$, $k, \ell \in \mathbb{Z}$, forms an orthonormal basis of $\bar{W}_k$. These arguments now establish

Theorem 7.19 *The collection $\{\bar{\phi}^a_{k\ell} \mid a = 0, 1 \wedge k, \ell \in \mathbb{Z}\}$ is an orthonormal basis for $\bar{V}_k = V_k \cap L^2([0,1])$, while $\{\bar{\psi}^a_{k\ell} \mid a = 0, 1 \wedge k, \ell \in \mathbb{Z}\}$ forms an orthonormal basis for $\bar{W}_k = \bar{V}_{k+1} \ominus \bar{V}_k$. Furthermore, $L^2([0,1]) = \bigoplus_{k \in \mathbb{Z}} \bar{W}_k$.* ∎

Remark. In the case $s_0 = s_1 = -1/5$, it is easy to see that $\bar{\psi}^0 = \psi_s|_{[0,1]}$ and

$$\bar{\psi}^1 = \begin{cases} \psi_a|_{[0,1]}, & \text{supp } \psi_a \subseteq [0,1], \\ 0, & \text{otherwise.} \end{cases}$$

7.4 N-refinable Scaling Functions

The results in the previous subsections can be modified to construct vector scaling functions and vector wavelets that satisfy the slightly more general matrix dilation equations

$$\phi = \sum_{\ell \in \mathbb{Z}} \mathbf{c}_\ell \phi(N \cdot -\ell) \quad \text{and} \quad \psi = \sum_{\ell \in \mathbb{Z}} \mathbf{d}_\ell \psi(N \cdot -\ell), \qquad (7.110)$$

with $2 < N \in \mathbb{N}$. In this setting the space $\mathfrak{C} = \theta \left(\prod_{j=0}^{N-1} \mathcal{P}^1 \right) \cap C_{\mathbb{R}}(I)$ has dimension $N + 1$. Let $\{ e_{y_i} \}_{i=0}^N$ be a basis of $\mathfrak{C}$ with $y_i := \varepsilon_{i+1}$, $i = 0, \dots, N-1$, and $y_N := (1, p_1, p_2, \dots, p_{N-1}, 0)^t$. Here $\{ \varepsilon_i \}_{i=0}^N$ denotes the canonical basis of $\mathbb{R}^{N+1}$. The parameters p_j and the scaling factors s_j, $j = 0, 1, \dots, N$, have to be adjusted in such a way that $e_{y_N} \neq 0$ and $\langle e_{y_N}, e_{y_j} \rangle = 0$, for all $j = 0, 1, \dots, N-1$. Once this has been accomplished, orthogonal scaling functions ϕ^{*a}, $a = 0, 1, \dots, N-1$ can be obtained by applying the Gram-Schmidt procedure to the set $\{ e_{y_j} \}_{j=0}^{N-2}$. Since the e_{y_j}, $j = 0, 1, \dots, N-2$, are continuous functions vanishing at $x = 0$ and $x = 1$, the scaling functions ϕ^{*a} will also be continuous and supported on $[0, 1]$. The scaling function ϕ^{*N-1} can be obtained by subtracting from $e_{y_{N-1}}$ its projection onto the subspace spanned by $\{ \phi^{*a} \}_{a=0}^{N-2}$ and piecing it together continuously with ϕ^{*N}, as in the case considered in the previous subsection. (That it suffices to only consider this special basis for $\mathfrak{C}$ follows from Proposition 7.9.)

Since there are N orthogonality relations and $2N - 1$ unknowns, p_j, $j = 1, \dots, N-1$, and s_j, $j = 0, 1, \dots, N-1$, it may be possible to impose other desirable conditions besides orthogonality and still obtain a basis $\{ \phi^{*a} \}_{a=0}^N$.

In order for the orthogonality relations $\langle e_{y_N}, e_{y_j} \rangle = 0$, $j = 0, 1, \dots, N-1$, to hold, the functions $\lambda_j^a = b_j^a \mathrm{id}_{\mathbb{R}} + c_j^a$, $a = 0, 1, \dots, N$, need to be calculated. Using Eqs. (5.16) — (5.19) yields

$$b_j^a = \delta_{aj} - \delta_{a+1,j}, \qquad c_j^a = \delta_{a+1,j}, \qquad a = 0, 1, \dots, N-2,$$

$$b_j^{N-1} = \delta_{j,N-1} - s_j, \qquad c_j^{N-1} = 0,$$

and

$$b_j^N = p_{j+1} - p_j + s_j, \qquad c_j^N = p_j - s_j,$$

for $j = 0, 1, \dots, N-1$. One proceeds as before by examining conditions on $\mathfrak{C}$ that guarantee the existence of compactly supported scaling functions. The wavelets are then constructed as in the previous section. The next result is a generalization of Proposition 7.9.

Proposition 7.14 *Let V be an N-dimensional linear subspace of $C^0([0,1])$ such that there does not exist an orthonormal basis of V with $N - 1$ of the basis functions vanishing at $x = 0$ or $x = 1$, or vice versa. Then there do not exist continuous and compactly supported functions composed of linear combinations of basis elements of V and their integer-translates so that the set $F := \{\phi^a(\cdot - \ell)|_{[0,1]} \mid a = 0, 1, \ldots, N-1 \wedge \ell \in \mathbb{Z}\}$ spans V and $\langle \phi^a, \phi^{a'} \rangle = \delta_{aa',0\ell}$.*

Proof. The proof uses induction on N and can be found in [54]. ∎

For illustrative purposes the case $N = 3$ and $s_0 = s_2$ is briefly considered. Applying essentially the same type of arguments as in the previous subsection gives the following result. (Here and in the remainder of this section the reader is referred to [54] for detailed but lengthy calculations and proofs.)

Proposition 7.15 *Continuous, compactly supported, and orthogonal scaling functions and wavelets can be constructed from elements of $\mathfrak{C}$ and its integer-translates iff $|s_i| < 1$, $i = 0.1$, and $\bar{\mathsf{p}}(s_0, s_1) = 0$, where*

$$
\begin{aligned}
\bar{\mathsf{p}}(s_0, s_1) \quad := \quad & 48s_0^2 - 256s_0^3 + 16s_0^2 s_1^2 + 192s_0^2 s_1 + 400s_0^2 \\
& + 16s_0 s_1^3 - 96s_0 s_1^2 - 272s_0 s_1 - 192s_0 + s_1^4 \\
& + 16s_1^3 + 70s_1^2 + 48s_1 + 9.
\end{aligned}
$$

Proof. See [54]. ∎

In the case $s_0 = s_2 = 3/7$ and $s_1 = -3/7$, the graphs of the three scaling functions ϕ^a, $a = 0, 1, 2$, and the six associated wavelets ψ^b, $b = 0, 1, \ldots, 5$, are depicted in Fig. 7.6 and Fig. 7.7, respectively.

Finally, the following setup is considered. Let $s_j = 0$, for all $j = 0, 1, \ldots, N - 2$, and $s_{N-1} =: s$.

Theorem 7.20 *Let $2 < N \in \mathbb{N}$ and let $s = (1/2)(-B + \sqrt{B^2 - N}$, where $B := (1/11)(N + 1)([-2\sqrt{3}]^{N-1}[1 - 4\sqrt{3}] + [-2\sqrt{3}]^{3-N}[1 + 4\sqrt{3}])$. Then there exist affine fractal scaling functions $\{\phi^a\}_{a=0}^N$ and affine fractal wavelets $\{\psi^a\}_{a=0}^N$ supported on $[0,2]$ satisfying the N-scale dilation equations*

$$
\phi = \sum_{\ell \in \mathbb{Z}} \mathbf{c}_\ell \phi(N \cdot -\ell) \quad and \quad \psi = \sum_{\ell \in \mathbb{Z}} \mathbf{d}_\ell \psi(N \cdot -\ell),
$$

where the $\{\mathbf{c}_\ell\}_{\ell \in \mathbb{Z}}$ and $\{\mathbf{d}_\ell\}_{\ell \in \mathbb{Z}}$ are $N \times N$ matrices.

Proof. The reader is referred to [54] for the proof. ∎

A similar result can be proven for $s_j := s$, for all $j = 0, 1, \ldots, N$. The interested reader is again referred to [54] for more details and complete proofs.

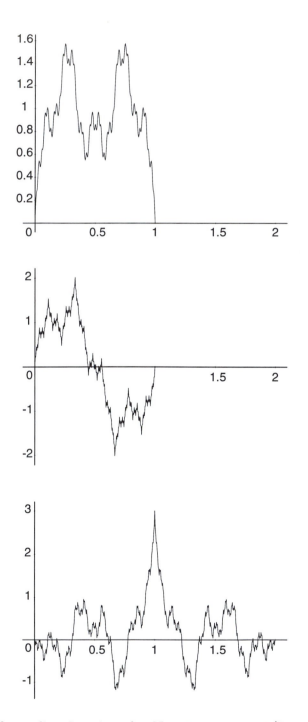

Figure 7.6: The scaling functions for $N = 3$, $s_0 = s_2 = 3/7$, and $s_1 = -3/7$.

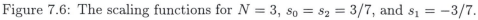

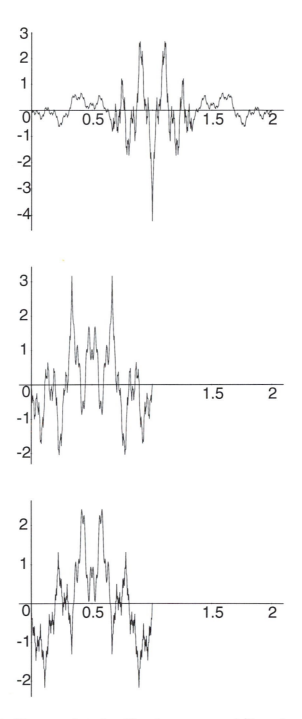

Figure 7.7: The wavelets for $N = 3$, $s_0 = s_2 = 3/7$, and $s_1 = -3/7$.

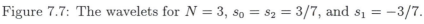

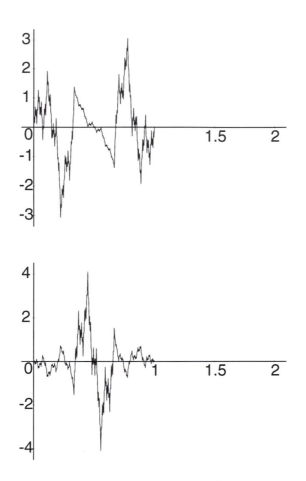

Figure 7.7: The wavelets for $N = 3$, $s_0 = s_2 = 3/7$, and $s_1 = -3/7$ (continued).

Chapter 8

Fractal Surfaces

This chapter deals with $\mathbb{R}^m$-valued multivariate continuous fractal functions $\boldsymbol{f} : X \subseteq \mathbb{R}^n \to \mathbb{R}^m$ and some of their properties. The existence and construction of these multivariate functions is already implicitly contained in Theorems 5.1, 5.2, and 5.3; simply choose X to be a non-empty compact subset of $\mathbb{R}^n$ and $Y := \mathbb{R}^m$. Here, however, the issues have to be re-addressed. There are two reasons for this: Firstly, the more complex geometry is hidden within the construction and needs to be investigated more closely. Secondly, the graphs of these $\mathbb{R}^m$-valued multivariate continuous fractal functions, the so-called *fractal surfaces*, can be used to construct wavelet bases in $\mathbb{R}^n$. This construction is based on results from the theory of *Coxeter groups*, and certain issues involving the geometry of fractal surfaces need to be clarified. The next chapter will deal exclusively with this last question.

Fractal surfaces constructed via iterated function systems were first introduced in [128]. A slightly more general construction of such fractal surfaces was later presented in [76], and in [86] $\mathbb{R}^m$-valued multivariate fractal functions $\boldsymbol{f} : X \subseteq \mathbb{R}^n \to \mathbb{R}^m$ were investigated. The latter two constructions use recurrent iterated function systems. Of course, it is always possible to construct fractal surfaces as *tensor products* of univariate continuous fractal functions. However, these tensor product fractal surfaces lack most of the exciting features of the aforementioned fractal surfaces. S. Dubuc and his co-workers also have constructed fractal surfaces using a multidimensional iterative interpolation process.

The first section in this chapter introduces tensor product fractal surfaces. The construction of non-tensor product fractal surfaces is presented next. This construction is based on recurrent iterated function systems and emphasizes the interpolatory nature of fractal surfaces in $\mathbb{R}^{n+m}$. The special

case $m = 1$ is then considered as an illustrative example. Finally, a few com-
ments are made about Dubuc's fractal surfaces. In the third section properties
of fractal surfaces such as Hölder continuity, oscillation, box dimension, and
regularity are discussed.

8.1 Tensor Product Fractal Surfaces

Here it is shown how the tensor product can be used to obtain fractal-like
surfaces from univariate $\mathbb{R}$-valued continuous fractal functions.

Before commencing with the construction, the definition of tensor product
is needed. The definition that will be given is not the most general one, but
it is more than adequate for our purposes.

Definition 8.1 Let X and Y be vector spaces over $\mathbb{K}$ of dimension n and
m, respectively. With the spaces X and Y is associated an nm-dimensional
vector space $X \otimes Y$ in the following way: Let $x \in X$ and $y \in Y$, and let
$\theta : X \times Y \to X \otimes Y$, $(x, y) \overset{\theta}{\longmapsto} x \otimes y$. This mapping θ is required to have the
following three properties:

(a) (Distributive Law) If $x, x_1, x_2 \in X$ and $y, y_1, y_2 \in Y$, then

$$x \otimes (y_1 + y_2) = x \otimes y_1 + x \otimes y_2,$$

$$(x_1 + x_2) \otimes y = x_1 \otimes y + x_2 \otimes y.$$

(b) (Associative Law) If $x \in X$, $y \in Y$, and $\alpha \in \mathbb{K}$, then

$$\alpha x \otimes y = x \otimes \alpha y = \alpha (x \otimes y).$$

(c) If $\{x_1, \ldots, x_n\}$ and $\{y_1, \ldots, y_m\}$ are bases of X and Y, respectively, then
the nm elements

$$x_i \otimes y_j, \quad i = 1, \ldots, n, \quad j = 1, \ldots, m,$$

are to form a basis of $X \otimes Y$.

If such a mapping exists, then the vector space $X \otimes Y$ is called the tensor
product of X with Y.

An example of a tensor product is provided by the linear spaces $\mathcal{P}^d(\mathbb{R})$.
Suppose that $d = 1$. Then $\{e_1 := 1, e_2 := \mathrm{id}_{\mathbb{R}}\}$ is a basis for the vector space

$\mathcal{P}^1(\mathbb{R})$. The tensor product $\mathcal{P}^1(\mathbb{R}) \otimes \mathcal{P}^1(\mathbb{R})$ is then the vector space of all bivariate real polynomials p of the form

$$p(x, y) = \sum_{i=0}^{1} \sum_{j=0}^{1} a_{ij} x^i y^j,$$

where $x^i y^j := e_i \otimes e_j(x, y)$ and the a_{ij}, $i, j = 0, 1$, are real numbers.

In what follows the notation and terminology of Section 7.2 is used. Suppose that N is an integer greater than one. Let $X := [x_0, x_N]$ be a non-empty compact interval of $\mathbb{R}$, and let ${}^1\Lambda$ and ${}^2\Lambda$ be finite-dimensional dilation-invariant subspaces of B^N. Denote by ${}^1\Lambda_r$, respectively ${}^2\Lambda_s$, the space $\theta \left({}^1\Lambda\right) \cap C_{\mathbb{R}}^r(X)$, respectively, $\theta \left({}^2\Lambda\right) \cap C_{\mathbb{R}}^s(X)$, $r, s \in \mathbb{N}_0$. Furthermore, let $\{f^1, \ldots, f^n\}$ and $\{g^1, \ldots, g^m\}$ be bases for these spaces, and let f and g be fractal functions generated by the maps $\lambda^f = (\lambda_0^f, \lambda_1^f, \ldots, \lambda_{N-1}^f) \in {}^1\Lambda$ and $\lambda^g = (\lambda_0^g, \lambda_1^g, \ldots, \lambda_{N-1}^g) \in {}^2\Lambda$, respectively, i.e.,

$$f = \sum_{i=1}^{n} \alpha_i \, f^i \qquad \text{and} \qquad g = \sum_{j=1}^{m} \beta_j \, g^j.$$

Then the tensor product of f with g, $f \otimes g$, is the bivariate fractal function

$$f \otimes g = \sum_{i=1}^{n} \sum_{j=1}^{m} \alpha_i \beta_j \, f^i \otimes g^j. \tag{8.1}$$

The graph of the fractal function $f \otimes g$ is called a *tensor product fractal surface*.

Example 8.1 Let $N := 2$, let $X := [0, 1]$, let ${}^1\Lambda := \prod_{k=0}^{1} \mathcal{P}^1 =: {}^2\Lambda$, and let $r := 0 =: s$. Then the fractal functions f and g, as defined earlier, are affine. The space $\theta \left(\prod_{k=0}^{1} \mathcal{P}^1\right) \cap C_{\mathbb{R}}(X)$ is then three-dimensional, and a basis is, for instance, given by $\{f^1, f^2, f^3\}$, where f^i is the affine fractal function that interpolates the points $(0, \delta_{1i})$, $(1/2, \delta_{2i})$, and $(1, \delta_{3i})$. If the scaling factors are denoted by s_0 and s_1, then the following expressions for the functions $\lambda \in B^2$ are obtained:

$$f^1: \qquad \lambda^1 = \left((s_0 - 1)(\mathrm{id}_{\mathbb{R}} - 1), s_1(\mathrm{id}_{\mathbb{R}} - 1)\right),$$

$$f^2: \qquad \lambda^2 = (\mathrm{id}_{\mathbb{R}} - s_0, -\mathrm{id}_{\mathbb{R}} + 1 - s_1),$$

and

$$f^3: \qquad \lambda^3 = \left(-s_0(\mathrm{id}_{\mathbb{R}} + 1), (1 - s_1)\mathrm{id}_{\mathbb{R}} - s_1\right).$$

The tensor products $f^i \otimes f^j$ satisfy the functional equations

$$f^i \otimes f^j = \sum_k^1 \left(\lambda_k^i \circ u_k^{-1} + s_k f^i \circ u_k^{-1} \right) \left(\lambda_k^j \circ u_k^{-1} + s_k f^j \circ u_k^{-1} \right) \chi_{u_k X},$$

with $u_k = (1/2)\, \mathrm{id}_{\mathbb{R}} + (k-1)$.

Note that if (x, y, z) is a coordinate system in $\mathbb{R}^3$ and if the graph of $f \otimes f$ is analytically represented by $z = f \otimes f(x, y)$, then the intersection of graph $f \otimes f$ with the hyperplane $y = c$, $c \in [0, 1]$, is a rescaled verson of graph $f \otimes f$.

8.2 Affine Fractal Surfaces in $\mathbb{R}^{n+m}$

In this section non-tensor product fractal surfaces defined on certain polyhedra in $\mathbb{R}^n$ are introduced. The construction of such surfaces is given in a general setting, that is, recurrent iterated function systems are used. Although some of the first results stated in this section follow from earlier theorems in Chapter 7, their proofs are repeated here. The reason for this repetition lies in the fact that these proofs explicitly use the underlying geometry of the construction, an issue that will be taken up again in more detail and more generality in the next chapter. But first a few definitions are needed.

Definition 8.2 Let $\{q_0, q_1, \ldots, q_n\}$ be a set of geometrically independent points in $\mathbb{R}^k$, $k \geq n$. The n-dimensional geometric simplex or n-simplex, denoted by σ^n, is defined as

$$\sigma^n = \left\{ x \in \mathbb{R}^n \,\middle|\, \exists\, \alpha_0, \alpha_1, \ldots, \alpha_n \in \mathbb{R}^+ : x = \sum_{j=0}^n \alpha_j q_j \wedge \sum_{j=0}^n \alpha_j = 1 \right\}. \quad (8.2)$$

The points $q_0, q_1, \ldots q_n$ are called the vertices of the n-simplex.

Definition 8.3 Two simplices σ^n and σ^m are called properly joined if either $\sigma^n \cap \sigma^m = \emptyset$ or $\sigma^n \cap \sigma^m$ is a face of both σ^n and σ^m.

Definition 8.4 A simplicial complex in $\mathbb{R}^n$ is a finite family τ^n of properly joined p-geometric simplices σ^p, $p \leq n$, such that each face a member of τ^n is also a member of τ^n.

A simplex σ^m is called a *face* of the simplex σ^n, $m \leq n$, if each vertex of σ^m is also a vertex of σ^n. The n-simplex σ^n endowed with the Euclidean subspace topology of $\mathbb{R}^k$ is denoted by $|\sigma^n|$ and is called the *geometric carrier of σ^n*.

8.2.1   The construction

Let X be a polyhedron made up of finitely many n-simplices $\sigma_i^n \subseteq \mathbb{R}^n$, $i = 1, \ldots, N$, where $1 \neq N \in \mathbb{N}$. Denote by Q the set of all vertices $q_j \in X$, $j = 1, \ldots, M$ of X. Let $\Delta := \{(q_j, z_j) \in X \times \mathbb{R}^m \mid j = 1, \ldots, M\}$ be an interpolating set. Let τ_k^n be an n-simplicial complex in X that is a union of some of the σ_k^n, $k = 1, \ldots, K$. After relabelling — if necessary — the vertices of τ_k^n are denoted by $q_1, \ldots, q_L$. A function $\ell : \{1, \ldots, M\} \to \{1, \cdots, L\}$ is called a *labelling map* if whenever $q_{j_1}, \ldots, q_{j_{n+1}}$ are the vertices of some σ_i^n, then $q_{\ell(j_1)}, \ldots, q_{\ell(j_{n+1})}$ are the vertices of some τ_k^n.

Now let $\boldsymbol{u}_i : \mathbb{R}^n \to \mathbb{R}^n$ be the unique affine map such that

$$\boldsymbol{u}_i(q_{\ell(j)}) = q_j, \qquad \text{for all } q_j \in \sigma_i^n, \tag{8.3}$$

$i = 1, \ldots, N$. The maps $\boldsymbol{u}_i$ can be represented as

$$\boldsymbol{u}_i(x) = A_i x + D_i, \tag{8.4}$$

where $A_i \in M_{nn}$, the algebra of all $n \times n$ matrices over $\mathbb{R}$, and $D_i \in \mathbb{R}^n$. Let $B \in M_{mm}$ and suppose that the spectral radius s of B is less than one. Note that there exists a norm $\| \cdot \|_B$ on $\mathbb{R}^m$ such that the induced matrix norm of B equals s. Let $\boldsymbol{v}_i : \mathbb{R}^n \times \mathbb{R}^m \to \mathbb{R}^m$ be the unique affine map of the form

$$\boldsymbol{v}_i(x, y) = C_i x + B y + E_i, \tag{8.5}$$

where $C_i \in M_{mn}$, $E_i \in \mathbb{R}^m$ are such that

$$\boldsymbol{v}_i(q_{\ell(j)}, z_{\ell(j)}) = z_j, \tag{8.6}$$

for all j such that $q_j \in \sigma_i^n$, and for all $i = 1, \ldots, N$. Let $C^*(X, \mathbb{R}^m) := \{ \boldsymbol{f} \in C(D, \mathbb{R}^m) \mid \boldsymbol{f}(q_j) = z_j \wedge j = 1, \ldots, M \}$. Define a norm $\| \cdot \|_\infty$ on $C^*(X, \mathbb{R}^m)$ by $\|\boldsymbol{f}\|_\infty := \sup\{\|\boldsymbol{f}(x)\|_B \mid x \in X\}$ and let $\Phi : C^*(X, \mathbb{R}^m) \to C(D, \mathbb{R}^m)$ be the Read-Bajraktarević operator defined by

$$(\Phi \boldsymbol{f})(x) = \sum_{i=1}^{N} \boldsymbol{v}_i(\boldsymbol{u}_i^{-1}(x), \boldsymbol{f} \circ \boldsymbol{u}_i^{-1}(x)) \, \chi_{\sigma_i^n}. \tag{8.7}$$

Theorem 8.1 *The mapping* Φ *in (8.7) is well-defined, maps the space* $C^*(X, \mathbb{R}^m)$ *into itself, and is contractive in* $\| \cdot \|_\infty$.

Proof. Clearly $\Phi(\boldsymbol{f})$ is continuous on each σ_i^n. Let $\phi_i := \Phi(\boldsymbol{f})\big|_{\sigma_i^n}$, $i = 1, \ldots, N$. Suppose σ_i^n and $\sigma_{i'}^n$ intersect along a face, i.e., $\sigma_i^n \cap \sigma_{i'}^n = F$, where F is a p-simplex with $p < n$. To prove that Φ is well-defined, it suffices to show that $\phi_i(F) = \phi_{i'}(F)$. Note that $\phi_i(q_j) = z_j = \phi_{i'}(q_j)$ for each vertex $q_j \in F$. But (8.4) and the fact that each $x \in F$ is a linear combination of the vertices of F imply that $\phi_i(x) = \phi_{i'}(x)$, for all $x \in F$. Equations (8.4) and (8.6) imply that $\Phi(\boldsymbol{f})(q_j) = z_j$, $j = 1, \ldots, M$. Therefore, Φ maps $C^*(X, \mathbb{R}^m)$ into itself.

Now let $\boldsymbol{f}, \boldsymbol{g} \in C^*(X, \mathbb{R}^m)$. Then

$$\|\Phi(\boldsymbol{f}) - \Phi(\boldsymbol{g})\|_\infty =$$

$$= \sup_{1 \le i \le N} \{\|\boldsymbol{v}_i(\boldsymbol{u}_i^{-1}(x), \boldsymbol{f}(\boldsymbol{u}_i^{-1}(x))) - \boldsymbol{v}_i(\boldsymbol{u}_i^{-1}(x), \boldsymbol{g}(\boldsymbol{u}_i^{-1}(x)))\|_\infty\}$$

$$= \sup_{1 \le i \le N} \{\|B(\boldsymbol{f}(\boldsymbol{u}_i^{-1}(x)) - \boldsymbol{g}(\boldsymbol{u}_i^{-1}(x)))\|_B \mid x \in X\} \le s\,\|\boldsymbol{f} - \boldsymbol{g}\|_\infty.$$

∎

The unique fixed point $\boldsymbol{f}^* \in C^*(X, \mathbb{R}^m)$ is called *an $\mathbb{R}^m$-valued multivariate (continuous) affine fractal function* and its graph an *affine fractal surface in $\mathbb{R}^{n+m}$*. Note that $\boldsymbol{f} \in \mathcal{RF}(X, \mathbb{R}^m)$. Furthermore, the graph G^* of $\boldsymbol{f}^*$ is the attractor of the recurrent IFS $(X \times \mathbb{R}^m, \mathbf{w})$, where

$$\boldsymbol{w}_i \begin{pmatrix} x \\ y \end{pmatrix} = \begin{pmatrix} \boldsymbol{u}_i(x) \\ \boldsymbol{v}_i(x, y) \end{pmatrix} \qquad (i = 1, \ldots, N). \tag{8.8}$$

The fractal surfaces constructed previously can be used to define *hidden variable fractal surfaces*. To this end, assume that $\boldsymbol{u}_i = A_i x + D_i$, where A_i is a similitude of norm $\|A_i\| = a_i$, and $\boldsymbol{v}_i(x, y) = C_i x + By + E_i$, where $B = s\Theta$ for an isometry Θ, $i = 1, \ldots, N$. Consider the components of $\boldsymbol{f}^* = (f_1^*, \ldots, f_m^*)^t$. The graph of f_j^* is the orthogonal projection of $\boldsymbol{f}^*$ onto $\mathbb{R}^n \times 0 \times \ldots \times \mathbb{R} \times \ldots \times 0$, where the factor $\mathbb{R}$ is in the jth position. Since f_j^* still depends continuously on all the variables, it is referred to as a *hidden variable multivariate fractal function*, and its graph is called a *hidden variable fractal surface*.

Example 8.2 The sequence of pictures in Fig. 8.1 and Fig. 8.2 displays the two projections of a fractal function $\boldsymbol{f}^*: \sigma^2 \to \mathbb{R}^2$ with $Q := \{(j/3, k/3) \mid j + k \le 3; j, k = 0, 1, 2, 3\}$, $z_{jk} := (\sin(\frac{j}{3}), \cos(\frac{k}{3}))$, and $s = 4/5$.

8.2.2 $\mathbb{R}$-valued affine fractal surfaces

Next $\mathbb{R}$-valued affine fractal surfaces are considered. For this purpose let $X := \sigma^2$ and let $(X \times \mathbb{R}, \mathbf{w})$ be simply an IFS.

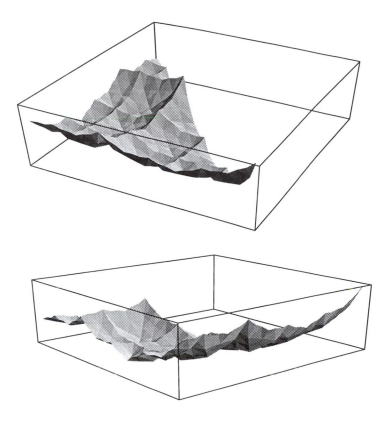

Figure 8.1: Two views of the projection f_1^* of f^*.

In this special setting, the 2-simplex σ^2 is to be subdivided into N non-degenerate subsimplices σ_i^2 such that $\bigcup \sigma_i^2 = \sigma^2$ and $\partial \sigma^2 = \sum_i^N \partial \sigma_i^2$, i.e., if $\sigma_i^2 \cap \sigma_{i'}^2 = F_{ii'} \subset \sigma^2$, $i \neq i'$, then $F_{ii'}$ has opposite orientations in σ_i^2 and $\sigma_{i'}^2$, $i, i' \in \{1, \ldots, N\}$. With the 2-simplex σ^2 and its collection of subsimplices $\{\sigma_i^2\}_{i=1}^N$ is associated a finite graph G whose vertices are the distinct vertices of $\bigcup_{i=1}^N \sigma_i^2$ and whose edges are the distinct (non-oriented) faces in this union. The main issue is to find conditions on G that guarantee that the Read-Bajraktarević operator $\Phi : C^*(\sigma^2, \mathbb{R}) \to \mathbb{R}^{\sigma^2}$,

$$f \xmapsto{\Phi} v_i(u_i^{-1}, f \circ u_i^{-1}),$$ (8.9)

is well-defined along adjacent faces $F_{ii'} = \sigma_i^2 \cap \sigma_{i'}^2$ in σ^2.

Proposition 8.1 *Suppose that the graph* G *associated with the 2-simplex* σ^2 *and its subdivision* $\{\sigma_i^2\}$ *has chromatic number three. Then there exists a*

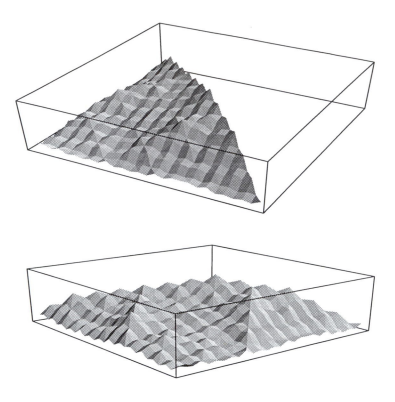

Figure 8.2: Two views of the projection f_2^* of f^*.

labelling map $\ell : \{1, \ldots, V\} \to \{1, \ldots, V\}$ *such that*

$$u_i(q_{\ell(j)}) = q_j \qquad and \qquad v_i(q_{\ell(j)}, z_{\ell(j)}) = z_j,$$

for all $q_j \in \mathsf{V}$, $i = 1, \ldots, N$. *Here* V *denotes the vertex set of* G *and* V *its cardinality.*

Remark. Recall that the *chromatic number* of a graph G is the the least number of symbols needed to label the vertices of G in such a way that any two vertices that are joined by an edge have distinct symbols.

Proof. The proof is easy and left to the reader. (See also [76].)   ∎

In Fig. 8.4 shows such a fractal surface. The scaling factor s was chosen as $1/2$. The domain of definition is the 2-simplex σ^2 which is shown together with its subdivision in Fig. 8.3.

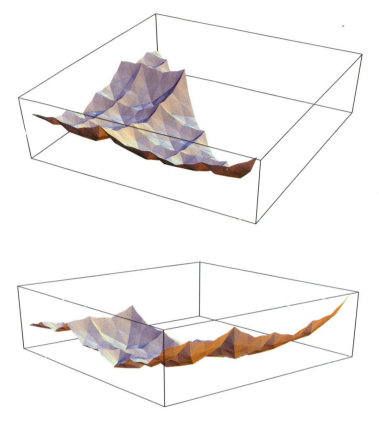

Figure 8.1.   Two views of the projection f_1^* of f^*.

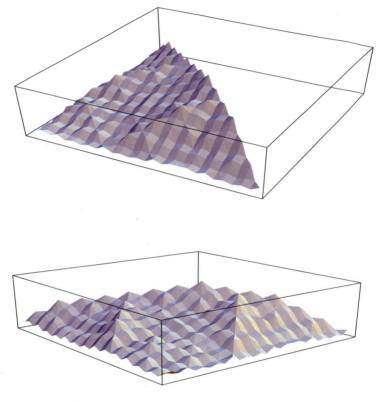

Figure 8.2.   Two views of the projection f_2^* of f^*.

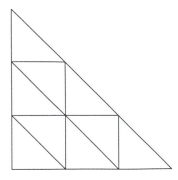

Figure 8.3: The simplex σ^2 and its subdivison.

The preceding construction shows that in order for the Read-Bajraktarević operator Φ to be well-defined, there can only be one free parameter $|s| < 1$. Next it is shown that there is another construction that allows more free parameters but, unfortunately, requires the boundary data of the interpolating set to be co-planar (also [128]).

Suppose that $m = 1$, $X = \sigma^2$, $(X \times \mathbb{R}, \mathbf{w})$ is an IFS, and Π is a non-vertical hyperplane in $\mathbb{R}^3$. Let $C^{**}(\sigma^2, \mathbb{R})$ denote the collection of all functions $f \in C(\sigma^2, \mathbb{R})$ such that $(x, f(x)) \in \Pi$ for all $x \in \partial\sigma^2$. Furthermore, suppose there exist affine contractions $u_i : \sigma^2 \to u_i \sigma^2 =: \sigma_i^2$ such that $\{\sigma_i^2 \mid i = 1, \ldots, N\}$ is a partition of σ^2. Let $\{q_j \mid j = 1, \ldots, M\}$ be the collection of all distinct vertices of $\bigcup_{i=1}^{N} \sigma_i^2$ with q_1, q_2, and q_3 being the vertices of σ^2. Define a labelling map $\ell : \{1, \ldots, N\} \times \{1, 2, 3\} \to \{1, \ldots, M\}$ by requiring that $\{q_{\ell(i,j)} \mid j = 1, 2, 3\}$ are the vertices of σ_i^2 given that $\{q_j \mid j = 1, 2, 3\}$ are the vertices of σ^2. Let $\Delta := \{(q_j, z_j) \in X \times \mathbb{R} \mid j = 1, \ldots, M\}$ be an interpolating set. Note that u_i is the unique affine contraction satisfying

$$u_i(q_j) = q_{\ell(i,j)}, \qquad j = 1, 2, 3; \; i = 1, \ldots, N. \qquad (8.10)$$

As before, define unique affine mappings $v_i : \sigma^2 \times \mathbb{R} \to \mathbb{R}$,

$$v_i(x, y, z) := b_i x + c_i y + s_i z + e_i,$$

where b_i, c_i, and e_i are uniquely determined by

$$v_i(q_j, z_j) = z_{\ell(i,j)}, \qquad j = 1, 2, 3; \; i = 1, \ldots, N, \qquad (8.11)$$

and $|s_i| < 1$, $i = 1, \ldots, N$.

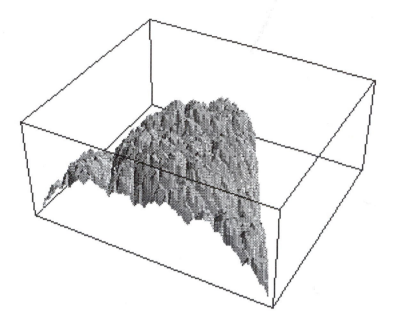

Figure 8.4: A fractal surface with $s = 1/2$.

Proposition 8.2 *Suppose* $\Delta := \{(q_j, z_j) \in X \times \mathbb{R} \mid j = 1, \ldots, M\}$ *is an interpolating set such that* $\{(q_j, z_j) \mid q_j \in \partial\sigma^2\} \subset \Pi$. *Then the Read-Bajraktarević operator* Φ, *as defined by Eq. (8.9), but with* v_i *as before, is well-defined and contractive in the norm* $\|\cdot\|_\infty$ *with contractivity factor* $\max\{|s_i| \mid i = 1, \ldots, N\}$. *Moreover,* $\Phi(f)(q_j) = z_j$, *for all* $j = 1, \ldots, M$, *and the unique fixed point* f^{**} *of* Φ *is an element of* $C^{**}(\sigma^2, \mathbb{R})$.

Proof. Let $f \in C^{**}(\sigma^2, \mathbb{R})$. Then graph f contains the three line segments $\overline{(q_j, z_j)(q_{j'}, z_{j'})}$, $j, j' = 1, 2, 3$ and $j \neq j'$. Therefore, graph Φf contains the line segments $\overline{(q_{\ell(i,j)}, z_{\ell(i,j)})(q_{\ell(i',j')}, z_{\ell(i',j')})}$. Hence,

$$v_i(u_i^{-1}(x), f \circ u_i^{-1}(x)) = v_{i'}(u_{i'}^{-1}(x), f \circ u_{i'}^{-1}(x)), \qquad (8.12)$$

for all $x \in \overline{q_j q_{j'}}$, whenever $\overline{q_j q_{j'}} = \sigma_i^2 \cap \sigma_{i'}^2$. The well-definedness and continuity of Φf now follows from the fact that $\Phi f|_{\overline{q_j q_{j'}}} = \overline{(q_j, z_j)(q_{j'}, z_{j'})}$. Furthermore, since $\phi f|_{\partial\sigma^2}$ consists of line segments that join points in $\{(q_j, z_j) \mid q_j \in \partial\sigma^2\} \subset \Pi$, graph $\Phi f|_{\partial\sigma^2} \subset \Pi$. Thus, $\Phi f \in C^{**}(\sigma^2, \mathbb{R})$. It is straightforward to show that Φ is contractive in $\|\cdot\|_\infty$ with contractivity $\max\{|s_i| \mid i = 1, \ldots, N\}$ and that Φf interpolates Δ. ∎

Remark. It is an immediate consequence of the proofs of Propositions 8.1 and 8.2 that more general mappings v_i can be considered, as long as as the "join-up" condition (8.12) is satisfied (also (B⁰) on page 139). This more general setup will be considered in detail in the next chapter, where it is used to define wavelet bases in $L^2(\mathbb{R}^n)$.

Fig. 8.5 shows an example of an affine fractal surface with co-planar boundary data and $s = 1/2$.

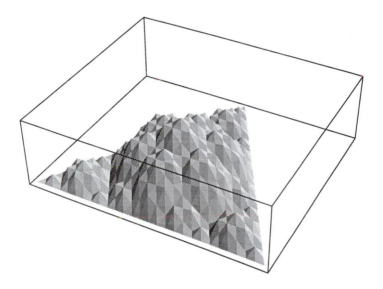

Figure 8.5: An affine fractal surface with co-planar boundary data.

Remark on multidimensional iterative interpolation functions.
In Section 5.3 the iterative interpolation process was introduced. The careful reader may have noticed that the results on the continuity of the iterative interpolation process do also hold for X to be a closed discrete subgroup of $\mathbb{R}^n$. Therefore, an iterative interpolation process (f, X, T, p) can also be used to define fractal-like surfaces. Some properties of such surfaces generated by iterative interpolation processes, including continuity and differentiability, are discussed in more detail in [50, 139]. The reader who would like to learn more about surfaces generated by iterative interpolation processes is referred to these references.

8.3   Properties of Fractal Surfaces

In this section some properties of fractal surfaces are discussed. Among those are oscillation, regularity and differentiability, Hölder continuity, and the box dimension. It also will be shown how moments of $\mathbb{R}$-valued multivariate fractal functions can be recursively calculated.

8.3.1   The oscillation of f^*

In what follows it is assumed that $X = \sigma^n$, i.e., $(X, \mathbf{w})$ is just an IFS on X, and that

$$\sigma^n = \bigcup_{i=1}^{N} \sigma_i^n, \tag{8.13}$$

for some $1 < N \in \mathbb{N}$, and where each σ_i^n is similar to σ^n. (That such a simplex exists follows from the theory of Coxeter groups. In the next chapter this claim will be verified.) This then implies that if $\boldsymbol{u}_i$ is of the form (8.4), A_i has to be a similitude with scaling factor $a_i < 1$, i.e., $A_i = a_i \Theta_i$, with Θ_i being an isometry on $\mathbb{R}^n$. Let us also assume that B is a similitude, i.e. $B = s\Theta$, where Θ is an isometry on $\mathbb{R}^m$. In order to calculate the oscillation of f^* over σ^n, a special class of covers of σ^n is needed.

Definition 8.5 Let $\varepsilon > 0$. An ε-cover $\mathcal{C}_\varepsilon$ of a bounded set $S \subseteq \mathbb{R}^n$ is called admissible if it is of the form

$$\mathcal{C}_\varepsilon = \{B_\varepsilon(r_\alpha) \,|\, r_\alpha, r_{\alpha'} \in S \text{ and } |r_\alpha - r_{\alpha'}| \geq \varepsilon/(2\sqrt{n}) \text{ for all } r_\alpha \neq r_{\alpha'}\}, \tag{8.14}$$

where $B_\varepsilon(r_\alpha)$ denotes the n-dimensional ball of radius ε centered at $\alpha \in S$.

At this point the definition of oscillation of a function over a set is recalled.

Definition 8.6 Let $X \subseteq \mathbb{R}^n$ be the domain of a function $f : \mathbb{R}^n \to \mathbb{R}^m$. The oscillation of f over $B \subseteq X$ is defined as

$$\operatorname{osc}(f; B) = \sup_{x, x' \in B} \|f(x) - f(x')\|, \tag{8.15}$$

and the ε-oscillation of f over X as

$$\operatorname{osc}_\varepsilon(f; X) = \inf \sum_{B \in \mathcal{C}_\varepsilon} \operatorname{osc}(f; B), \tag{8.16}$$

where the infimum is taken over all admissible ε-covers $\mathcal{C}_\varepsilon$ of X.

The next lemma states how fast the ε-oscialltion of $\boldsymbol{f}^*$ increases as ε decreases to zero. Recall that M is the number of interpolation points (q_j, z_j).

Lemma 8.1 *Suppose the set of interpolation points* $\Delta = \{(q_j, z_j) : j = 1, \ldots, M\}$ *is not contained in any n-dimensional hyperplane of* IR^{n+m}, *and* $\sum_{i=1}^{N} s a_i^{n-1} > 1$. *Then*

$$\lim_{\varepsilon \to 0+} \varepsilon^{n-1} \mathrm{osc}_\varepsilon (f^*; \sigma^n) = +\infty. \tag{8.17}$$

Proof. The first hypothesis implies that there exists an $\tilde{x} \in \overset{\circ}{\sigma}{}^n$ such that $V := \|\boldsymbol{f}^*(\tilde{x}) - \pi(\tilde{x})\| > 0$, where $\pi : \mathrm{IR}^n \to \mathrm{IR}^m$ is the unique affine map such that $\pi(q_j) = z_j$, for each of the $n+1$ vertices $q_1, \ldots, q_{n+1}$ of σ^n. Let $\tilde{\sigma}^n$ be a closed and connected subset of $\overset{\circ}{\sigma}{}^n$ such that whenever $x \in \sigma^n$, there is an $x' \in \tilde{\sigma}^n$ with $\|\boldsymbol{f}^*(x) - \boldsymbol{f}^*(x')\| \le V/2$. Let $\eta > 0$ be the distance between $\tilde{\sigma}^n$ and $\partial \sigma^n$. Let $\underline{a} := \min\{a_i : i = 1, \ldots, N\}$, let $0 < \varepsilon < \eta \underline{a}/2$, and let Σ_ε be the collection of all finite codes $\mathbf{i} \in \Sigma$ such that

$$2\varepsilon \le \eta \, a_{\mathbf{i}} \le 2\varepsilon/\underline{a} \tag{8.18}$$

holds for $\mathbf{i}$ but no curtailment of it. Since $\pi(\tilde{x})$ is in the convex hull of $\{\boldsymbol{f}^*(q_j) : j = 1, \ldots, n+1\}$, it follows that $\pi(\tilde{x}) = \Sigma_{j=1}^{n+1} \alpha_j \boldsymbol{f}^*(q_j)$, where $\alpha_j \ge 0$ and $\Sigma_{j=1}^{n+1} \alpha_j = 1$. Let $x_j \in \tilde{\sigma}^n$ be such that $\|\boldsymbol{f}^*(q_j) - \boldsymbol{f}^*(x_j)\| \le V/2$, for $j = 1, \ldots, n+1$. Then

$$\mathrm{osc}(\boldsymbol{f}^*; \tilde{\sigma}^n) \ge \Sigma_{j=1}^{n+1} \alpha_j \|\boldsymbol{f}^*(\tilde{x}) - \boldsymbol{f}^*(x_j)\|$$

$$\ge V \|\boldsymbol{f}^*(\tilde{x}) - \pi(\tilde{x})\| - \|\Sigma_{j=1}^{n+1} \alpha_j (\boldsymbol{f}^*(q_j) - \boldsymbol{f}^*(x_j))\|$$

$$\ge V/2.$$

This inequality together with the fixed-point property of $\boldsymbol{f}^*$ gives

$$\mathrm{osc}(\boldsymbol{f}^*; \boldsymbol{u_i}\tilde{\sigma}^n) \ge \left\| \boldsymbol{f}^*(\boldsymbol{u_i}(\tilde{x})) - \sum_{j=1}^{n+1} \alpha_j \boldsymbol{f}^*(\boldsymbol{u_i}(x_j)) \right\|$$

$$\ge \|(\Phi_{\mathbf{i}} \boldsymbol{f}^*)(\boldsymbol{u_i}(\tilde{x})) - (\Phi_{\mathbf{i}} \pi)(\boldsymbol{u_i}(\tilde{x}))\|$$

$$- \left\| (\Phi_{\mathbf{i}} \pi)(\boldsymbol{u_i}(\tilde{x})) - \sum_{j=1}^{n+1} \alpha_j (\Phi_{\mathbf{i}} \boldsymbol{f}^*)(\boldsymbol{u_i}(x_j)) \right\|$$

$$= \|B_{\mathbf{i}}\| \left(\|f^*(\tilde{x}) - \pi(\tilde{x})\| - \|\pi(\tilde{x}) - \sum_{j=1}^{n+1} \alpha_j f^*(x_j)\| \right)$$

$$\geq s_{\mathbf{i}}(V/2).$$

Here the notation $(\Phi_{\mathbf{i}} f)(x) = v_{\mathbf{i}}(u_{\mathbf{i}}^{-1}(x), f \circ u_{\mathbf{i}}^{-1}(x))$ was used, where $v_{\mathbf{i}}(x, y)$ is such that $w_{\mathbf{i}}(x, y) = (u_{\mathbf{i}}(x), v_{\mathbf{i}}(x, y))$, for all finite codes $\mathbf{i}$ in the associated code space Σ. Note that $\mathrm{osc}_\varepsilon(f^*; u_{\mathbf{i}} \tilde{\sigma}^n) \geq \mathrm{osc}(f^*; u_{\mathbf{i}} \tilde{\sigma}^n)$, since f^* is continuous and $u_{\mathbf{i}} \tilde{\sigma}^n$ connected. Therefore, using (8.18), one has

$$\mathrm{osc}_\varepsilon(f^*; \sigma^n) \geq \mathrm{osc}_\varepsilon(f^*; \bigcup_{\mathbf{i}} u_{\mathbf{i}} \tilde{\sigma}^n) = \sum_{\mathbf{i}} \mathrm{osc}_\varepsilon(f^*; u_{\mathbf{i}} \tilde{\sigma}^n))$$

$$\geq \left(\sum_{\mathbf{i}} s_{\mathbf{i}} \right) (V/2). \tag{8.19}$$

From (8.18) it also follows that $(a_{\mathbf{i}}/\varepsilon)^{n-1} \leq (2/\eta \underline{a})^{n-1} =: \gamma^{-1}$. Thus,

$$\frac{1}{2} \sum_{\mathbf{i}} s_{\mathbf{i}} V \geq \frac{\gamma}{2} \sum_{\mathbf{i}} (a_{\mathbf{i}}^{n-1} s_{\mathbf{i}}) V \varepsilon^{-n+1} = \frac{\gamma}{2} \sum_{\mathbf{i}} \left(a_{\mathbf{i}}^{d-1} s_{\mathbf{i}} \right) \left(a_{\mathbf{i}}^{n-d} V \right) \varepsilon^{-n+1}, \tag{8.20}$$

where d is the unique positive solution of $\sum_{i=1}^{N} s\, a_i^{d-1} = 1$. Note that the second hypothesis implies $d > n$.

Now define a probability measure μ on Σ by $\mu(\mathbf{i}) = s_{\mathbf{i}}\, a_{\mathbf{i}}^{d-1}$, for any cylinder set $\mathbf{i}$. Since Σ_ε partitions Σ, one has $1 = \sum_{i \in \Sigma_\varepsilon} \mu(\mathbf{i}) = \sum_{i \in \Sigma_\varepsilon} s_{\mathbf{i}}\, a_{\mathbf{i}}^{d-1}$. Therefore,

$$\mathrm{osc}_\varepsilon(f^*; \sigma^n) \geq (\gamma/2)(\overline{a}^{k(n-d)} V) \varepsilon^{-n+1},$$

where $\overline{a} = \max\{a_i \mid i = 1, \dots, N\}$ and $k = \min\{|\mathbf{i}| \mid \mathbf{i} \in \Sigma_\varepsilon\}$. Since $k \to \infty$ as $\varepsilon \to 0+$, the result follows.   ∎

The preceding lemma will now be used to derive upper and lower bounds on the ε-oscillation of f^* over σ^n.

Theorem 8.2 *Assume that the hypotheses of Lemma 8.1 are satisfied. Then there exist positive constants ε_0, k_1, and k_2 such that*

$$k_1 \varepsilon^{-\delta} \leq \mathrm{osc}_\varepsilon(f^*; \sigma^n) \leq k_2 \varepsilon^{-\delta}, \tag{8.21}$$

for all $0 < \varepsilon < \varepsilon_0$, where δ is the unique positive solution of $\sum_{i=1}^{N} s\, a_i^\delta = 1$.

Proof. Let $i \in \{1, \ldots, N\}$, let $0 < \varepsilon < 1$, and suppose $B_{\varepsilon/a_i}(r) \subseteq \sigma^n$. The fixed-point property of $\boldsymbol{f}^*$ implies that

$$
\mathrm{osc}(\boldsymbol{f}^*; \boldsymbol{u}_i B_{\varepsilon/a_i}(r)) = \sup_{x, x' \in B_\varepsilon(\boldsymbol{u}_i(r))} \|\boldsymbol{v}_i(x, \boldsymbol{f}^*(x)) - \boldsymbol{v}_i(x', \boldsymbol{f}^*(x))\|
$$

$$
\leq \|B\| \, \mathrm{osc}(\boldsymbol{f}^*; B_{\varepsilon/a_i}(r)) + \frac{2\varepsilon}{a_i} \|C_i\|. \tag{8.22}
$$

Let $\sigma_i^n = \{x \in \boldsymbol{u}_i \sigma^n \mid \mathrm{dist}(x, \partial \boldsymbol{u}_i \sigma^n) \geq 2\varepsilon\}$. (Here $\mathrm{dist}(A, B)$ denotes the distance between the sets A and B.) Note that $\sigma_i^n \neq \emptyset$ for ε small enough. To proceed, the following observation is needed: If $\mathcal{C}_\varepsilon$ is an admissible ε-cover of a set S and x is a point not covered by $\mathcal{C}_\varepsilon$, then $\mathcal{C}_\varepsilon \cup B_\varepsilon(x)$ is an admissible ε-cover of $S \cup \{x\}$. Thus, any admissible cover of a set S may be extended to an admissible cover of a superset of S.

Let $\mathcal{C}_{\varepsilon/a_i}$ be an admissible ε/a_i-cover of σ^n. Applying $\boldsymbol{u}_i$ to this cover yields an admissible ε-cover $\mathcal{C}_\varepsilon^i$ of $\boldsymbol{u}_i \sigma^n$. Let $\tilde{\mathcal{C}}_\varepsilon^i = \{B_\varepsilon \in \mathcal{C}_\varepsilon^i \mid B_\varepsilon \cap \sigma_i^n \neq \emptyset\}$. Note that $\tilde{\mathcal{C}}_\varepsilon := \bigcup_i \tilde{\mathcal{C}}_\varepsilon^i$ is an admissible ε-cover of $\bigcup_i \sigma_i^n$ and may be extended to an admissible ε-cover $\mathcal{C}_\varepsilon$ of σ^n by adding ε-balls with centers in $\sigma^n \setminus \bigcup_i \sigma_i^n$, as described earlier. Therefore,

$$
\mathrm{osc}_\varepsilon(\boldsymbol{f}^*; \sigma^n) \leq \sum_{B_\varepsilon \in \mathcal{C}_\varepsilon} \mathrm{osc}(\boldsymbol{f}^*, B_\varepsilon)
$$

$$
= \sum_{B_\varepsilon \in \tilde{\mathcal{C}}_\varepsilon} \mathrm{osc}(\boldsymbol{f}^*, B_\varepsilon) + \sum_{B_\varepsilon \in \mathcal{C}_\varepsilon \setminus \tilde{\mathcal{C}}_\varepsilon} \mathrm{osc}(\boldsymbol{f}^*, B_\varepsilon).
$$

It follows from (8.14) and $\mathrm{vol}_n(\sigma^n \setminus \bigcup_i \sigma_i^n) \leq 4\varepsilon \, \mathrm{vol}_{n-1}(\bigcup_i \partial \boldsymbol{u}_i \sigma^n)$ that there exists a positive constant c_0 such that $\mathcal{C}_\varepsilon \setminus \tilde{\mathcal{C}}_\varepsilon$ contains at most $c_0 \varepsilon^{-n+1}$ ε-balls. Thus,

$$
\sum_{B_\varepsilon \in \mathcal{C}_\varepsilon \setminus \tilde{\mathcal{C}}_\varepsilon} \mathrm{osc}(\boldsymbol{f}^*, B) \leq 2 \|\boldsymbol{f}^*\| c_0 \varepsilon^{-n+1}.
$$

Furthermore, if $B_\varepsilon \in \tilde{\mathcal{C}}_\varepsilon^i$, then $B_\varepsilon = \boldsymbol{u}_i B_{\varepsilon/a_i}$, for some $B_{\varepsilon/a_i} \in \mathcal{C}_{\varepsilon/a_i}$. Hence

$$
\sum_{B_\varepsilon \in \tilde{\mathcal{C}}_\varepsilon} \mathrm{osc}(\boldsymbol{f}^*; B_\varepsilon) = \sum_i \sum_{B_{\varepsilon/a_i} \in \mathcal{C}_{\varepsilon/a_i}} \mathrm{osc}(\boldsymbol{f}^*; B_{\varepsilon/a_i})
$$

$$
\leq \sum_i \sum_{B_{\varepsilon/a_i} \in \mathcal{C}_{\varepsilon/a_i}} \left(s \, \mathrm{osc}(\boldsymbol{f}^*; B_{\varepsilon/a_i}) + \frac{2\varepsilon}{a_i} \|C_i\| \right).
$$

Note that by (8.14), $\sum_{B_{\varepsilon/a_i} \in \mathcal{C}_{\varepsilon/a_i}} (2\varepsilon/a_i) \|C_i\| \leq c_1 \varepsilon^{-n+1}$, for some $c_1 > 0$. Therefore,

$$
\mathrm{osc}_\varepsilon(\boldsymbol{f}^*; \sigma^n) \leq \sum_i \sum_{B_{\varepsilon/a_i} \in \mathcal{C}_{\varepsilon/a_i}} s \, \mathrm{osc}(\boldsymbol{f}^*; B_{\varepsilon/a_i}) + c_2 \varepsilon^{-n+1},
$$

for some $c_2 > 0$.

Since the preeding inequality holds for any admissible ε/a_i-cover, one has

$$\mathrm{osc}_\varepsilon(\boldsymbol{f}^*; \sigma) \leq \sum_i s\,\mathrm{osc}_{\varepsilon/a_i}(\boldsymbol{f}^*; \sigma) + c_2\varepsilon^{-n+1}. \tag{8.23}$$

On the other hand, if $\mathcal{C}_\varepsilon$ is an admissible cover of $\sigma_i^n := \boldsymbol{u}_i\sigma^n$, $i = 1, \ldots, N$, then again by the fixed-point property of $\boldsymbol{f}^*$ (assuming $s \neq 0$),

$$\mathrm{osc}(\boldsymbol{f}^*, B_{\varepsilon/a_i}(\boldsymbol{u}_i^{-1}(r_\alpha)) =$$

$$= \sup_{x,x' \in B_\varepsilon(r_\alpha)} \|(\boldsymbol{v}_i^{-1}(x, \boldsymbol{f}^*(x)) - \boldsymbol{v}_i^{-1}(x', \boldsymbol{f}^*(x'))\|$$

$$\leq \|B^{-1}\|\,\mathrm{osc}(\boldsymbol{f}^*; B_\varepsilon(r_\alpha)) + 2\|B^{-1}C_iA_i\|\,\varepsilon,$$

for all $B_\varepsilon(r_\alpha) \in \mathcal{C}_\varepsilon$, where the mapping $\boldsymbol{v}_i^{-1}$ is such that $\boldsymbol{w}_i^{-1}(\,\cdot\,, *) = (\boldsymbol{u}_i^{-1}(\,\cdot\,), \boldsymbol{v}_i^{-1}(\,\cdot\,, *))$. Thus,

$$\sum_i (\varepsilon/a_i) - \mathrm{osc}(\boldsymbol{f}^*; \sigma^n) \;\leq\; \sum_i s^{-1}\,\mathrm{osc}_\varepsilon(\boldsymbol{f}^*; \sigma_i^n)$$

$$+ c_0\left(\sum_i \|B^{-1}C_iA\|\right)\varepsilon^{-n+1},$$

that is,

$$\mathrm{osc}_\varepsilon(\boldsymbol{f}^*; \sigma^n) \geq \sum_i s\,\mathrm{osc}_{\varepsilon/a_i}(\boldsymbol{f}^*; \sigma^n) - c_1\varepsilon^{-n+1}, \tag{8.24}$$

for $c_1 = c_0(\sum_i \|B^{-1}C_iA_i\|) > 0$. Note that (8.24) holds trivially for $s = 0$. Hence, combining (8.23) and (8.24) gives

$$\sum_i s\,\mathrm{osc}_{\varepsilon/a_i}(\boldsymbol{f}^*, \sigma^n) - c_1\varepsilon^{-n+1} \leq \mathrm{osc}_\varepsilon(\boldsymbol{f}^*; \sigma^n)$$

$$\leq \sum_i s\,\mathrm{osc}_{\varepsilon/a_i}(\boldsymbol{f}^*; \sigma^n) + c_2\,\varepsilon^{-n+1}.$$

Now let $\gamma := \sum_{i=1}^N sa_i^{n-1}$ and let $\bar{a} := \max\{a_i : i = 1, \cdots, N\}$. By Lemma 8.1, an $\varepsilon_0 > 0$ can be chosen small enough so that

$$\mathrm{osc}_\varepsilon(\boldsymbol{f}^*; \sigma^n) \geq [2c_1/(\gamma - 1)]\,\varepsilon^{-n+1},$$

for $\varepsilon_0 < \varepsilon \leq \varepsilon_0/\underline{a}$. Choose $K_1 > 0$ small enough so that $K_1\varepsilon_0^{-\delta} \leq [c_1/(\gamma - 1)]\,\varepsilon^{-n+1}$ and $K_2 > 0$ large enough so that

$$\mathrm{osc}_\varepsilon(\boldsymbol{f}^*, \sigma^n) \leq [c_2/(1 - \gamma)]\,\varepsilon^{-n+1} + K_2\varepsilon^{-\delta}$$

for $\varepsilon_0 \leq \varepsilon \leq \varepsilon_0/\underline{a}$.

Now define functions $\underline{\varphi}, \overline{\varphi} : (0, \varepsilon_0] \to \mathbb{R}$ by

$$\underline{\varphi}(\varepsilon) := \left(\frac{c_1}{\gamma - 1}\right) \varepsilon^{-n+1} + K_1 \varepsilon^{-\delta}$$

and

$$\overline{\varphi}(\varepsilon) := \left(\frac{c_2}{1 - \gamma}\right) \varepsilon^{-n+1} + K_2 \varepsilon^{-\delta},$$

respectively. It follows that, for all $\varepsilon_0 \leq \varepsilon \leq \varepsilon_0 / \underline{a}$,

$$\underline{\varphi}(\varepsilon) \leq \mathrm{osc}_\varepsilon(\boldsymbol{f}^*; \sigma^n) \leq \overline{\varphi}(\varepsilon).$$

Note that

$$\underline{\varphi}(\varepsilon) = \sum_i s\, \underline{\varphi}\,(\varepsilon / a_i) - c_1 \varepsilon^{-n+1}$$

and

$$\overline{\varphi}(\varepsilon) = \sum_i s\, \overline{\varphi}\,(\varepsilon / a_i) + c_2 \varepsilon^{-n+1}.$$

If $\overline{a}\, \varepsilon_0 \leq \varepsilon \leq \varepsilon_0$, then $\varepsilon_0 \leq \varepsilon / a_i \leq \varepsilon / \underline{a}$ and

$$\mathrm{osc}_\varepsilon(\boldsymbol{f}^*; \sigma^n) \leq \sum_i s\, \mathrm{osc}_{\varepsilon/a_i}(\boldsymbol{f}^*; \sigma^n) + c_2 \varepsilon^{-n+1}$$

$$\leq \sum_i s\overline{\varphi}\,(\varepsilon / a_i) + c_2 \varepsilon^{-n+1} = \overline{\varphi}(\varepsilon).$$

Similarly,

$$\underline{\varphi}(\varepsilon) \leq \mathrm{osc}_\varepsilon(\boldsymbol{f}; \sigma^n)$$

for $\overline{a}\, \varepsilon_0 \leq \varepsilon \leq \varepsilon_0$. Now, if

$$\underline{\varphi}(\varepsilon) \leq \mathrm{osc}_\varepsilon(\boldsymbol{f}^*; \sigma^n) \leq \overline{\varphi}(\varepsilon)$$

holds for $(\overline{a})^\ell \varepsilon_0 \leq \varepsilon \leq \varepsilon_0$, it must hold for $(\overline{a})^{\ell+1} \varepsilon_0 \leq \varepsilon \leq \varepsilon_0$. Therefore,

$$\underline{\varphi}(\varepsilon) \leq \mathrm{osc}_\varepsilon(\boldsymbol{f}^*; \sigma^n) \leq \overline{\varphi}(\varepsilon)$$

for all $0 < \varepsilon \leq \varepsilon_0$. Since $\delta > n - 1$, there exist positive constants k_1 and k_2 such that

$$k_1 \varepsilon^{-\delta} \leq \mathrm{osc}_\varepsilon(\boldsymbol{f}^*; \sigma^n) \leq k_2 \varepsilon^{-\delta}.$$

∎

Remark. The inequalities in (8.21) imply that

$$\delta = \lim_{\varepsilon \to 0+} \frac{\log \mathrm{osc}_\varepsilon(\boldsymbol{f}^*; \sigma^n)}{-\log \varepsilon}.$$

8.3.2   Box dimension of the projections f_j^*

In this subsection a formula for the box dimension of a hidden variable fractal surface is presented.

For this purpose let f_j^* be the orthogonal projection of $\boldsymbol{f}^*$ onto $\mathbb{R}^n \times 0 \times \ldots \times \mathbb{R} \times \ldots \times 0$, where the factor $\mathbb{R}$ is in the jth position. Recall that $\boldsymbol{u}_i = A_i \cdot + D_i$, where A_i is a similitude of norm a_i, and that $\boldsymbol{v}_i(\cdot, *) = C_i \cdot + s\Theta * + E_i$, with Θ being an isometry on $\mathbb{R}^m$ and $s < 1$. Denote by $\lambda_1, \ldots, \lambda_m$ and by $h_1, \ldots, h_m$, the eigenvalues and orthonormal eigenvectors of Θ, respectively. Let the eigenvalues of Θ be ordered in such a way that $\lambda_1, \ldots, \lambda_\mu$ are all the distinct eigenvalues of Θ, $1 \le \mu \le m$. The canonical basis of $\mathbb{R}^m$ is denoted by $\{e_1, \ldots, e_m\}$. Define

$$c_{kj} = \langle h_k, e_j \rangle_E \qquad (8.25)$$

($\langle \cdot, \cdot \rangle_E$ denotes the Euclidean inner product in $\mathbb{R}^m$). Then $\boldsymbol{f}^*$ can be written as

$$\boldsymbol{f}^*(x) = \sum_{k=1}^m b_k(x) h_k, \qquad \text{where} \quad b_k(x) = \langle \boldsymbol{f}^*(x), h_k \rangle_E. \qquad (8.26)$$

Also, $f_j^*(x) = \langle \boldsymbol{f}^*(x), e_j \rangle_E$. Let $I(\kappa)$ be the set of all indices from $\{1, \ldots, m\}$ indexing the same eigenvalue of Θ, $\kappa \in \{1, \ldots, \mu\}$, and let

$$d_\kappa(x) = \sum_{k \in I(\kappa)} b_k(x) c_{kj}. \qquad (8.27)$$

Theorem 8.3 *Suppose that* (i) $d_\nu \not\equiv 0$, *for some* $\nu \in \{1, \ldots, \mu\}$ *and* (ii) $\sum_{i=1}^n a_i^{n-1} s > 1$. *Then the box dimension* d *of graph* f_j^* *is the unique positive solution of*

$$\sum_{i=1}^N a_i^{d-1} s = 1; \qquad (8.28)$$

otherwise, $d = n$.

Proof. To prove the theorem, use is made of a special class $\mathcal{K}_\varepsilon$, $\varepsilon > 0$, of covers of graph f_j^*. The covers $G_\varepsilon \in \mathcal{K}_\varepsilon$ are defined as follows:

$$G_\varepsilon = \{B_\varepsilon(r_\alpha) \times [y_\alpha + (k-1)\varepsilon, y_\alpha + k\varepsilon] \mid y_\alpha \le \inf_{B_\varepsilon} f_j^* \wedge$$

$$y_\alpha + n_\alpha \varepsilon \ge \sup_{B_\varepsilon} f_j^* \wedge k = 1, \ldots, n_\alpha \wedge n_\alpha \in \mathbb{N}\},$$

where $\{B_\varepsilon(r_\alpha)\}$ is an admissible ε-cover of σ^n. Let $\mathcal{N}_\varepsilon(f_j^*; \sigma^n) = \inf\{\|G_\varepsilon\|_c \mid G_\varepsilon \in \mathcal{K}_\varepsilon\}$.

Next it is shown that it suffices to consider covers from $\mathcal{K}_\varepsilon$ to calculate the box dimension of graph F_j^*. Let $\tilde{G}_\varepsilon$ be an arbitrary minimal cover of graph f_j^* consisting of sets of the form $B_\varepsilon(r) \times [z, z + \varepsilon]$, $z \in \mathbb{R}$. Denote by $\tilde{\mathcal{N}}_\varepsilon(f_j^*; D)$ the cardinality of this minimal cover. Then, obviously,

$$\tilde{\mathcal{N}}_\varepsilon(f_j^*; \sigma^n) \leq \mathcal{N}_\varepsilon(f_j^*; \sigma^n). \tag{8.29}$$

But since $\{B_\varepsilon(r_\alpha)\}$ is an admissible cover of σ^n, one also has that $B_{\varepsilon/2\sqrt{n}}(r_\alpha)$ $\cap B_{\varepsilon/2\sqrt{n}}(r_\beta) = \emptyset$, $r_\alpha \neq r_\beta$, and that the number of balls $B_{\varepsilon/2\sqrt{n}}(r_\alpha)$ contained in the ball $B_{\varepsilon/2\sqrt{n}}(r_\alpha)$ is less than or equal to $\xi := (4\sqrt{n} + 1)^n$. Thus, any $B_\varepsilon(r)$ meets at most ξ elements of any admissible ε-cover of σ^n. Hence,

$$\mathcal{N}_\varepsilon(f_j^*; \sigma^n) \leq 3\,\xi\,\tilde{\mathcal{N}}_\varepsilon(f_j^*; \sigma^n). \tag{8.30}$$

Note that by (8.14) there exists a constant $c > 0$ such that $|\mathcal{C}_\varepsilon| \leq c\varepsilon^{-n}$ for any admissible ε-cover of σ^n. Furthermore, (8.15), (8.16), and the definition of $\mathcal{N}_\varepsilon(f_j^*; \sigma^n)$ give

$$\varepsilon^{-1}\mathrm{osc}_\varepsilon(f_j^*; \sigma^n) - c_1\varepsilon^{-n} \leq \mathcal{N}_\varepsilon(f_j^*; \sigma^n) \leq \varepsilon^{-1}\mathrm{osc}_\varepsilon(f_j^*; \sigma^n) + c_2\varepsilon^{-n}, \tag{8.31}$$

for some positive constants c_1 and c_2.

Since $\mathrm{osc}_\varepsilon(f_j^*; \sigma^n) \leq \mathrm{osc}_\varepsilon(f^*; \sigma^n)$, Theorem 8.5 and (8.30) imply that

$$\mathcal{N}_\varepsilon(f_j^*; \sigma^n) \leq k_2\varepsilon^{-d}, \tag{8.32}$$

where $k_2 > 0$ and d is given by (8.28).

Next a lower bound for $\mathcal{N}_\varepsilon(f_j^*; \sigma^n)$ needs to be derived. In order to do this two technical lemmas have to be stated and proved. The first of these lemmas gives estimates on osc_ε. For this purpose, a definition is needed.

Definition 8.7 *For $f : \sigma^n \to \mathbb{R}$, define the upper ε-oscillation of f over σ^n by*

$$\overline{\mathrm{osc}}_\varepsilon(f; \sigma^n) = \sup\left\{ \sum_{B_\varepsilon \in \mathcal{C}_\varepsilon} \mathrm{osc}(f; B_\varepsilon) \right\}, \tag{8.33}$$

where the supremum is taken over all admissible ε-covers $\mathcal{C}_\varepsilon$ of σ^n.

Lemma 8.2 *Let $f : \sigma^n \to \mathbb{R}$ and let $\varepsilon > 0$ be arbitrary. Then there exist positive constants K, k_1, and k_2 so that*
(i) $\mathrm{osc}_\varepsilon(f; \sigma^n) \leq \overline{\mathrm{osc}}_\varepsilon(f; \sigma^n) \leq K\mathrm{osc}_\varepsilon(f; \sigma^n)$ and
(ii) $k_1\mathrm{osc}_\varepsilon(f; \sigma^n) \leq \mathrm{osc}_{c\varepsilon}(f; \sigma^n) \leq k_2\mathrm{osc}_\varepsilon(f; \sigma^n)$, for any $c > 0$. Furthermore, k_1 and k_2 depend only on f and c.

Proof of Lemma 8.2. (i) Let $\mathcal{C}_\varepsilon$ be an arbitrary admissible ε-cover of σ^n and let $\mathcal{C}'_\varepsilon$ be an admissible ε-cover of σ^n such that $\sum_{B' \in \mathcal{C}'_\varepsilon} \omega(f; B') \leq 2\Omega_\varepsilon(f; \sigma^n)$. For each $B \in \mathcal{C}_\varepsilon$ let $S(B) := \{B' \in \mathcal{C}'_\varepsilon \mid B \cap B' \neq \emptyset\}$. Then any B' meets at most $\xi = (4\sqrt{n}+1)^n$ elements of $\mathcal{C}_\varepsilon$. Furthermore, since $S(B) \supset B \cap \sigma^n$, one has $\mathrm{osc}(f; B) \leq \sum_{B' \in S(B)} \mathrm{osc}(f; B')$ for any $B \in \mathcal{C}_\varepsilon$. Thus,

$$\sum_{B \in \mathcal{C}_\varepsilon} \mathrm{osc}(f; B) \;\leq\; \sum_{B \in \mathcal{C}_\varepsilon} \sum_{B' \in S(B)} \mathrm{osc}(f; B') \leq \xi \sum_{B' \in \mathcal{C}'_\varepsilon} \mathrm{osc}(f; B')$$

$$\leq\; 2\xi \mathrm{osc}_\varepsilon(f; \sigma^n).$$

(ii) As in the proof of Theorem 8.2, it follows that any $B_{c_1 \varepsilon}(r)$ meets at most $[2(c_1/c_2 + 1)\sqrt{n} + 1]^n$ elements of an admissible $c_2 \varepsilon$-cover of σ^n for any $c_1, c_2 > 0$. Now the result follows as in (i).   ∎

Next a lower bound for $\mathrm{osc}_\varepsilon(d_\nu; \sigma^n)$ using a functional inequality as in the proof of Theorem 8.2 is derived. In order to get the induction started, the next lemma is needed.

Lemma 8.3 *Suppose that conditions (i) and (ii) of Theorem 8.3 are satisfied. Then*

$$\lim_{\varepsilon \to 0^+} \varepsilon^n \, \mathcal{N}_\varepsilon(f^*_j; \sigma^n) = +\infty.$$

Proof of Lemma 8.3. Since Θ is an isometry $|\lambda^n_\nu| = 1$, for all $n \in \mathbb{N}$. For α, $\beta \in \mathrm{span}\,\{(\lambda^n_\nu)^\infty_{n=0} \mid \nu = 1, \cdots, \mu\}$, define

$$(\alpha, \beta) := \lim_{T \to \infty} \frac{1}{T} \sum_{n=0}^{T} \alpha_n \overline{\beta}_n.$$

Now it follows from hypothesis (i) in Theorem 8.3 and

$$((\lambda^n_\nu), (\lambda^n_{\nu'})) = \delta_{\nu\nu'} x$$

that there exists an $n_0 \in \mathbb{N}$ with $\sum_\nu d_\nu(x)\lambda^{n_0}_\nu \neq 0$. Assume without loss of generality that $n_0 = 0$. Moreover, it can be assumed that $z_1 = \ldots = z_{n+1} = 0$ in $\mathbb{R}^m$, with one of the vertices being the origin. To simplify notation, the following definition is introduced.

Definition 8.8 Let $a, b \in \mathbb{R}^{n+1}$ and let π be a plane perpendicular to b containing the point $P \in \mathbb{R}^{n+1}$. Then

$$\mathrm{Proj}_{a;b,P}(x) := \frac{x - \langle x - P, b \rangle_E}{\langle a, b \rangle_E}\, a, \qquad \text{for all } x \in \mathbb{R}^{n+1}.$$

Since $f_j^* \not\equiv 0$ and continuous, there exists a $b \in \mathbb{R}^n$ such that $\mathrm{Proj}_{(b,0);(b,0),O}$ graph f_j^* contains an n-dimensional cube $C_{2\rho}$ of side 2ρ, for some $\rho > 0$. (Here O denotes the origin in $\mathbb{R}^{n+1}$.) By the Poincaré Recurrence Theorem there exists a subsequence $\{n_k\} \subseteq \mathbb{N}$ such that

$$|\lambda_\nu - \lambda_\nu^{n_k}| < \rho/2c$$

for all $\nu = 1, \ldots, \mu$, and $c = \max\{\sum_\nu |d_\nu(x)| \,|\, x \in \sigma^n\}$. Hence,

$$|f_j^*(x) - (\Theta^{n_k} f_j^*)(x)| = \left| \sum_\nu d_\nu(x)(\lambda_\nu - \lambda_\nu^{n_k}) \right| < \sum_\nu d_\nu(x) \cdot \frac{\rho}{2c} \le \frac{\rho}{2}.$$

Let $\mathbf{i} \in \Sigma$, $|\mathbf{i}| = n_k$, and $a := \boldsymbol{w}_{\mathbf{i}}(b,0) - \boldsymbol{w}_{\mathbf{i}}(0,0)$. Then

$$\mathrm{Proj}_{a;(b,0),P} \text{ graph } f_j^*|_{\sigma_{\mathbf{i}}^n} = \mathrm{Proj}_{a;(b,0),P} \left\{ (\boldsymbol{u}_{\mathbf{i}}(x), f_j^* \circ \boldsymbol{u}_{\mathbf{i}}(x)) \,|\, x \in \sigma^n \right\}.$$

After some algebra, this reduces to

$$T^{(\mathbf{i})}\left(x - \frac{\langle x, b \rangle_E}{\langle a, b \rangle_E} b, \langle \Theta^{n_k} f^*(x), e_j \rangle_E \right),$$

where $T^{(\mathbf{i})} : \mathbb{R}^n \times \mathbb{R}^m \to \mathbb{R}^n \times \mathbb{R}^m$ is defined by

$$T^{(\mathbf{i})}(x, y); = (\boldsymbol{u}_{\mathbf{i}}(x), s^{n_k} y + C_{\mathbf{i}} x + E_{\mathbf{i}}).$$

Note that the Jacobian of $T^{(\mathbf{i})}$ is given by

$$\mathrm{Jac}\, T^{(\mathbf{i})} = a_{\mathbf{i}}^{n-1} s^{|\mathbf{i}|}, \quad \mathbf{i} \in \Sigma.$$

Thus,

$$\mathrm{vol}\, T^{n_k}(C_{2p}) = \mathrm{Jac}\, T^{n_k}\, \mathrm{vol}\, C_{2p}.$$

Now, if $\mathcal{N}_\varepsilon(f_j^*; \sigma^n)$ is the cardinality of a minimal ε-cover $G_\varepsilon \in \mathcal{K}_\varepsilon$ of graph f_j^*, then, for $\varepsilon < 2\rho \underline{a}^{n_k}$,

$$\varepsilon^n \mathcal{N}_\varepsilon(f_j^*; \sigma^n) \ge \sum_{\mathbf{i}} \mathrm{Jac}\, T^{(\mathbf{i})}\, \mathrm{vol}\, C_{2\rho} \ge \left(\sum_{i=1}^N a_i^{n-1} s \right)^{|\mathbf{i}|} \mathrm{vol}\, C_{2\rho},$$

and this last term tends to infinity as $\varepsilon \to 0+$. ∎

Continuation of the proof of Theorem 8.3. Observe that

$$f_j^* = \sum_{\kappa=1}^\mu d_\kappa. \tag{8.34}$$

Lemma 8.2(i) and (8.34) give

$$\operatorname{osc}_\varepsilon(f_j^*;\sigma^n) \leq \sum_{\kappa=1}^{\mu} \overline{\operatorname{osc}}_\varepsilon(d_\kappa;\sigma^n) \leq K \sum_{\kappa=1}^{\mu} \operatorname{osc}_\varepsilon(d_\kappa;\sigma^n). \tag{8.35}$$

Lemma 8.3, (8.31), and (8.35) and possibly re-indexing yield

$$\limsup_{\varepsilon\to 0^+} \varepsilon^{n-1}\operatorname{osc}_\varepsilon(d_\nu;\sigma^n) = +\infty. \tag{8.36}$$

Taking the inner product of (8.26) with h_k and using (8.27), the following functional equation for d_ν is obtained:

$$d_\nu(\boldsymbol{u}_i(x)) = \left\langle c_i x + E_i, \; \sum_{k\in I(\kappa)} h_k c_{kj} \right\rangle + s\lambda_\nu d_\nu(x) \tag{8.37}$$

for $x \in \sigma^n$, $i \in \{1,\ldots,N\}$. Arguments similar to those in the derivation of (8.24) lead to

$$\operatorname{osc}_\varepsilon(d_\nu;\sigma^n) \geq \sum_i s\operatorname{osc}_{\varepsilon/a_i}(d_\nu;\sigma^n) - c\varepsilon^{-n+1}, \tag{8.38}$$

for some $c > 0$. By Lemmas 8.2(ii) and 8.3, there exists an $\varepsilon_0 > 0$ such that $\operatorname{osc}_\varepsilon(d_\nu;\sigma^n) \geq 2c/(\gamma-1)$ for $\varepsilon_0 \leq \varepsilon \leq \varepsilon_0/\underline{a}$, where $\gamma := \sum_{i=1}^{N} sa_i^{n-1}$ and $\underline{a} := \min\{|a_i| \,|\, i = 1,\ldots,N\}$. Thus, a small enough constant $K_1 > 0$ can be chosen so that

$$\operatorname{osc}_\varepsilon(d_\nu;\sigma^n) \geq \left(\frac{c}{\gamma-1}\right)\varepsilon^{-n+1} + K_1\varepsilon^{-d+1}, \tag{8.39}$$

for $\varepsilon_0 \leq \varepsilon \leq \varepsilon_0/\underline{a}$. Using induction and (8.38) it is easily seen that (8.39) holds for all $0 < \varepsilon \leq \varepsilon_0$. In particular,

$$\operatorname{osc}_\varepsilon(d_\nu;\sigma^n) \geq K_1\varepsilon^{-d+1} \tag{8.40}$$

for $0 < \varepsilon \leq \varepsilon_0$. Let $\mathbf{i} \in \Sigma$, be a finite code of length ℓ. Note that by (8.26)

$$f^* \circ \boldsymbol{u}_\mathbf{i}(x) = v_\mathbf{i}(x, f^*(x)) = c_\mathbf{i} x + B^\ell f^*(x) + E_\mathbf{i},$$

for $x \in \sigma^n$. Therefore, Eqs. (8.26) and (8.27) imply

$$f_j^* \circ \boldsymbol{u}_\mathbf{i}(x) - f_j^* \circ \boldsymbol{u}_\mathbf{i}(y) = \langle C_\mathbf{i}(x-y), e_j \rangle_E + \sum_{\kappa-1}^{\mu} s^\ell \lambda_\kappa^\ell(d_\kappa(x) - d_\kappa(y)),$$

for $x, y \in \sigma^n$. Hence,

$$\operatorname{osc}_\varepsilon(f_j^*; \boldsymbol{u_i}\sigma^n) \geq -\frac{2\|C_i\|}{a_i}\varepsilon^{n+1} + s^\ell \operatorname{osc}_{\varepsilon/a_i}\left(\sum_\kappa \lambda_\kappa^\ell d_\kappa; \sigma^n\right),$$

and, by Lemma 8.2(ii),

$$\operatorname{osc}_\varepsilon(f_j^*; \boldsymbol{u_i}\sigma^n) \geq -\frac{2\|C_i\|}{a_i}\varepsilon^{-n+1} + k_1 s^\ell \operatorname{osc}_\varepsilon\left(\sum_\kappa \lambda_\kappa^\ell d_\kappa; \sigma^n\right), \qquad (8.41)$$

where $k_1 = k_1(a_i)$. Therefore,

$$\operatorname{osc}_\varepsilon(f_j^*; \sigma^n) \geq -c_\ell \varepsilon^{-n+1} + c_\ell' \operatorname{osc}_\varepsilon\left(\sum_\kappa \lambda_\kappa^\ell d_\kappa; \sigma^n\right), \qquad (8.42)$$

where $c_\ell, c_\ell' > 0$.

To obtain a lower bound for $\operatorname{osc}_\varepsilon(f_j^*; \sigma^n)$, we relate the preceding lower bound for $\operatorname{osc}_\varepsilon(d_\nu; \sigma^n)$ to a lower bound for $\operatorname{osc}_\varepsilon(\sum_\kappa \lambda_\kappa^\ell d_\kappa; \sigma^n)$. Note that, since the Vandermonde determinant $V(\lambda_1, \ldots, \lambda_\mu)$ of $\lambda_1, \ldots, \lambda_\mu$,

$$V(\lambda_1, \ldots, \lambda_\mu) = \begin{pmatrix} 1 & 1 & \cdots & 1 \\ \lambda_1 & \lambda_2 & \cdots & \lambda_\mu \\ \lambda_1^2 & \lambda_2^2 & \cdots & \lambda_\mu^2 \\ \vdots & \vdots & \cdots & \vdots \\ \lambda_1^{\mu-1} & \lambda_2^{\mu-1} & \cdots & \lambda_\mu^{\mu-1} \end{pmatrix},$$

is non-zero, there exist constants $\alpha_0, \ldots, \alpha_{\mu-1}$ such that

$$d_\nu = \sum_{\ell=0}^{\mu-1} \alpha_\ell \left(\sum_\kappa d_\kappa\right). \qquad (8.43)$$

Therefore, by (8.40) and Lemma 2(i), the following inequalities hold:

$$
\begin{aligned}
0 < K_1 \;\;\leq\;\; & \liminf_{\varepsilon \to 0+} \varepsilon^{-d+1}\operatorname{osc}_\varepsilon(d_\nu; \sigma^n) \\
= \;\;& \liminf_{\varepsilon \to 0+} \varepsilon^{-d+1}\operatorname{osc}_\varepsilon\left(\sum_\ell \alpha_\ell \left(\sum_\kappa \lambda_\kappa^\ell d_\kappa\right); \sigma^n\right) \\
\leq \;\;& \liminf_{\varepsilon \to 0+} \sum_\ell \varepsilon^{-d+1}|\alpha_\ell|\,\overline{\operatorname{osc}}_\varepsilon\left(\sum_\kappa \lambda_\kappa^\ell d_\kappa; \sigma^n\right) \\
\leq \;\;& \liminf_{\varepsilon \to 0+} \sum_\ell \varepsilon^{-d+1}|\alpha_\ell|\, K \operatorname{osc}_\varepsilon\left(\sum_\kappa \lambda_\kappa^\ell d_\kappa; \sigma^n\right) \\
\leq \;\;& \liminf_{\varepsilon \to 0+} c'' \sum_\ell \varepsilon^{-d+1}\operatorname{osc}_\varepsilon\left(\sum_\kappa \lambda_\kappa^\ell d_\kappa; \sigma^n\right),
\end{aligned}
$$

where $c'' = \max\{K\,|\alpha_\ell|\,|\,\ell = 0,\ldots,\mu-1\}$. Thus, there exists an $\varepsilon_0 > 0$ such that, for each $0 < \varepsilon \le \varepsilon_0$, there is some $\ell \in \{0,\ldots,\mu-1\}$ with

$$\varepsilon^{-d+1}\operatorname{osc}_\varepsilon\left(\sum_\kappa \lambda_\kappa^\ell d_\kappa;\sigma^n\right) \ge \frac{K_1}{2\mu c''}. \qquad (8.44)$$

Hence, using (8.42) and (8.44), there exist $K_2, K_3 > 0$ such that

$$\operatorname{osc}_\varepsilon(f_j^*;\sigma^n) \ge K_2\,\varepsilon^{-n+1} + K_3\,\varepsilon^{-d+1},$$

for $0 < \varepsilon \le \varepsilon_0$. The result now follows from Theorem 8.2. ∎

8.3.3 Box dimension of $f^* \in \mathcal{RF}(X \subset \mathbb{R}^2, \mathbb{R})$

Now a formula for the box dimension of a fractal surface in $\mathbb{R}^2 \times \mathbb{R}$ is presented. At this point the reader should recall the notation and terminology introduced in Section 5.4 and the construction of a multivariate recurrent $\mathbb{R}$-valued affine fractal function $f^* : X \subset \mathbb{R}^2 \to \mathbb{R}$. Throughout this subsection it is assumed that the mappings u_i, as defined in Equation (8.3), are similitudes whose contractivity factors are given by a_i, $i = 1,\ldots,N$, and that the connectivity matrix C of the underlying recurrent IFS is irreducible. Before the main result can be stated a few remarks about covers have to be made and some technical lemmas must be proven.

To this end, let $G^* := \operatorname{graph} f^*$ and $G_i^* := \operatorname{graph} f^*|_{\sigma_i^2}$, $i = 1,\ldots,N$. Let $Z := B_\varepsilon(r) \times [z, z+\varepsilon]$, $\varepsilon > 0$, and let $\mathcal{C}_\varepsilon(G^*)$ denote the class of all covers of G^* whose covering elements are of the form Z. That only this class of covers needs to be considered in the calculation of the box dimension of G^* should be fairly clear. Let $\mathcal{N}(\varepsilon) = \mathcal{N}(\varepsilon;G^*) := \min\{|C_\varepsilon(G^*)|\,|\,C_\varepsilon(G^*) \in \mathcal{C}_\varepsilon\}$. In order to calculate $\dim_\beta G^*$, an even more special class of covers needs to be introduced. A cover $K_\varepsilon(G^*)$ of G^* is called *admissible* iff it is of the form

$$K_\varepsilon(G^*) := \{B_\varepsilon(r_\alpha) \times [z + (k-1)\varepsilon, z + k\varepsilon]\,|\,k = 1,\ldots,n\}$$

and $|r_\alpha - r_\beta| > \varepsilon$ whenever $\alpha \ne \beta$. The class of all admissible covers of G^* is denoted by $\mathcal{K}_\varepsilon(G^*)$. Define $N^*(\varepsilon) = N^*(\varepsilon;G^*) := \min\{|K_\varepsilon(G^*)|\,|\,K_\varepsilon(G^*) \in \mathcal{K}_\varepsilon\}$.

Lemma 8.4 $\mathcal{N}(\varepsilon) \le N^*(\varepsilon) \le 48\,\mathcal{N}(\varepsilon)$.

Proof. Lemma 3.2 implies that at most 16 balls $B_\varepsilon(r_\alpha)$ from an admissible cover meet any ball $B_\varepsilon(r)$. Thus, any set of the form Z meets at most $3 \cdot 16$ of the elements of $K_\varepsilon(G^*)$. ∎

Now let $\mathcal{N}_i(\varepsilon) := \min\{\|K_\varepsilon(G_i^*)\|_c \mid K_\varepsilon(G_i^*) \in \mathcal{K}_\varepsilon\}$, $i = 1, \ldots N$. The next result is follows directly from the proofs of Lemma 6.2 and Theorem 6.1.

Lemma 8.5 *There exist positive constants A and B such that for $0 < \varepsilon < 1$,*

$$\frac{|s|}{a_i} \left(\sum_{j \in \mathbf{I}(i)} \mathcal{N}_i(\varepsilon/a_i) \right) - \frac{A}{\varepsilon^2} \leq \mathcal{N}_i(\varepsilon) \leq \frac{|s|}{a_i} \left(\sum_{j \in \mathbf{I}(i)} \mathcal{N}_i(\varepsilon/a_i) \right) + \frac{B}{\varepsilon^2}. \quad (8.45)$$

Here $\mathbf{I}(i) := \{j \in \{1, \ldots, N\} \mid \sigma_j^2 \subset \tau_{k(i)}\}$, where $k(i)$ is that index for which $u_i \tau_{k(i)}^2 = \sigma_i^2$. ∎

Lemma 8.6 *Let $B := \operatorname{diag}(|s|a_i)C$ have spectral radius $r < 1$. If there exists a $k \in \{1, \ldots, N\}$ such that $\{(q_j, z_j) \mid q_j \in \tau_k^2\}$ is not contained in a hyperplane of $\mathbb{R}^3$, then*

$$\lim_{\varepsilon \to 0+} \varepsilon^2 \mathcal{N}_i(\varepsilon) = +\infty.$$

Proof. Note that the irreducibility of the connection matrix C implies that G_i^* is not contained in any hyperplane of $\mathbb{R}^2$, $i = 1, \ldots, N$. Let Π_i denote the hyperplane through the points $\{(q_{i_j}, z_{i_j}) \mid j = 1, 2, 3\}$, where $\{q_{i_j} \mid j = 1, 2, 3\}$ are the vertices of σ_i^2. Denote by $\sigma_i^2(\eta)$ that 2-simplex contained in σ_i^2 whose η-body is σ_i^2. Now let $v \in \mathbb{R}^3$ and let P be a vertical plane in $\mathbb{R}^3$. The projection onto P in the direction of v is denoted by $\operatorname{proj}_{v,P}$. By the continuity of f^* and the non-coplanarity of G^*, one can choose an $\eta > 0$, a $v_i \in \mathbb{R}^3$ parallel to Π_i and P_i such that $\operatorname{proj}_{v_i,P_i}$ graph $f^*|_{\sigma_i^2(\eta)}$ has area $A_i > 0$. Denote by $\tilde{A}_i$ the area of $\operatorname{proj}_{v_i,P_i} B_1(r) \times [0, 1]$. Then it is immediate that, for any $\varepsilon > 0$, $\mathcal{N}_i(\varepsilon) \geq (A_i/\tilde{A}_i)^{-2}$. Let $e = (e_1, \ldots, e_N)^t$ be a positive eigenvector of B, and choose $\theta > 0$ sufficiently small so that $A_i/\tilde{A}_i \geq \theta e_i$, $i = 1, \ldots, N$. Next apply the map w_i to graph $f^*|_{\sigma_j^2(\eta)}$ for each $j \in \mathbf{I}(i)$. Since w_i maps vertical planes onto vertical planes and has Jacobian $\operatorname{Jac}(w_i) = sa_i^2$, it follows that $\operatorname{proj}_{w_i(v_j),w_i P_j}$ graph $f^*|_{\sigma_j^2(\eta)}$ has area equal to $|s|a_i A_i$. Hence, for $0 < \varepsilon < \underline{a}\,\eta$, with $\underline{a} := \min\{a_i \mid i = 1, \ldots, N\}$, one has

$$\mathcal{N}_i(\varepsilon) \geq |s|a_i \sum_{j \in \mathbf{I}(i)} \frac{A_j}{\tilde{A}_j} \varepsilon^{-2} \geq \theta \varepsilon^{-2} |s|a_i \sum_{i=1}^{N} c_{ij} e_j = \theta \varepsilon^{-2} r e_i.$$

Induction then gives

$$\mathcal{N}_i(\varepsilon) \geq \theta \varepsilon^{-2} r^n e_i,$$

for $0 < \varepsilon < \underline{a}^n \eta$, $n \in \mathbb{N}$, and $i = 1, \ldots, N$. ∎

Now the dimension theorem can finally be stated.

Theorem 8.4 *Suppose that the connection matrix C is irreducible and that there exists a k such that $\{(q_j, z_j) \mid q_j \in \tau_k^2\}$ is not contained in a hyperplane of $\mathbb{R}^3$ and that the spectral radius r of $\mathrm{diag}(|s|a_i)$ is greater than one. Then $\dim_\beta$ graph f^* is the unique d such that*

$$r(\mathrm{diag}(|s|a_i^{d-1})\, C) = 1; \qquad (8.46)$$

otherwise, $\dim_\beta$ graph $f^* = 2$.

Proof. Using the preceding lemmas as well as the arguments given in the proof of Theorem 6.2 gives

$$A_1 \varepsilon^{-d} + A_2 \varepsilon^{-2} \leq \mathcal{N}_i(\varepsilon) \leq A_3 \varepsilon^{-d} + A_4 \varepsilon^{-2},$$

for positive constants $A_1, \ldots, A_4$, and $\varepsilon > 0$. Hence,

$$\dim_\beta \text{ graph } f^* = \max\{2, d\}.$$

Note that if $\{(q_j, z_j) \mid q_j \in \tau_{k(i)}^2\}$ is contained in a hyperplane for all $i = 1, \ldots, N$, then graph f^* is piecewise co-planar and and thus has box dimension equal to two. ∎

In the case $(X, \mathbf{w})$ is an IFS rather than a recurrent IFS, i.e., the connection matrix C is given by $(c_{ij}) = (1)$, for all $i, j \in \{1, \ldots, N\}$, Eq. (8.46) reduces to the following formula.

Corollary 8.1 *Suppose that $\{(q_j, z_j) \mid j = 1, \ldots, M\}$ is not contained in any hyperplane of $\mathbb{R}^3$ and that $\sum_{i=1}^{N} |s|a_i > 1$. Then the box dimension of graph f^* is the unique positive solution d of*

$$\sum_{i=1}^{N} |s|a_i^{d-1} = 1; \qquad (8.47)$$

otherwise, $\dim$ graph $f^* = 2$.

Finally, a dimension result for affine fractal surfaces in $\mathbb{R}^3$ with co-planar interpolation points is stated. For this purpose, let Δ_{xy} be a set of $N + 1$ equally spaced points on $\partial \sigma^2$ from which N^2 subsimplices σ_i^2 are constructed. The maps $u_i : \sigma^2 \twoheadrightarrow \sigma_i^2$ are then of the form

$$u_i = \begin{pmatrix} \pm\frac{1}{N} & 0 \\ 0 & \pm\frac{1}{N} \end{pmatrix} \cdot + D_i,$$

for $i = 1, \ldots, N^2$. The mappings v_i are given as in Eq. (8.11). Denote by f^{**} the bivariate fractal function generated by these maps and by G^{**} its graph. First, however, a lemma is needed.

Lemma 8.7 *Let B_ε denote a ball in $\mathbb{R}^3$ of radius $\varepsilon > 0$. For $\varepsilon > 0$, let Π_ε be the ε-prismatoidal set $\varepsilon\,\sigma^2 \times \varepsilon\,[0,1]$, where — as usual — εA denotes the set $\{\varepsilon\,a \,|\, a \in A\}$. Let S be a bounded set in $\mathbb{R}^3$ and let $\mathcal{N}(B_\varepsilon)$ denote the minimum number of ε-balls B_ε necessary to cover S. Then*

$$\mathcal{N}(B_\varepsilon) \leq \mathcal{N}(\Pi_\varepsilon) \leq 8\,\mathcal{N}(B_\varepsilon),$$

where $\mathcal{N}(\Pi_\varepsilon)$ is the minimum number of ε-prismatoidal sets needed to cover S.

Proof. Every B_ε can be covered by at most eight Π_ε, and every Π_ε can be covered by one B_ε. ∎

Theorem 8.5 *Assume that $\sum_{i=1}^{N^2} |s_i| > N$ and that $\bigcup_{i=1}^{N^2} |\sigma_i^2|$ is not contained in any hyperplane of $\mathbb{R}^3$. Then the box dimension of graph f^* is given by*

$$\dim_\beta \text{graph } f^{**} = 1 + \log_N \sum_{i=1}^{N^2} |s_i|; \tag{8.48}$$

otherwise, $\dim_\beta \text{graph } f^{**} = 2$.

Proof. It is fairly clear that it suffices to prove the theorem for $\varepsilon = \varepsilon_m := N^{-m}$, $m \in \mathbb{N}$.

Let $m \in \mathbb{N}$ and let $\mathcal{U}_m$ be a collection of covers of G^{**} by ε_m-prismatoidal sets $\Pi = \Pi_m(p, q)$ of the form $\text{proj}_{z=0}\Pi = |\sigma_p^2|$, $p = 1, \ldots, N^{2m}$, $\sigma_p^2 := \mathfrak{w}_{i(p)}\sigma^2$, with $i(p) \in \Sigma$ denoting a finite code of length p, and $q = 1, \ldots, \mathcal{N}(m, p)$, where $\mathcal{N}(m, p) \in \mathbb{N}$ is the number of such ε_m-prismatoidal sets over $|\sigma_p^2|$. Furthermore, it is required that $\Pi_m(p, q)$ and $\Pi_m(p, q+1)$ intersect along their respective faces. Now let $U_m \in \mathcal{U}_m$ be a cover of G^{**} of minimal cardinality $\mathcal{N}(m) \in \mathbb{N}$. Observe that $\mathcal{N}(m) = \sum_p \mathcal{N}(m, p)$. Let $G_p^{**} := \bigcup_{q=1}^{\mathcal{N}(m,p)} \Pi_m(p, q)$, $p = 1, \ldots, N^{2m}$. It is immediate that G_p^{**} is compact and connected.

The image G_{pi}^{**} of G_p^{**} under the map w_i, $i = 1, \ldots, N^2$, is a compac set in $\sigma_p^2 \times \mathbb{R}$ above the ith subsimplex σ_{pi}^2 of σ_p^2. Since $G^{**} = \bigcup_{i=1}^{N^2} w_i G^{**}$, it follows that $G^{**} \subseteq \bigcup_i w_i \left(\bigcup_p G_p^{**} \right)$. The compact set G_{pi}^{**} is contained entirely in the prismatoidal set $(N^{m+1})^{-1} \sigma^2 \times (\mathcal{N}(m, p)|s_i| + |c_i| + |d_i|)(N^{-m})\,[0, 1]$. Hence, if $\mathcal{N}(m+1, p, i)$ denotes the number of ε_{m+1}-prismatoidal sets above σ_{pi}^2, then

$$\mathcal{N}(m + 1, p, i) \leq N\,[\mathcal{N}(m, p)|s_i| + |c_i| + |d_i|] + 1.$$

Note that $\mathcal{N}(m+1) = \sum_i \sum_p \mathcal{N}(m,p,i)$. Summing the preceding inequality over p and i yields

$$\mathcal{N}(m+1) \le \left[N \sum_{i=1}^{N^2} |s_i| \right] \mathcal{N}(m) + c_1 N^{2m+1},$$

with $c_1 := \sum_i (|c_i| + |d_i| + N) > 0$. If $\sum_i |s_i| \le N$, then induction on m gives

$$\mathcal{N}(m) \le \left[\mathcal{N}(1) + c_1 m N^{-1} \right] N^{2m},$$

and thus,

$$\limsup_{m \to \infty} \frac{\log \mathcal{N}(m)}{\log N^m} \le 2.$$

Hence, $\dim_\beta G^{**} = 2$ in this case. Also, if $\bigcup_i |\sigma_i^2|$ is contained in a hyperplane P of $\mathbb{R}^3$, then $G^{**} = P$, and thus, $\dim_\beta G^{**} = 2$.

If $\gamma := \sum_i |s_i| > N$ then, again by induction on m,

$$\begin{aligned} \mathcal{N}(m) &\le (N\gamma) \left[\mathcal{N}(1) + \frac{c_1}{\gamma} \left(1 + \frac{N}{\gamma} + \cdots + \left(\frac{N}{\gamma} \right)^{m-1} \right) \right] \\ &\le (N\gamma) \left[\mathcal{N}(1) + \frac{c_1}{\gamma - N} \right], \end{aligned}$$

which implies that

$$\overline{\dim}_\beta G \le 1 + \log_N \gamma.$$

To obtain a lower bound for $\dim_\beta G^{**}$ when $\gamma > N$ and $\bigcup_i |\sigma_i^2|$ is not contained in any hyperplane of $\mathbb{R}^3$, one proceeds as follows: Choose $i \in \{1, \ldots, N^2\}$ and assume that $s_i \ne 0$. The inverse image of G_p^{**} under w_i^{-1} is contained in a prismatoidal set $(N^{-m+1}) \sigma_i^2 \times N^{-m} (\mathcal{N}(m,p)|s_i^{-1}| + |c_i s_i^{-1}|N)$ $[0,1]$. Hence, in the same notation as before,

$$\mathcal{N}(m-1,p,i) \le |s_i^{-1}| \mathcal{N}(m,p,i) + (|s_i^{-1}|N)(|c_i| + |d_i|) + 1.$$

Summing over p and i yields

$$\mathcal{N}(m) \ge (N\gamma) \mathcal{N}(m-1) - c_2 N^{2m-1},$$

with $c_2 := N \sum_i |s_i^{-1}|(|c_i| + |d_i|) + N > 0$. Note that the preceding inequality holds trivially for $s_i = 0$. Induction on m for all $m_0 \in \{1, \ldots, m\}$ gives

$$\mathcal{N}(m) \ge (N\gamma)^{m-m_0} (\mathcal{N}(m_0) - c_2 N^{2m_0} (\gamma - N)^{-1}).$$

By Lemma 8.8 below, one can choose m_0 large enough to guarantee that

$$\mathcal{N}(m_0) > c_2 N^{2m_0}(\gamma - N)^{-1}.$$

Let $c_3 := (N\gamma)^{-m_0}(\mathcal{N}(m_0) - c_2 N^{2m_0}(\gamma - N)^{-1}) > 0$. Then

$$\mathcal{N}(m) \geq c_3(N\gamma)^m.$$

Hence,

$$\liminf_{m \to \infty} \frac{\log \mathcal{N}(m)}{\log N^m} \geq 1 + \log_N \gamma.$$

Thus, $\dim_\beta G^{**} = 1 + \log_N \gamma$ for $\gamma > N$ and $\bigcup_i |\sigma_i^2|$ not contained in any hyperplane of $\mathbb{R}^3$. ∎

The proof of the preceding theorem used the following lemma, which is a special case of Lemma 8.6, and it is therefore left to the reader to establish its validity.

Lemma 8.8 *If $\gamma > N$ and $\bigcup_i |\sigma^2|$ is not co-planar, then*

$$\lim_{m \to \infty} \frac{N^{2m}}{\mathcal{N}(m)} = 0.$$

8.3.4   Hölder continuity

Next it is shown that the bivariate fractal function f^{**} defined at the end of the previous section is Hölder continuous with exponent $\alpha = 3 - \dim_\beta G^{**}$.

Theorem 8.6 *Let f^{**} be defined as before. Then, if $0 \leq h < 1$,*

$$|f^{**}(x + h, y + h) - f^{**}(x, y)| \leq c h^\alpha, \qquad (x, y) \in \sigma^2, \qquad (8.49)$$

for a positive constant c and $\alpha = 3 - \dim_\beta G^{**} = 2 - \log_N \sum_i^{N^2} |s_i|$.

Proof. Let $n \in \mathbb{N}$. For every finite code $\mathbf{i}(n) \in \Sigma := \{1, \ldots, N^2\}^{\mathbb{N}}$ of length n denote by σ_p^2 the simplex $w_{\mathbf{i}(n)}\sigma^2$, $p = 1, \ldots, N^{2n}$. Let $\Pi(m, p) := \bigcup_{k=0}^{\mathcal{N}}(m, p)\{N^{-m}\sigma^2 \times N^{-m}[k, k + N^{-m}]$ be the collection of all sets of the above form that cover graph $f^{**}|_{\sigma_p^2}$. Let $0 \leq h < 1$ be given. Let m be the least integer such that there exists a finite code $\mathbf{i}(m) \in \Sigma$ of length m so that (x, y) and $(x + h, y + h)$ are both in the interior of $u_{\mathbf{i}(m)}\sigma^2$. Then it follows from Theorem 8.5 that

$$|f^{**}(x + h, y + h) - f^{**}(x, y)| \leq N^{-m} \max\{\mathcal{N}(m, p) \,|\, p = 1, \ldots, N^{2m}\}.$$

However, since $\mathcal{N}(m,p) \geq cN^{-2m}(N^{-m}N^{-d})$, where $d = \dim_\beta G^{**}$, it follows that

$$\frac{\log |f^{**}(x+h,y+h) - f^{**}(x,y)|}{\log h} \geq \frac{\log cN^{-2m}(N^{-m}N^{-d})}{\log h}$$

$$\geq \frac{\log c}{\log h} + \frac{\log N^{-3m+md}}{\log N^{-m}}$$

$$= \frac{\log c}{\log h} + 3 - d,$$

because $h \leq N^{-m}$.   ∎

The fact that f^{**} has Hölder exponent $\alpha = 3 - \dim_\beta G^{**}$ has the following consequence.

Proposition 8.3 *Suppose that f^{**} is Hölder continuous with exponent $\alpha = 3 - \dim_\beta G^{**}$. Then $\mathcal{H}^{\dim_\beta G^{**}}(G^{**}) < +\infty$.*

Proof. Let $0 \leq h < 1$ and let m be the least integer so that $h \leq N^{-2m}$. Let σ_p^2 be any of the N^{2m} subsimplices of σ^2. Then graph $f^{**}|_{\sigma_p^2}$ can be covered by at most $N^{2m}(ch^\alpha + 1)$ sets of the form $N^{-m}\sigma^2 \times N^{-m}[k, k + N^{-m}]$, $k \in \mathbb{N}$. Thus,

$$\mathcal{H}^{\dim_\beta G^{**}}_{N^{-2m}}(G^{**}) \leq N^{2m}\left(ch^{3-\dim_\beta G^{**}} + 1\right)\left(N^{-m}\right)^{\dim_\beta G^{**}}$$

$$\leq N^{2m}\left(cN^{(-m)(3-\dim_\beta G^{**})} + 1\right)\left(N^{-m}\right)^{\dim_\beta G^{**}}$$

$$= cN^{-m} + \left(N^{-m}\right)^{\dim_\beta G^{**} - 2}$$

$$\leq \frac{c}{N} + \left(\frac{1}{N}\right)^{\dim_\beta G^{**} - 2} < +\infty.$$

Therefore, $\mathcal{H}^{\dim_\beta G^{**}}(G^{**}) < +\infty$.   ∎

8.3.5   p-balanced measures and moment theory for f^{**}

The projection $(\sigma^2, \mathbf{u})$ of the IFS $(\sigma^2 \times \mathbb{R}, \mathbf{w})$ that generates graph f^{**} onto $\mathbb{R}^2$ is also an IFS whose attractor is clearly σ^2. Let μ, respectively, $\tilde{\mu}$, denote the **p**-balanced measure of $(\sigma^2 \times \mathbb{R}, \mathbf{w})$, respectively $(\sigma^2, \mathbf{u})$. Recall that $G^{**} = \text{graph } f^{**}$. Let $\mathcal{M}(\sigma^2)$ and $\mathcal{M}(G^{**})$ denote the measure spaces of σ^2 and G^{**}, respectively, and let $H : \sigma^2 \to G^{**}$ be the homeomorphism defined by $(x,y) \mapsto (x,y,f^{**}(x,y))$. Then the following relation between measures on σ^2 and G^{**} holds.

Theorem 8.7 *The homeomorphism $H : \sigma^2 \to G^{**}$, as defined earlier, induces a contravariant homeomorphism $\mathcal{M}(H) : \mathcal{M}(G^{**}) \to \mathcal{M}(\sigma^2)$. Furthermore,*

$$\forall \, \tilde{E} \in \tilde{\mathcal{B}}(\sigma^2) : \tilde{\mu}\tilde{E} = \mu H(\tilde{E}).$$

If $g \in L^1(\sigma^2 \times \mathbb{R}, \mu)$, then

$$\int_{G^{**}} g \, d\mu = \int_{\sigma^2} g \circ H \, d\tilde{\mu} = \int_{H^{-1}\sigma^2} g \, d\tilde{\mu}.$$

Proof. The proof is similar to that of Theorem 5.8 and is therefore omitted. ∎

The preceding theorem implies the following integral relation:

Corollary 8.2 *If the probabilities in the IFS $(\sigma^2 \times \mathbb{R}, \mathbf{w})$ are chosen according to $p_i := \text{area}\,(\sigma_i^2)$, $i = 1, \ldots, N^2$, and if dA denotes two-dimensional Lebesgue measure on $\mathbb{R}^2$, then*

$$\int_{G^{**}} g \, d\mu = \int_{\sigma^2} g \circ H \, dA.$$

Proof. With the preceding choice of probabilities, $dA(u_i^{-1}\tilde{B}) = dA(\tilde{B})$, for all $\tilde{B} \in \mathcal{B}(\sigma^2)$. ∎

This corollary, together with the stationarity of the **p**-balanced measure μ, can be used to calculate moments. Let $g \in L^1(\sigma^2 \times \mathbb{R}, \mu)$ and let $p_i = \text{area}\,(\sigma_i^2)$, $i = 1, \ldots, N^2$. Then

$$\int_{G^{**}} g(x,y,z)\, d\mu(x,y,z) = \sum_{i=1}^{N^2} p_i \int_{G^{**}} g \circ w_i(x,y,z)\, d\mu(x,y,z)$$

$$= \int_{\sigma^2} g(x,y,f^{**}(x,y))\, dA.$$

Also,

$$\int_{G^{**}} g(x,y,z)\, d\mu(x,y,z) = \sum_{i=1}^{N^2} p_i \int_{\sigma^2} g(u_i(x,y), v_i(x,y,f^{**}(x,y)))\, dA.$$

Hence,

$$\int_{\sigma^2} g(x, y, f^{**}(x, y))\, dA = \sum_{i=1}^{N^2} p_i \int_{\sigma^2} g(u_i(x, y), v_i(x, y, f^{**}(x, y)))\, dA. \quad (8.50)$$

Recalling the definition of moments (Definition 5.4), a moment theorem for affine fractal surfaces defined by functions f^{**} can now be stated and proven. To this end, let $\alpha, \beta, \gamma \in \mathbb{N}_0$.

Theorem 8.8 *Let* G^{**} *be the graph of an affine fractal function of the form* f^{**}. *Then the moments* $f^{**}_{\alpha,\beta,\gamma}$ *can be recursively and explicitly calculated from the lower order moments.*

Proof. Let $N_\pm := \pm N$.

$$
\begin{aligned}
f_{\alpha,\beta,\gamma} &= \int_{\sigma^2} x^\alpha y^\beta f^{**\gamma}\, dA \\
&= \sum_{i=1}^{N^2} (2N^{-2})^{-1} \int_{\sigma^2} \left(\frac{x}{N_\pm} + x_i\right)^\alpha \left(\frac{y}{N_\pm} + y_i\right)^\beta \\
&\qquad\qquad\qquad \times (L_i(x, y) + s_i z)^\gamma\, dA,
\end{aligned}
$$

with $L_i(x, y) := k_i x + \ell_i y + z_i$, and where the x_i, y_i, z_i, k_i, and ℓ_i are uniquely determined from Eqs. (8.3) and (8.11). To simplify notation the following abbreviations are introduced:

$$\sum_a := \sum_{a=0}^{\alpha-1} \binom{\alpha}{a} \left(\frac{x}{N_\pm}\right)^a x_i^{\alpha-a}, \qquad \sum_b := \sum_{b=0}^{\beta-1} \binom{\beta}{b} \left(\frac{y}{N_\pm}\right)^b y_i^{\beta-b},$$

and

$$\sum_c := \sum_{c=0}^{\gamma-1} \binom{\gamma}{c} (s_i f^{**})^c L_i^{\gamma-c}, \qquad \sum_i := \sum_{i=1}^{N^2} (2N^{-2})^{-1}.$$

Then

$$
\begin{aligned}
f_{\alpha,\beta,\gamma} &= \sum_i \int_{\sigma^2} \left(\sum_a + \left(\frac{x}{N_\pm}\right)^\alpha\right) \left(\sum_b + \left(\frac{y}{N_\pm}\right)^\beta\right) \\
&\qquad\qquad \times \left(\sum_c + (s_i f^{**})^\gamma\right) dA
\end{aligned}
$$

$$= \sum_i f^{**}_{\alpha,\beta,\gamma} + \sum_i \left[\sum_a \sum_b \sum_c f^{**}_{a,b,c} + \sum_a \sum_b f^{**}_{a,b,\gamma} \right.$$

$$+ \sum_b \sum_c f^{**}_{\alpha,b,c} + \sum_c \sum_a f^{**}_{a,\beta,c} + \sum_a f^{**}_{a,\beta,\gamma}$$

$$\left. + \sum_b f^{**}_{\alpha,b,\gamma} + \sum_c f^{**}_{a,b,\gamma} \right].$$

Or, upon transposing the first term on the right-hand side and substituting for $\sum_i$,

$$f^{**}_{\alpha,\beta,\gamma} = \frac{\left[\begin{array}{c} \sum_a \sum_b \sum_c f^{**}_{a,b,c} + \sum_a \sum_b f^{**}_{a,b,\gamma} + \sum_b \sum_c f^{**}_{\alpha,b,c} \\ + \sum_c \sum_a f^{**}_{a,\beta,c} + \sum_a f^{**}_{a,\beta,\gamma} + \sum_b f^{**}_{\alpha,b,\gamma} + \sum_c f^{**}_{a,b,\gamma} \end{array} \right]}{1 - \sum_{i=1}^{N^2}(s_i^\gamma/2N^{\alpha+\beta+\gamma+2})}. \tag{8.51}$$

Note that the numerator in the preceding equation is strictly less than 1. ∎

Remark. In the case $\beta := 0$ and $\gamma := 1$, the moment formula for f^{**} reduces to something more appealing, namely,

$$f^{**}_\alpha = \frac{\sum_{i=1}^{N^2} \sum_{a=0}^{\alpha-1} \binom{\alpha}{a}(x_i^{\alpha-a} s_i/2N^{a+3}) f^{**}_a + \mathfrak{P}_\alpha}{1 - \sum_{i=1}^{N^2} s_i/2N^{\alpha+3}}, \tag{8.52}$$

where

$$\mathfrak{P}_\alpha := \sum_{i=1}^{N^2}(2N^{-2})^{-1} \int_{\sigma^2} \left(\frac{x}{N_\pm} + x_i \right)^\alpha L_i(x,y)\,dA.$$

8.4 Fractal Surfaces of Class C^k

In this section two methods for constructing smooth fractal surfaces are presented. The first one is an extension of the method outlined in Section 5.8; the second one defines smooth fractal surfaces as indefinite integrals of C^0-fractal surfaces.

8.4.1 Construction via IFSs

Let $Q = [0,1] \times [0,1]$, let $e_1 = (1,0)^t$, $e_2 = (0,1)^t$, and let N be a fixed integer greater than one. Let $\Gamma = \{(m/N)e_1 + (n/N)e_2 : m,n \in \mathbb{Z}\}$ be a lattice in $\mathbb{R}^2$. Suppose that for each lattice point $(x_j, y_j) \in \Gamma \cap Q$ a real number z_{ij}, $i,j \in \{0,1,\ldots,N\}$ is given. The set $\Delta := \{(x_j, y_j, z_{ij}) \mid i,j = 0,1,\ldots,N\}$

can be thought of as a given set of data or interpolation points on Q. Smooth affine fractal surfaces containing this interpolating set will be constructed. To this end, let $u_{ij} : Q \to Q$ be given by

$$u_{ij}(x, y) = \begin{pmatrix} \frac{1}{N} & 0 \\ 0 & \frac{1}{N} \end{pmatrix} \begin{pmatrix} x \\ y \end{pmatrix} + \begin{pmatrix} \frac{i-1}{N} \\ \frac{j-1}{N} \end{pmatrix}, \tag{8.53}$$

and let $v_{ij} : Q \times \mathbb{R} \to \mathbb{R}$ be defined as

$$v_{ij}(x, y, z) = A_{ij}x^2 + B_{ij}y^2 + C_{ij}z^2 + D_{ij}xy + E_{ij}yz + F_{ij}zx + G_{ij}, \tag{8.54}$$

such that

$$v_{ij}(0, 0, z_{0,0}) = z_{i-1,j-1}, \qquad v_{ij}(0, 1, z_{0,N}) = z_{i-1,j},$$
$$\tag{8.55}$$
$$v_{ij}(1, 0, z_{N,0}) = z_{i,j-1}, \qquad v_{ij}(1, 1, z_{N,N}) = z_{i,j},$$

$i, j = 1, \ldots, N$, and that the following join-up conditions are satisfied:
For $j = 1, \ldots, N$ and $y \in [(j-1)/N, j/N]$,

$$v_{ij}(0, y, \varphi(0, y)) = v_{i-1,j}(1, y, \varphi(1, y)), \qquad i = 2, \ldots, N,$$
$$\tag{8.56}$$
$$v_{ij}(1, y, \varphi(1, y)) = v_{i+1,j}(0, y, \varphi(0, y)), \qquad i = 1, \ldots, N-1,$$

and for $i = 1, \ldots, N$ and $x \in [(j-1)/N, j/N]$

$$v_{ij}(x, 0, \varphi(x, 0)) = v_{i,j-1}(x, 1, \varphi(x, 1)), \qquad j = 2, \ldots, N,$$
$$\tag{8.57}$$
$$v_{ij}(x, 1, \varphi(x, 1)) = v_{i,j+1}(x, 0, \varphi(x, 0)), \qquad j = 1, \ldots, N-1.$$

Here φ denotes any C^0-function interpolating Δ.

Conditions (8.55), (8.56), and (8.57) uniquely determine some of the coefficients $A_{ij}, \ldots, G_{ij}$. For instance, if $z_{0,0} = z_{0,N} = z_{N,0} = z_{N,N} = 0$, then

$$A_{ij} = z_{i,j-1} - z_{i-1,j-1}, \qquad\qquad B_{ij} = z_{i-1,j} - z_{i-1,j-1}$$
$$D_{ij} = (z_{ij} - z_{i,j-1}) - (z_{i-1,j} - z_{i-1,j-1}), \qquad G_{ij} = z_{i-1,j-1},$$

for $i, j = 1, \ldots, N$. If $\varphi \equiv 0$ on ∂Q, then $A_{ij} = B_{ij} = D_{ij} \equiv 0$, and the join-up conditions are automatically satisfied. If $\varphi(0, y) \equiv \varphi(1, y)$ and $\varphi(x, 0) \equiv \varphi(x, 1)$, then in addition to Eq. (8.53) one also needs that

$$A_{ij} = A_{i,j-1}, \qquad\qquad B_{ij} = B_{i-1,j},$$
$$C_{ij} = C_{i-1,j} = C_{i,j-1}, \qquad E_{ij} = F_{ij} = 0,$$

in order for the join-up conditions to be satisfied.

Now let $\mathfrak{F}(Q) := \{\varphi \in C(Q, \mathbb{R}) | \varphi(x_j, y_j) = z_{ij}, i, j = 0, 1, \ldots, N\}$. Define a Read-Bajraktarević operator $\Phi : \mathfrak{F}(Q) \to \mathbb{R}^Q$ by

$$(\Phi\varphi)(x, y) := v_{ij}(u_{ij}^{-1}(x, y), \varphi \circ u_{ij}^{-1}(x, y))), \qquad (x, y) \in u_{ij}(Q). \qquad (8.58)$$

Suppose, without loss of generality, that $\|\varphi\|_\infty \leq 1$ on Q and that $s := 2 \max_{i,j} |C_{ij}| + \max_{i,j} |E_{i,j}| + \max_{i,j} |F_{ij}| < 1$.

Theorem 8.9 Φ *maps $\mathfrak{F}(Q)$ into itself, is well-defined, and is contractive in the* sup-*norm with contractivity s.*

Proof. The results follow inmmediately from the definition of Φ, conditions (8.55), (8.56), and (8.57) and the assumption on s. ∎

The unique fixed point of Φ is the graph of a C^0-function $f : Q \to \mathbb{R}$ that interpolates Δ. The graph of f is again called a fractal surface. Fig. 8.6 displays the fifth-level approximation of one of these surfaces with $z_{00} = z_{02} = z_{20} = z_{22} = 0$, $z_{01} = z_{10} = z_{12} = z_{21} = 1/2$, $z_{11} = 1$.

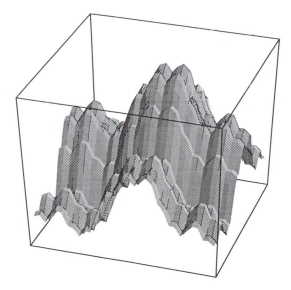

Figure 8.6: A fractal surface with $C_{ij} = 9/20$.

Imposing the following C^1 join-up conditions, one is guaranteed that f is a C^1-function: Suppose that $\varphi \in \mathfrak{F}^1(Q) := \{\psi \in C^1(Q, \mathbb{R}) | \psi(x_j, y_j) = z_{ij}, i, j = 0, 1, \ldots, N\}$. Let

$$\nabla v_{ij}(u_{ij}^{-1}(\cdot, \cdot), \varphi \circ u_{ij}^{-1}(\cdot, \cdot)) = \nabla v_{i,j-1}(u_{i,j-1}^{-1}(\cdot, \cdot), \varphi \circ u_{i,j-1}^{-1}(\cdot, \cdot)), \qquad (8.59)$$

for all $(x, y) \in [i/N, (i+1)/N] \times \{j/N\}$, and similarly for the three other edges. (Here $\nabla \in \mathcal{L}(\mathbb{R}^2, \mathcal{L}(\mathbb{R}^2, \mathbb{R}))$ denotes the differential operator $\nabla := (\partial/\partial x, \partial/\partial y)^t$.) In the case where $z_{0,0} = z_{0,N} = z_{N,0} = z_{N,N} = 0$, this implies that $E_{ij} = F_{ij} = 0$ and that $\nabla \varphi|_{\partial Q} \equiv 0$. If the class $\tilde{\mathfrak{F}}^1(Q) := \{\varphi \in \mathfrak{F}^1(Q) : \nabla \varphi|_{\partial Q} \equiv 0\}$ is considered, then the following result holds.

Theorem 8.10 *Let the mapping* $\Phi : \mathfrak{F}(Q) \to \mathbb{R}^Q$ *be defined as in (8.58). Suppose that condition (8.59) is satisfied. Then* Φ *maps* $\tilde{\mathfrak{F}}^1(Q)$ *into itself, is well-defined, and is contractive in the* C^1*-topology with contractivity* s.

Proof. This follows directly from Theorem 8.9 and the preceding considerations. ∎

Fig. 8.7 shows the fifth-level approximation of such a C^1 fractal surface with $C_{ij} := 1/4$.

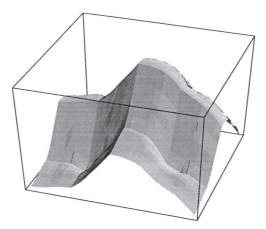

Figure 8.7: A C^1 fractal surface.

8.4.2   Smooth fractal surfaces via integration

In this subsection fractal surfaces defined on Q are considered that are generated by choosing u_{ij} as in the previous subsection, but the mappings v_{ij} are to be of the following form: (i) $v_{ij}(\cdot, \cdot, z) - v_{ij}(0, 0, z_0)$ is a symmetric quadratic form for all $z, z_0 \in \mathbb{R}$, and (ii) $v_{ij}(x, y, \cdot) - v_{ij}(x, y, z_0)$ is a linear form for all $(x, y) \in Q$ and $z_0 \in \mathbb{R}$. Furthermore, it is required that conditions (8.55), (8.56), and (8.57) also hold for this particular choice of v_{ij}. In the special case $z_{0,0} = z_{N,0} = z_{0,N} = z_{N,N} = 0$, the same expressions for A_{ij},

B_{ij}, D_i, and G_{ij} as before are obtained, if

$$v_{ij}(x, y, z) = A_{ij}x^2 + B_{ij}y^2 + C_{ij}z + D_{ij}xy + G_{ij}.$$

Note that (8.56) and (8.57) follow whether $\varphi \equiv 0$ on ∂Q or $\varphi|_{[0,1]\times\{0\}} \equiv \varphi|_{[0,1]\times\{1\}}$ and $\varphi|_{\{0\}\times[0,1]} \equiv \varphi|_{\{1\}\times[0,1]}$, $\varphi \in \mathfrak{F}(Q)$.

Defining an operator Φ as in (8.58) and assuming $s = \max_{i,j} |C_{ij}| < 1$, the following theorem is obtained, whose straightforward proof is left for the reader.

Theorem 8.11 *The unique fixed point of Φ is a C^0-function $f : Q \to \mathbb{R}$ such that $f(x_j, y_j) = z_{ij}$, for all $i, j = 0, 1, \ldots, N$.*

The reason for choosing this particular form of the v_{ij} will become clear shortly. Let

$$\tilde{f}(x, y) := \tilde{z}_{0,0} + \int_0^x \int_0^y f(s, t)\, dt\, ds,$$

for some $\tilde{z}_{0,0} \in \mathbb{R}$. Denote the integral operator $\int_0^x \int_0^y (\cdot)\, dt\, ds$ by $I_{(0,0)}^{(x,y)}(\cdot)$, and let $u_{ij}(x, y) = (\xi_i(x), \eta_j(y))$, where $\xi_i(x) := (1/N)x + (i-1)/N$ and $\eta_j(y) := (1/N)y + (j-1)/N$, $i, j = 0, 1, \ldots, N$. Then

$$\tilde{f} \circ u_{ij}(x, y) \;=\; \tilde{z}_{0,0} + I_{(0,0)}^{(x_{i-1}, y_{j-1})}(f) + I_{(x_{i-1},0)}^{(\xi_i(x), y_{j-1})}(f)$$

$$+ I_{(0, y_{i-1})}^{(x_{i-1}, \eta_j(y))}(f) I_{(x_{i-1}, y_{j-1})}^{(\xi_i(x), \eta_j(y))}(f)$$

$$=\; \tilde{z}_{0,0} + I_{(0,0)}^{(x_{i-1}, y_{j-1})}(f) + I_{(x_{i-1},0)}^{(\xi_i(x), y_{j-1})}(f)$$

$$+ I_{(0, y_{j-1})}^{(x_{i-1}, \eta_j(y))}(f) + \frac{1}{N^2} I_{(0,0)}^{(x_{i-1}, y_{j-1})}(f \circ u_{ij}).$$

Since $f \circ u_{ij} = v_{ij}(\cdot, f)$, this gives

$$\tilde{f} \circ u_{ij}(x, y) \;=\; \tilde{z}_{0,0} + I_{(0,0)}^{(x_{i-1}, y_{j-1})}(f) + I_{(x_{i-1},0)}^{(\xi_i(x), y_{j-1})}(f)$$

$$+ I_{(0, y_{j-1})}^{(x_{i-1}, \eta_j(y))}(f) + \frac{C_{ij}}{N^2} I_{(0,0)}^{(x,y)}(f)$$

$$+ \frac{1}{N^2} I_{(0,0)}^{(x,y)}(v_{ij}|_{z=0})$$

$$=:\; \frac{C_{ij}}{N^2} \tilde{f}(x, y) + R_{ij}(x, y).$$

Hence, $\tilde{f}$ is the unique fixed point of the operator $\Psi : C^1(Q, \mathbb{R}) \to C^1(Q, \mathbb{R})$,

$$\Psi\varphi := \tilde{v}_{ij}(u_{ij}^{-1}(\cdot, \cdot), \varphi \circ u_{ij}^{-1}(\cdot, \cdot)), \tag{8.60}$$

where $\tilde{v}_{ij}(x, y, z) = R_{ij}(x, y, z) + \frac{C_{ij}}{N^2} z$, or equivalently, graph f is the unique attractor of the IFS $(Q \times \mathbb{R}, \tilde{\mathbf{w}})$ with $\tilde{\mathbf{w}} = \{\tilde{w}_{ij} : Q \times \mathbb{R} \to Q \times \mathbb{R} \,|\, \tilde{w}_{ij} = (u_{ij}, \tilde{v}_{ij}), \; i, j = 1, \ldots, N\}$. Since the operator $I_{(0,0)}^{(x,y)}(\cdot)$ is continuous, $\tilde{f}$ is continuous at its interpolating set $\tilde{\Delta} := \{(x_j, y_j, \tilde{z}_{ij}) : \; i, j = 0, 1, \ldots, N\}$. To determine the $\tilde{z}_{ij}$, notice that $u_{ij}(0, 0) = (x_{i-1}, y_{j-1}) = z_{i-1, j-1}$, and thus $\tilde{f} \circ u_{ij}(0, 0) = \tilde{z}_{0,0} + I_{(0,0)}^{(x_{i-1}, y_{j-1})}(f) =: \tilde{z}_{i-1, j-1}$. Therefore,

$$
\begin{aligned}
\tilde{z}_{ij} &= \tilde{z}_{i-1, j-1} + \frac{C_{ij}}{N^2} I_{(0,0)}^{(1,1)}(f) + \frac{1}{N^2} I_{(0,0)}^{(1,1)}(v_{ij}\,|_{z=0}) + I_{(x_{i-1},0)}^{(x_i, y_{j-1})}(f) \\
&\quad + I_{(0, y_{j-1})}^{(x_{i-1}, y_j)}(f) \\[2mm]
&= \frac{C_{ij}}{N^2}(\tilde{z}_{N,N} - \tilde{z}_{0,0}) + \frac{1}{N^2} I_{(0,0)}^{(1,1)}(v_{ij}\,|_{z=0}) + (\tilde{z}_{i-1,j} - \tilde{z}_{i-1,j-1}) \\
&\quad + \tilde{z}_{i, j-1}.
\end{aligned}
$$

Hence the $\tilde{z}_{ij}$ can be expressed in terms of $\tilde{z}_{0,0}$, C_{ij}, and $I_{(0,0)}^{(1,1)}(v_{ij}\,|_{z=0})$, $(i, j) \neq (0, 0)$. These results are now summarized in a theorem.

Theorem 8.12 *Let graph f be a fractal surface generated by the IFS $(Q \times \mathbb{R}, \mathbf{w})$, where $w_{ij} = (u_{ij}, v_{ij})$ with*

$$u_{ij}(x, y) = \begin{pmatrix} \frac{1}{N} & 0 \\ 0 & \frac{1}{N} \end{pmatrix} \begin{pmatrix} x \\ y \end{pmatrix} + \begin{pmatrix} \frac{i-1}{N} \\ \frac{j-1}{N} \end{pmatrix}$$

and

$$v_{ij}(x, y, z) = A_{ij} x^2 + B_{ij} y^2 + C_{ij} z + D_{ij} xy + G_{ij},$$

such that $\max_{i,j} |C_{ij}| < 1$. Let

$$\tilde{f}(x, y) := \tilde{z}_{0,0} + \int_0^x \int_0^y f(s, t)\, dt\, ds, \qquad \text{for some } \tilde{z}_{0,0} \in \mathbb{R}.$$

Then graph $\tilde{f}$ is the attractor of the IFS $(Q \times \mathbb{R}, \tilde{\mathbf{w}})$ with $\tilde{w}_{ij} = (u_{ij}, \tilde{v}_{ij})$, where

$$\tilde{v}_{ij}(x, y, z) = \tilde{z}_{i-1, j-1} + \frac{C_{ij}}{N^2} z + \frac{1}{N^2} \int_0^x \int_0^y v_{ij}(s, t, 0)\, dt\, ds,$$

$i, j = 1, \ldots N$. *Furthermore, the* $\tilde{z}_{ij}$, $(i,j) \neq (0,0)$, *are recursively and uniquely determined by the free parameter* $\tilde{z}_{0,0}$, *the constants* C_{ij}, *and* $\int_0^x \int_0^y v_{ij}(s,t,0)\, dt\, ds$. *Also,* $\nabla \tilde{f}(x,y) = (g_y(x), h_x(y))$, *where*

$$g_y(x) = \int_0^y f(x,t)\, dt \quad and \quad h_x(y) = \int_0^x f(s,y)\, ds.$$

Moreover, $\frac{\partial}{\partial x}\frac{\partial}{\partial y}\tilde{f} = \frac{\partial}{\partial y}\frac{\partial}{\partial x}\tilde{f} = f$.

Proof. The last part of the theorem follows from calculus. ∎

It should now be clear how one can construct C^n-interpolating fractal surfaces, $n \in \mathbb{N}$: the foregoing procedure can be iterated an arbitrary number of times.

Chapter 9

Fractal Wavelets in $\mathbb{R}^n$

In this chapter multivariate $\mathbb{R}$-valued fractal functions are used to construct wavelet bases in $\mathbb{R}^n$. In order to obtain nested subspaces, a prerequisite for the existence of a multiresolution analysis, *foldable figures* have to be chosen as the domains of the multivariate fractal functions. These foldable figures are in one-to-one correspondence with so-called *crystallographic Coxeter groups*. It is therefore necessary to briefly review some concepts from the theory of Coxeter groups, or more generally, the theory of buildings. This is done in the first section. Unfortunately, only the most rudimentary concepts of this fascinating subject can be introduced. The reader who is interested in pursuing the subject further is referred to an — albeit incomplete — list of references in the bibliography.

In the second section a multiresolution analysis on $L^2(\mathbb{R}^n)$ is presented and the corresponding wavelet basis constructed.

9.1 Brief Review of Coxeter Groups

This section is a compendium of definitions and theorems from the theory of Coxeter groups, or more generally, the theory of buildings. Because of the limited scope of the presentation, only definitions and results relevant to the developments in this chapter are presented.

Throughout this section, $\mathbb{E}$ denotes a fixed *Euclidean space*, i.e., a finite-dimensional vector space over $\mathbb{R}$ endowed with a positive definite symmetric bilinear form $\langle \cdot, \cdot \rangle$. A *(linear) hyperplane in* $\mathbb{E}$ is a co-dimension one linear subspace of $\mathbb{E}$ and an *affine hyperplane* of $\mathbb{E}$ is a subset of $\mathbb{E}$ of the form $x + H_0$ with $x \in \mathbb{E}$ and H_0 a linear hyperplane of $\mathbb{E}$. Affine hyperplanes may be defined via linear equations of the form $f = c$, where $f : \mathbb{E} \to \mathbb{R}$ is a non-zero linear mapping and $c \in \mathbb{R}$. An *affine isometry* is a norm-preserving affine map. Let H_0 be a linear hyperplane. A linear transformation $r_{H_0} : \mathbb{E} \to \mathbb{E}$ is called a *reflection in* $\mathbb{E}$ *with respect to* H_0 iff $r_{H_0}|_{H_0} = id_{H_0}$ and $r_{H_0}(h^\perp) = (-1)h^\perp$, for all $h^\perp \in H_0^\perp$, the one-dimensional orthogonal complement of H_0 in $\mathbb{E}$. Now let H be an affine hyperplane and H_0 the linear hyperplane parallel to it. Then there exists an $x \in \mathbb{E}$ such that $H = x + H_0$. The reflection r_H with respect to H is defined by $r_H = \tau_x \circ r_{H_0} \circ \tau_{-x}$, where r_{H_0} is the orthogonal reflection with respect to H_0 and $\tau_x : \mathbb{E} \to \mathbb{E}$ is the translation $y \mapsto y + x$. Clearly, r_H is an (affine) isometry. Also, any non-zero vector $x \in \mathbb{E}$ determines a reflection r_x, with *reflecting hyperplane* $H_x := \{y \in \mathbb{E} \,|\, \langle y, x \rangle = 0\}$. It is easy to derive an explicit formula for r_x:

$$r_x(y) = y - \frac{2\langle y, x \rangle}{\langle x, x \rangle}\, x =: y - (y, x)\, x.$$

The *general linear group of* $\mathbb{E}$, $\mathrm{GL}(\mathbb{E})$, consists of all the invertible endomorphisms of $\mathbb{E}$. The next result characterizes reflections to a certain extent.

Proposition 9.1 *Let B be a finite subset of the set $\mathbb{E}$ with the property that* $\mathrm{span}(B) = \mathbb{E}$. *Suppose that all reflections r_x, $x \in B$, leave B invariant. If $r \in \mathrm{GL}(\mathbb{E})$ leaves B invariant, fixes pointwise a hyperplane H of $\mathbb{E}$, and satisfies $r(x) = -x$, for some $x \in B$, then $r = r_x$ and $H = H_x$.* ∎

The following definition introduces the concept the *root system of* $\mathbb{E}$, a concept that is also of great importance in the theory of Lie groups.

Definition 9.1 A subset $\mathcal{R}$ of the set $\mathbb{E}$ is called a root system in $\mathbb{E}$ iff the following axioms are satisfied:

(a) $|\mathcal{R}| < \infty$, $\mathrm{span}(\mathcal{R}) = \mathbb{E}$, and $0 \notin \mathcal{R}$.

(b) $x \in \mathcal{R} \wedge \alpha x \in \mathcal{R} \implies \alpha = \pm 1$.

(c) $\forall\, x \in \mathcal{R} : r_x \mathcal{R} = \mathcal{R}$.

(d) $x, y \in \mathcal{R} \implies (y, x) \in \mathbb{Z}$.

Let $\mathcal{R}$ be a root system of $\mathbb{E}$ and let $\overline{W}$ be the subgroup of $\mathrm{GL}(\mathbb{E})$ that is generated by the reflections r_x, $x \in \mathcal{R}$. The invariance of $\mathcal{R}$ under the reflections r_x, $x \in \mathcal{R}$ implies that $\overline{W}$ permutes the set $\mathcal{R}$, and thus can be identified with a subgroup of the symmetric group on $\mathcal{R}$. Note that $\overline{W}$ is finite. The group $\overline{W}$ is called a *Weyl group for* $\mathrm{GL}(\mathcal{R})$.

Suppose $\mathcal{H}$ is a collection of affine hyperplanes of $\mathbb{E}$. Denote by $\mathcal{W}$ the group of affine isometries generated by reflections r_H with $H \in \mathcal{H}$. A collection $\mathcal{H}$ is called $\mathcal{W}$-invariant iff $r_H \mathcal{H} = \mathcal{H}$, for all $r_H \in \mathcal{W}$. The group $\mathcal{W}$ is said to be an *affine reflection group* if there exists a $\mathcal{W}$-invariant family of hyperplanes $\mathcal{H}$ that is locally finite, that is, every point in $\mathbb{E}$ has a neighborhood which intersects only finitely many $H \in \mathcal{H}$. Shortly, a relationship between the Weyl group $\overline{\mathcal{W}}$ and the group $\mathcal{W}$ will be exhibited.

A given collection $\mathcal{H}$ of hyperplanes partitions $\mathbb{E}$ into convex *cells*, these cells being defined by $f = c$, $f > c$, or $f < c$, where $f = c$ defines a hyperplane $H \in \mathcal{H}$, $c \in \mathbb{R}$. The *support* of a cell is the linear subspace L defined by the inequalities $f = 0$ that occur in its description. (If there are no such equalities, then $L := \mathbb{E}$.) The dimension of a cell is the dimension of its support. A *chamber* is a cell of maximal dimension $\dim \mathbb{E}$. The chambers are the connected components of the complement of $\mathcal{H}$ in $\mathbb{E}$. The *walls* of a chamber are the supports of its co-dimension one faces.

Now choose a chamber C and denote by R the set of all reflections with respect to the walls of C. Then the following facts hold:

Theorem 9.1 1. *R generates $\mathcal{W}$.*

2. *$\mathcal{W}$ is simply-transitive on the chambers.*

3. *$\mathcal{H}$ is the set of all hyperplanes $H \in \mathbb{E}$ with $r_H \in \mathcal{W}$.*

4. *The closure $\overline{C}$ of C is a fundamental domain for the action of $\mathcal{W}$ on $\mathbb{E}$, i.e., no $r_H \in \mathcal{W}$ maps a point of $\overline{C}$ to another point of $\overline{C}$, and for all $x \in \mathbb{E}$ there exists an $r_H \in \mathcal{W}$ such that $r_H(x) \in \overline{C}$.*

The next theorem addresses some finiteness questions.

Theorem 9.2 1. *C has only finitely many walls, and thus R is finite.*

2. *There are only finitely many linear hyperplanes H_0 such that $\mathcal{H}$ contains a translate of H_0.*

1. *Denote by $\overline{\mathcal{W}}$ the set of linear parts of all elements in $\mathcal{W}$. Then $\overline{\mathcal{W}}$ is a finite reflection group.*

Remark. A finite reflection group, such as $\overline{\mathcal{W}}$, is called a *Coxeter group.*

The following theorem deals with the structure of C. But first a few more definitions.

The group $\mathcal{W}$ is called *essential* if its associated finite reflection group $\overline{\mathcal{W}}$ is essential, that is, if the origin is the only point fixed by all $r \in \overline{\mathcal{W}}$. A reflection group is *irreducible* iff it cannot be expressed as a product of reflection groups.

Theorem 9.3 *Suppose that $\mathcal{W}$ is essential and irreducible. Then the chamber C has exactly $1 + \dim \mathbb{E}$ walls and is a simplex in $\mathbb{E}$. Furthermore, $\mathcal{W}$ is infinite.*

An essential irreducible infinite affine reflection group is called a *Euclidean reflection group* or an *affine Weyl group*.

Finally, consider the structure of $\mathcal{W}$. Since $\mathcal{W}$ acts on $\mathbb{E}$, the *stabilizer of* $x \in \mathbb{E}$ is the subgroup $\mathcal{W}_x$ of $\mathcal{W}$ given by $\{r_H \in \mathcal{W} : r_H x = x\}$. Before the next theorem can be stated, the *semi-direct product* of groups has to be introduced.

Definition 9.2 Let G be a group and H a normal subgroup of G. If there exists a subgroup K of G so that $G = HK$ and $H \cap K = \{e\}$ then G is said to be the semi-direct product of H and K, in symbols, $G = H \ltimes K$.

Theorem 9.4 1. *There exist points $x \in \mathbb{E}$ such that the stabilizer $\mathcal{W}_x$ is isomorphic to $\overline{\mathcal{W}}$.*

2. *Let $\Gamma := \{x \in \mathbb{E} \mid \tau_x \in \mathcal{W}\}$. By an appropriate choice of the origin, it may be assumed that $0 \in \Gamma$. Then Γ is a lattice in $\mathbb{E}$, that is, a subgroup of $\mathbb{E}$ of the form $\mathbb{Z}e_1 \oplus \cdots \oplus \mathbb{Z}e_n$, for some $\mathbb{R}$-basis $\{e_1, \ldots, e_n\}$ of $\mathbb{E}$. Furthermore, $\mathcal{W}$ is isomorphic to the semi-direct product $\Gamma \ltimes \overline{\mathcal{W}}$.*

Remark. It follows from the second statement in Theorem 9.4 that the finite reflection group $\overline{\mathcal{W}}$ leaves the lattice Γ invariant. Such groups are called *crystallographic*. There is a classification of these crystallographic Coxeter groups in terms of so-called *Coxeter or Dynkin diagrams*. For each $n = \dim \mathbb{E}$ there is only a finite number of such groups. Each such diagram not only represents the underlying group but also gives the explicit geometry of the chambers which, in the case of an essential and irreducible group $\mathcal{W}$, are simplices in $\mathbb{E}$. These simplices have to be used as domains for multivariate fractal functions in order to generate nested subspaces of $L^2(\mathbb{R}^n)$.

Definition 9.3 A compact and connected subset F of $\mathbb{E}$ is called a *foldable figure* iff there exists a finite set of hyperplanes in $\mathbb{E}$ that cuts F into finitely many congruent subfigures $F_1, \ldots, F_M$ each similar to F, so that the reflection in any of these hyperplanes bounding F_m takes it into some $F_{m'}$.

The following results relate foldable figures to Coxeter groups ([95]).

Theorem 9.5 *A foldable figure in* $\mathbb{E}$ *is a convex polytope that tessellates* $\mathbb{E}$ *by reflections in hyperplanes. Furthermore, foldable figures are in one-to-one correspondence with crystallographic Coxeter groups.*

Example 9.1 The standard simplex σ^2 is an example of a foldable figure in $\mathbb{E} := \mathbb{R}^2$.

The following are examples of foldable figures in $\mathbb{R}^3$ whose associated (crystallographic) Coxeter groups are reducible.

Example 9.2 (i) The three-dimensional unit cube $[0,1] \times [0,1] \times [0,1]$.

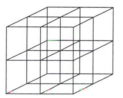

(ii) The three-dimensional prism $\sigma^2 \times [0,1]$.

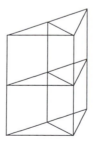

9.2   Fractal Functions on Foldable Figures

Fractal surfaces defined over a given foldable figure are now constructed. Let F be a foldable figure in $\mathbb{E} := \mathbb{R}^n$ with the property that 0 is one of its vertices. Let $\mathfrak{S}$ be the tesselation of $\mathbb{R}^n$ and $\mathcal{H}$ be the set of hyperplanes associated with F, and let $\mathcal{W}$ be the affine reflection group generated by $\mathcal{H}$, i.e., $\mathcal{W}$ is the group of affine isometries in $\mathbb{R}^n$ generated by the reflections r_H for $H \in \mathcal{H}$. Then the results in the previous section give the following properties of $\mathcal{H}$ and $\mathcal{W}$:

1. $\mathcal{H}$ consists of the translates of a finite set of linear hyperplanes.

2. $\mathcal{W}$ is simply transitive on $\mathfrak{S}$, i.e., for any $\mathfrak{s}, \mathfrak{s}' \in \mathfrak{S}$ there exists a unique element $r_{\mathfrak{s}'\mathfrak{s}} \in \mathcal{W}$ mapping $\mathfrak{s}$ onto $\mathfrak{s}'$.

3. $\varkappa\mathcal{H} \subseteq \mathcal{H}$ for any $\varkappa \in \mathbb{N}$, where $\varkappa\mathcal{H} := \{\varkappa H : H \in \mathcal{H}\}$.

Let $\varkappa$ be a fixed positive integer and let $X := \varkappa F$. Clearly, X is also a foldable figure whose N subfigures X_i are in $\mathfrak{S}$. Note that the tessellation and the set of hyperplanes associated with X are $\varkappa\mathfrak{S}$ and $\varkappa\mathcal{H}$, respectively. Furthermore, the affine reflection group generated by $\varkappa\mathcal{H}$ is an isomorphic subgroup of $\mathcal{W}$. The fact that the group $\mathcal{W}$ is simply transitive on $\mathfrak{S}$ yields a set of similitudes $u_i : X \to X_i$, $i = 1, \ldots, N$, $N = \varkappa^n$, in the following way. First, let $u_1 : X \to X_1 := F$ be given by $u_1 := (1/\varkappa)\,\mathrm{id}_{\mathbb{R}^n}$. Then define $u_j : X \to X_j$, by

$$u_j := r_{X_j X_1} \circ u_1, \tag{9.1}$$

for all $j \in \{2, \ldots, N\}$.

Let C^N consist of all N-tuples $\lambda := (\lambda_1, \ldots, \lambda_N)$ of continuous $\mathbb{R}$-valued functions λ_i on X satisfying the following property:

(W) If $u_i X$ and $u_j X$ have a common face E_{ij}, then $\lambda_i(x) = \lambda_j(x)$ for all $x \in u_i^{-1} E_{ij} = u_j^{-1} E_{ij}$, $i, j = 1, \ldots, N$.

For $\lambda = (\lambda_1, \ldots, \lambda_N) \in C^N$ and a fixed $-1 < s < 1$, let $\Phi_\lambda : C(X, \mathbb{R}) \to C(X, \mathbb{R})$ be the Read-Bajraktarević operator

$$\Phi_\lambda f := \sum_{i=1}^{N} \left[\lambda_i \circ u_i^{-1} + s f \circ u_i^{-1}\right] \chi_{u_i X}. \tag{9.2}$$

Theorem 9.6 *The mapping Φ_λ given by (9.2) is well-defined and contractive with contractivity factor s in the sup-norm. Hence, it possesses a unique fixed point $^*f_\lambda \in C(X, \mathbb{R})$. Furthermore, the mapping $\theta : C^N \to C(X, \mathbb{R})$,*

$$\lambda \mapsto {}^*f_\lambda, \tag{9.3}$$

is linear.

Proof. Suppose E_{ij} is a common face of $u_i X$ and $u_j X$. Since $\mathcal{W}$ is simply transitive on $\mathfrak{S}$, it follows that $u_j := r_{X_j X_i} \circ u_i$. As $r_{X_j X_i}$ is the identity on E_{ij}, one has $u_i^{-1}(x) = u_j^{-1}(x)$, for all $x \in E_{ij}$. Thus $\lambda_i \circ u_i^{-1}(x) + s f \circ u_i^{-1}(x) = \lambda_j \circ u_j^{-1}(x) + s f \circ u_j^{-1}(x)$, for all $x \in u_i X \cap u_j X$. Hence, for any $f \in C(X, \mathbb{R}^m)$, $\Phi_\lambda(f)$ is well-defined and continuous on X.

Now let $f, g \in C(X, \mathbb{R}^m)$. Then

$$\|\Phi_\lambda(f) - \Phi_\lambda(g)\|_\infty = \sup_{\substack{x \in X \\ i=1,\ldots,N}} \left\{ \left| s(f \circ u_i^{-1}(x) - g \circ u_i^{-1}(x)) \right| \right\}$$

$$\leq s \|f - g\|_\infty.$$

Therefore, by the Banach Fixed-Point Theorem, Φ_λ has a unique fixed point $^*f_\lambda \in C(X, \mathbb{R})$. The linearity of θ follows directly from Eq. (9.2). ∎

The fixed point $^*f_\lambda \in C(X, \mathbb{R})$ is again called a *multivariate fractal function* and, for $n > 1$, its graph a *fractal surface*.

Let $\prod C := \prod_{r \in \mathcal{W}} C^N$ be the direct product of C^N over $\mathcal{W}$ and let $B_c(\mathbb{R}^n)$ denote the linear space of real-valued functions bounded on compact subsets of $\mathbb{R}^n$. For $\lambda \in \prod C$, let $^*f_\lambda \in B_c(\mathbb{R}^n)$ be defined by

$$^*f_\lambda|_{r \overset{\circ}{X}} := {}^*f_{\lambda(r)} \circ r^{-1}, \qquad r \in W, \tag{9.4}$$

where $\lambda(r) := (\lambda(r)_1, \ldots, \lambda(r)_N)$ is the "rth component" of λ.

Remark. The values of $^*f_\lambda$ on $\varkappa \mathcal{H}$ will be left unspecified so that f_λ actually represents an equivalence class of functions.

Fig. 9.1 shows the graph of a univariate fractal function $^*f_\lambda$ defined on the interval $[0, 1] = [0, 1/2] \cup [1/2, 1] = r[0, 1/2]$, where r denotes the reflection about the line $x = 1/2$, and Fig. 9.2 that of a bivariate fractal function $^*g_\lambda$ defined on the standard simplex σ^2 made up of four congruent subsimplices, as in Example 9.1.

9.3 Interpolation on Foldable Figures

In this section it is shown that continuous fractal functions constructed on a given foldable figure X can be used to interpolate a given set of data points. For this purpose let $\overline{W}$ be the Coxeter group associated with the foldable figure X. If $\overline{W}$ is reducible, let $\overline{W}_1, \ldots, \overline{W}_M$ denote its irreducible factors. Let X_m be the foldable figure corresponding to $\overline{W}_m$, $m = 1, \ldots, M$.

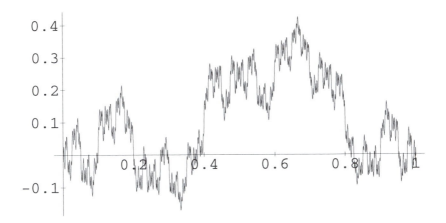

Figure 9.1: The graph of $^*f_\lambda$.

It is clear that X is isometrically isomorphic to $\times_{m=1}^{M} X_m$. Without loss of generality, it may be assumed that $X = \times_{m=1}^{M} X_m$ and that the origin is in the set V of vertices of X. By Theorem 9.3, each X_m is a simplex in $\mathbb{R}^{n_m}$ with $n_m + 1$ vertices. Let Λ_m be the collection of affine maps $\lambda : \mathbb{R}^{n_m} \to \mathbb{R}$ and let $\Lambda := \prod_{m=1}^{M} \Lambda_m$.

Let $\varkappa$ be a positive integer, let $n := n_1 + \cdots + n_m$, and let $N = \varkappa^n$. Define $u_1, \ldots, u_N$ and C^N as in the previous section. Denote the set of vertices of $u_i X$ by V_i.

Theorem 9.7 *Let* $\Delta := \{(v, z_v) \,|\, v \in \bigcup \mathsf{V}_i\}$ *be a given interpolating set in* $\mathbb{R}^n \times \mathbb{R}$. *Then there exists a unique* $\lambda := (\lambda_1, \ldots, \lambda_N)$ *such that each* $\lambda_i \in \Lambda$ *and* $\lambda \in C^N$. *Moreover,* $f_\lambda(v) = z_v$, *for all* $v \in \bigcup \mathsf{V}_i$.

Proof. First, suppose that $M = 1$, i.e., $\overline{W}$ is irreducible. Hence, X is a simplex and so the vertices of X are geometrically independent. Thus, given real numbers w_v, $v \in \mathsf{V}$, there exists a unique (affine) $\lambda \in \Lambda$ satisfying

$$\lambda(v) = w_v. \tag{9.5}$$

Using properties of the tensor product, it follows inductively that there is a unique $\lambda \in \Lambda$ satisfying Eq. (9.5) in the $M > 1$ case.

Now let λ_i, $i = 1, \ldots, N$, be the unique $\lambda_i \in \Lambda$ with the property

$$\lambda_i(v) = z_{u_i(v)} - s z_v, \qquad v \in \bigcup \mathsf{V}_i, \tag{9.6}$$

and define $\lambda := (\lambda_1, \ldots, \lambda_N)$.

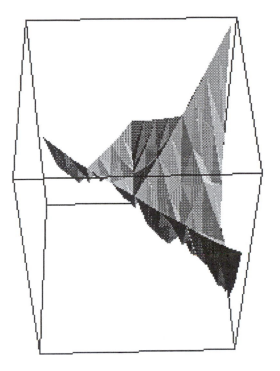

Figure 9.2: The graph of $^*g_\lambda$.

If E_{ij} is a common face of $u_i X$ and $u_j X$, then λ_i and λ_j agree on the vertices of $X_{ij} := u_i^{-1} E_{ij} = u_j^{-1} E_{ij}$. Note that X_{ij} is a foldable figure and so, by the unique interpolation property, $\lambda_i \equiv \lambda_j$ on X_{ij}. Hence, $\lambda \in C^N$.

Equations (9.2) and (9.6) imply that $f_\lambda(v) = z_v$, for all $v \in \bigcup V_i$. ∎

9.4 Dilation and $\mathcal{W}$ Invariant Spaces

Let V be a linear space of real-valued functions on $\mathbb{E} := \mathbb{R}^n$, $\varkappa$ a positive integer greater than one, and $\mathcal{W}$ a crystallographic Coxeter group.

Definition 9.4 The linear space V is said to be dilation-invariant (with scale $\varkappa$) iff $f \in V$ implies $U_\varkappa f := f(\,\cdot\,/\varkappa) \in \mathbb{V}$. It is called $\mathcal{W}$-invariant iff $f \in V$ implies $f \circ r \in V$ for all $r \in \mathcal{W}$.

Remark. Since the translations along the lattice Γ are contained in $\mathcal{W}$, any $\mathcal{W}$-invariant linear space V is also translation-invariant with respect to the lattice Γ, i.e., $f \circ \tau_\gamma \in V$ for any $\gamma \in \Gamma$.

Definition 9.5 *Let* $\Delta_\varkappa : \prod C \to \prod C$ *be defined by*

$$\Delta_\varkappa \lambda(\varkappa r \circ u_j)_i := \lambda(r)_j \circ u_i + s(\lambda(r)_i - \lambda(r)_j), \tag{9.7}$$

for $r \in \mathcal{W}$, $i, j = 1, \ldots, N$.

Remark. Equation (9.1) implies that $\varkappa r \circ u_j \in \mathcal{W}$ and, furthermore, that for any $r' \in \mathcal{W}$ there is some $r \in \mathcal{W}$ and some $j \in \{1, \ldots, N\}$ such that $r' = \varkappa r \circ u_j$.

Theorem 9.8 *Let* $\lambda \in \prod C$. *Then*

$$U_\varkappa f_\lambda = f_{\Delta_\varkappa \lambda}. \tag{9.8}$$

Proof. First it is shown that for any $r \in \mathcal{W}$ and $j = 1, \ldots, N$,

$$f_{\lambda(r)} \circ u_j = f_{\Delta_\varkappa \lambda(\varkappa r \circ u_j)}. \tag{9.9}$$

To see this, note that

$$
\begin{aligned}
f_{\lambda(r)} \circ u_j \circ u_i &= \lambda(r) \circ u_i + s f_{\lambda(r)} \circ u_i \\
&= \lambda(r) \circ u_i + s(\lambda(r) \circ u_i + s f_{\lambda(r)}) \\
&= \lambda(r) \circ u_i + s(\lambda(r) \circ u_i - \lambda(r) \circ u_j) \\
&\quad + s(\lambda(r) \circ u_j - s f_{\lambda(r)}) \\
&= \Delta_\varkappa \lambda(\varkappa r u_j) \circ u_i + s f_{\lambda(r)} \circ u_j.
\end{aligned}
$$

Equation (9.9) now follows from the uniqueness of the fixed point of $\Phi_{\Delta_\varkappa \lambda}$. Finally,

$$
\begin{aligned}
U_\varkappa f_\lambda|_{\varkappa r \circ u_j(X)} &= U_\varkappa f_\lambda|_{r \circ u_j X} = U_\varkappa f_{\lambda(r)} \circ r^{-1}|_{r \circ u_j X} \\
&= f_{\lambda(r)}|_{u_j X} = f_{\lambda(r)} \circ u_j \circ u_j^{-1} \circ (\varkappa \circ r)^{-1} \\
&= f_{\Delta_\varkappa \lambda(\varkappa r \circ u_j)} \circ (\varkappa r \circ u_j)^{-1}.
\end{aligned}
$$

■

Theorem 9.8 may be used to construct $U_\varkappa$-invariant function spaces from $\Delta_\varkappa$-invariant subspaces of $\prod C$. For example, let Λ_d be the collection of all $\lambda \in C^N$ such that each component λ_i is a polynomial of degree at most d. It is easy to verify that $\prod_{r \in \mathcal{W}} \Lambda_d$ is $\Delta_\varkappa$ invariant.

9.5 Multiresolution Analyses

Next, it is shown that the function spaces constructed previously form multiresolution analyses of $L^2(\mathbb{R}^n)$. To this end, let Λ be a finite-dimensional subspace of C^N such that $\prod_{r \in \mathcal{W}} \Lambda$ is $\Delta_{\varkappa}$ invariant. Define

$$V_0 := \left\{ f_\lambda \,\middle|\, \lambda \in \prod_{r \in \mathcal{W}} \Lambda \right\} \tag{9.10}$$

and V_k, for $0 \neq k \in \mathbb{Z}$, by $f \in V_k$ iff $U_{\varkappa^{-k}} f \in V_0$. Let $A := \dim \Lambda$. Clearly, $V_0|_X$ also has dimension A. Therefore, an orthonormal basis $\{\phi^1, \ldots, \phi^A\}$ of $V_0|_X$ can be constructed using, for instance, the Gram-Schmidt orthonormalization algorithm. Then $\mathcal{B}_\phi := \{\phi^a \circ r \,|\, a = 1, \ldots, A \wedge r \in \mathcal{W}\}$ is an orthonormal basis of V_0. Let $\phi := (\phi^1, \ldots, \phi^A)^t$. Since $V_1 \subset V_0$, there exists a sequence of $A \times A$ matrices $\{\mathbf{c}_r\}_{r \in \mathcal{W}}$, only a finite of which are different from the zero matrix $\mathbf{0}$ such that

$$\phi(x/\varkappa) = \sum_{r \in \mathcal{W}} \mathbf{c}_r (\phi \circ r)(x). \tag{9.11}$$

Remark. Since $\mathcal{W}$ is the semi-direct product of Γ with $\overline{\mathcal{W}}$, the basis $\mathcal{B}_\phi$ can also be generated by translates along the lattice Γ of the enlarged set of scaling functions $\{\phi \circ r \,|\, r \in \overline{\mathcal{W}}\}$.

Theorem 9.9 *The sequence $\{V_k\}_{k \in \mathbb{Z}}$ defined earlier is a multiresolution analysis of $L^2(\mathbb{E})$ with respect to $\mathcal{W}$.*

Proof. Conditions (M1), (M4), and (M5) required for a multiresolution analysis follow directly from the preceding construction. Conditions M2, M3, and M6 are a consequence of Theorem 2.1 in [138]. For the sake of completeness this theorem is restated next using the notation and terminology developed in this monograph. ∎

Theorem 9.10 ([138]) *Let $\phi \in L^2(\mathbb{R})$ and let $\phi_\mathbf{p} := \sum_{\ell \in \mathbb{Z}} |\phi(\cdot - \ell)|$ be its periodization. Denote by $\mathcal{L}^2(\mathbb{R})$ the set of all Lebesgue measurable functions ϕ for which $\|\phi_\mathbf{p}\|_{L^2([0,1]^n)} < \infty$.*

Suppose that $\phi \in \mathcal{L}^2(\mathbb{R})^A$ and that there exists a positive constant m such that

$$m \, \|\mathbf{s}\|_{\ell^2(\mathbb{Z}^n)^A} \leq \|\mathbf{s} \circledast \phi\|_{L^2(\mathbb{R})^A} .$$

If there exists a bi-infinite sequence $\mathbf{c}$ of $A \times A$ matrices whose elements are in $\ell^2(\mathbb{Z}^n)$ so that

$$\phi = \sum_{\ell \in \mathbb{Z}^n} \mathbf{c}_\ell \, \phi(2 \cdot -\ell),$$

then ϕ admits multiresolution.

Next, the wavelets associated with this multiresolution analysis are defined. For this purpose denote the orthogonal complement of V_k in V_{k-1} by W_k. First, a basis for W_0 needs to be found. Note that the function space $V_{-1}|_X$ has dimension $\varkappa^n A$, while $V_0|_X$ is an A-dimensional subspace of V_{-1}. Thus, again using the Gram-Schmidt orthonormalization algorithm, an orthonormal basis of $B = (\varkappa^n - 1)A$ functions $\psi^1, \ldots, \psi^B$ can be constructed. Then, obviously, $\mathcal{B}_\psi := \{\psi^b \circ r \,|\, b = 1, \ldots, B; \; r \in \mathcal{W}\}$ is an orthonormal basis of W_0. The functions $\psi^1, \ldots, \psi^B$ are now the wavelets, and $\mathcal{B}_\psi$ is the wavelet basis. Note that the scaling functions $\phi^1, \ldots, \phi^A$, as well as the wavelets $\psi^1, \ldots, \psi^B$, are orthogonal and compactly supported, but *not* continuous. Also, W_k is orthogonal to W_ℓ for $k \neq \ell$. Clearly, the W_k's form an orthogonal direct sum decomposition of $L^2(\mathbb{R}^n)$.

Let $\boldsymbol{\psi} := (\psi^1, \ldots, \psi^B)^t$. Since $W_1 \subset V_0$, there exists a sequence of $B \times A$ matrices $\{\mathbf{d}_r\}_{r \in \mathcal{W}}$ such that

$$\boldsymbol{\psi}(x/\varkappa) = \sum_{r \in \mathcal{W}} \mathbf{d}_r (\phi \circ r)(x). \tag{9.12}$$

As $\boldsymbol{\psi}$ is compactly supported, only a finite number of the $\mathbf{d}_r$'s are non-zero. The orthonormality of $\mathcal{B}_\phi$ and $\mathcal{B}_\psi$ leads to simple decomposition and reconstruction algorithms. Let $f_0 \in V_0$. Then $f_0 = f_1 + g_1$, where f_1 and g_1 denote the orthogonal projections of f_0 onto V_1 and W_1, respectively. Expressing $f_0 =, f_1$, and g_1 in terms their respective bases yields

$$f_0 = \sum_{r \in \mathcal{W}} \mathbf{a}_r^t(0) \, \phi \circ r,$$

$$f_1 = \sum_{r \in \mathcal{W}} \mathbf{a}_r^t(1) \, \phi \circ r,$$

and

$$g_1 = \sum_{r \in \mathcal{W}} \mathbf{b}_r^t(1) \, \psi \circ r,$$

where the coefficients are row matrices of the appropriate size. Equations (9.11) and (9.12) now give the decomposition algorithm

$$\mathbf{a}_r(1) = \sum_{r' \in \mathcal{W}} a_{r \varkappa r \varkappa - 1}(0)\mathbf{c}_{r'}^t,$$

and

$$\mathbf{b}_r(1) = \sum_{r' \in \mathcal{W}} b_{r \varkappa r \varkappa - 1}(0)\mathbf{d}_{r'}^t,$$

as well as the reconstruction algorithm

$$\mathbf{a}_r(0) = \sum_{r' \in \mathcal{W}} \mathbf{c}_{r'}^t a_{r \varkappa r \varkappa - 1}(1) + \mathbf{d}_{r'}^t b_{r \varkappa r \varkappa - 1}(1).$$

List of Symbols

Π_ε, 331

Σ, 25

(Σ, d_F), 26

$(\Sigma, \mathcal{F}, \mu)$, 26

Υ, 76

Φ, 136

Ψ, 196

$(\Omega, \mathcal{F}, \mu)$, 18

$(\Omega, \mathcal{F}, \pi)$, 18

∂, 88

X, 62

$\preceq$, 7

$\rightarrowtail$, 29

$\twoheadrightarrow$, 29

$\hookrightarrow$, 35

$*$, 225

$\otimes$, 306

$\ltimes$, 348

$\prod B^N$, 260

$\prod B_{\mathbb{R}}(I)$, 262

$\prod C^r$, 263

$\prod \mathcal{P}^d$, 264

$\prod \mathcal{P}_r^d$, 264

$\prod \theta$, 261

$\prod \tau$, 262

$(\prod_{i \in I} G_i, \theta)$, 33

$(\bigoplus_{i \in I} G_i, \theta)$, 33

$|\cdot|$, 43

$\langle \cdot, \cdot \rangle$, 4

$\ll$, 123

$[\![\cdot]\!]$, 185

$^\#f_i$, 195

$\|\cdot\|$, 4

$\|\cdot\|_c$, 27

$\|\cdot\|_p$, 22

$\|\cdot\|_\infty$, 22

$\|\cdot\|_{B_p^{\alpha,q}}$, 223

$\|\cdot\|_{\dot{B}_p^{\alpha,q}}$, 222

$\|\cdot\|_{C^k}$, 195

$\|\cdot\|_{C^{k,\alpha}}$, 219

$\|\cdot\|_{V^\alpha}$, 231

$\|\cdot\|_{W_q^s}$, 220

$\|\cdot\|_{X_0 \cap X_1}$, 224

$\|\cdot\|_{X_0 + X_1}$, 224

$\|\cdot\|_{\Lambda_\alpha}$, 231

$\|\cdot\|_{\dot{\Lambda}_\alpha}$, 231

$\|\cdot\|_{b_p^{\alpha,q}}$, 228

$\|\cdot\|_{\dot{b}_p^{\alpha,q}}$, 226

$\|\cdot\|_{b_p^{\alpha,q}(\mathbb{T}^n)}$, 229

$\|\cdot\|_{\dot{b}_p^{\alpha,q}(\mathbb{T}^n)}$, 229

Bibliography

[1] M. Abramowitz and I. Stegun, *Handbook of Mathematical Functions*, Dover Publications, Inc., New York, 1972.

[2] R. A. Adams, *Sobolev Spaces*, Pure and applied mathematics series, Vol. 65, Academic Press, New York, 1975.

[3] B. Alpert, *Sparse Representation of Smooth Operators*, Ph.D. Thesis, Yale University, 1990.

[4] M. Bajraktarević, "Sur une équation fonctionelle," *Glasnik Mat.-Fiz. Astr. Ser. II* **12**(1956), 201—205.

[5] Ch. Bandt, "Self-similar sets. I. Topological Markov chains and mixed self-similar sets". *Math. Nachr.* **142**(1989), 107—123.

[6] Ch. Bandt, "Self-similar sets. III. Constructions with sofic systems," *Mh. Math.* **108**(1989), 89—102.

[7] M. F. Barnsley, *Fractals everywhere*, Academic Press, Orlando, Florida, 1988.

[8] M. F. Barnsley, "Fractal functions and interpolation," *Constr. Approx.* **2**(1986), 303—329.

[9] M. F. Barnsley and S. Demko, "Iterated function systems and the global construction of fractals," *Proc. R. Soc. Lond. A* **399**(1985), 243—275.

[10] M. F. Barnsley, J. H. Elton, and D. P. Hardin, "Recurrent iterated function systems," *Constr. Approx.* **5**(1)(1989), 3—31.

[11] M. F. Barnsley, J. Elton, D. P. Hardin, and P. R. Massopust, "Hidden variable fractal interpolation functions," *SIAM J. Math. Anal.* **20**(5) (1989), 1218—1242.

[12] M. F. Barnsley and A. N. Harrington, "The calculus of fractal interpolation functions," *J. Approx. Th.* **57**(1989), 14—34.

[13] M. F. Barnsley, P. R. Massopust, H. Strickland and A. D. Sloan, "Fractal modeling of biological structures," *Ann. New York Acad. Sci.* **504**(1987), 179—194.

[14] T. Bedford, *Crinkly Curves, Markov Partitions and Dimension*, Ph.D. Thesis, University of Warwick, U.K., 1984.

[15] T. Bedford, "Dimension and dynamics for fractal recurrent sets," *J. Lond. Math. Soc.* **33**(2)(1986), 89—100.

[16] T. Bedford, "The box dimension of self-affine graphs and repellers," *Nonlinearity* **2**(1989), 53—71.

[17] T. Bedford, "On Weierstrass-like functions and random recurrent sets," *Proc. Cambr. Phil. Soc.* **106**(2) (1989), 14—34.

[18] T. Bedford, "Hölder exponents and box dimension for self-affine fractal functions," *Constr. Approx.* **5**(1)(1989), 33—48.

[19] T. Bedford, M. Dekking and M. Keane, "Fractal image coding techniques and contraction operators," *Delft University of Technology Report* **92-93**(1992).

[20] T. Bedford and M. Urbański, "The box and Hausdorff dimension of self-affine sets," *Erg. Th. and Dynam. Sys.* **10**(4)(1990), 627—644.

[21] R. Bellman, *Introduction to Matrix Analysis*, McGraw-Hill, New York, 1960.

[22] S. K. Berberian, *Lectures in Functional Analysis and Operator Theory*, Springer Verlag, New York, 1974.

[23] J. Bergh and J. Peetre, "On the spaces V_p, $0 < p \le \infty$," *Bolletino U. M. I.* **10**(4) (1974), 632—648.

[24] M. V. Berry and Z. V. Lewis, "On the Weierstrass-Mandelbrot fractal function," *Proc. R. Soc. Lond. A* **370** (1980), 459—484.

[25] A. S. Besicovitch and H. D. Ursell, "Sets of fractional dimension," *J. Lond. Math. Soc.* **12**(1937), 18—25.

[26] P. Billingsley, "Hausdorff dimension in probability theory," *Ill. J. Math.* **4**(1960), 187—209.

[27] P. Billingsley, "Hausdorff dimension in probability theory II.," *Ill. J. Math.* **5**(1961), 291—298.

[28] N. Bourbaki, *Groupes et Algèbres de Lie*, Chapitres IV, V, VI, Hermann, Paris, 1968.

[29] D. G. Bourgin, *Modern Algebraic Topology*, The MacMillan Company, New York, 1963.

[30] K. S. Brown, *Buildings*, Springer Verlag, New York, 1989.

[31] R. J. Buck, "A generalized Hausdorff dimension for functions and sets," *Pacific J. Math.* **44**(1973), 69—78.

[32] H. Cajar, *Billingsley Dimension in Probability Spaces*, Lecture Notes in Mathematics, No. 892, Springer-Verlag, Heidelberg, 1981.

[33] B. C. Carlson, *Special Functions of Applied Mathematics*, Academic Press, New York, 1977.

[34] D. M. Casesnoves, "Ensayo de teoría unitaria de las ondas relativistas de materia y radiación. Las funciones fractales y los objetos fractales en mecánica y en probabilidad," *Rev. Real Acad. Cienc. Natur. Fis. Madrid* **82**(1)(1988), 27—43.

[35] C. K. Chui, *An Introduction to Wavelets*, Academic Press, Boston, 1992.

[36] K. L. Chung, *A Course in Probability Theory*, 2nd ed., Academic Press, New York, 1974.

[37] H. S. M. Coxeter, *Regular Polytopes*, 3rd ed., Dover, New York, 1973.

[38] W. Dahmen and C. A. Micchelli, "Statistical encounters with *B*-splines," *Contemp. Math.* **59**(1986), 17—48.

[39] G. David, *Wavelets and Singular Integrals on Curves and Surfaces*, Springer Lecture Notes, No. 1465, Springer Verlag, Berlin, 1991.

[40] I. Daubechies, "Orthonormal bases of compactly supported wavelets," *Commun. Pure and Applied Math.*, Vol. XLI(1988), 909—996.

[41] I. Daubechies, *Ten Lectures on Wavelets*, CBMF Conference Series in Applied Mathematics, Vol. 61, SIAM, Philadelphia, 1992.

[42] I. Daubechies, *personal communication*.

[43] C. de Boor, "Splines as linear combinations of B-splines: A survey," in G. G. Lorentz, C. K. Chui, and L. L. Schumaker, Eds. , *Approximation Theory II*, Academic Press, New York, 1976.

[44] C. de Boor, R. A. DeVore, and A. Ron, "Approximation from shift-invariant subspaces of $L^2(\mathbb{R})$," *preprint*.

[45] M. F. Dekking, "Recurrent sets," *Adv. in Math.* **44**(1982), 78—104.

[46] M. F. Dekking, "Replicating superfigures and endomorphisms of free groups," *J. Comb. Th., Series A* **32**(1982), 315—320.

[47] M. F. Dekking, "Recurrent sets: A fractal formalism," Report 82-32, Technische Hogeschool, Delft, 1982.

[48] A. Deliu and B. Jawerth, "Geometrical dimension versus smoothness," *Constr. Approx.* **8**(1992), 211—222.

[49] A. Deliu, J. S. Geronimo, R. Shonkwiler and D. Hardin, "Dimensions associated with recurrent self-similar sets," *Math. Proc. Camb. Phil. Soc.* **110**(1991), 327—336.

[50] G. Deslauriers, J. Dubois and S. Dubuc, "Multidimensional iterative interpolation," *Can. J. Math.* **43**(2)(191), 297—312.

[51] R. Devaney, *Introduction to Chaotic Dynamical Systems*, Benjamin/ Cummings Publishing Co., Melo Park, New Jersey,1986.

[52] R. Devaney, "Overview: Dynamics of simple maps," in AMS Short Course Lecture Notes, Vol. 39, R. Devaney and Linda Keen, eds., Providence, Rhode Island, 1988.

[53] J. Dieudonné, *Éléments d'Analyse*, Fascicule XXXI, Gauthier-Villars, Paris, 1969.

[54] G. Donovan, J. S. Geronimo, D. P. Hardin and P. R. Massopust, "Construction of orthogonal wavelets using fractal interplation functions," *to appear in SIAM J. Math. Anal.*.

[55] S. Dubuc, "Interpolation through an iterative scheme," *J. Math. Anal. Appl.* **114**(1)(1986), 185—204.

[56] S. Dubuc, "Interpolation fractale," in *Fractal Geometry and Analysis*, J. Bélais and S. Dubuc, eds., Kluwer Academic Publishers, Dordrecht, The Netherlands, 1989.

[57] S. Dubuc and A. Elqortobi, "Valeurs extremes de fonctions fractales," *Cahiers du C.E.R.O.* **30**(1)(1988), 3—12.

[58] J.-M. Dumont and A. Thomas, "Systèmes de numeration et fonctions fractales aux substitutions," *Theoret. Comput. Sci.* **65**(2) (1987), 153—169.

[59] J. P. Eckmann and D. Ruelle, "Ergodic theory of chaos and strange attractors," *Rev. Mod. Phys.* **57**(1)(1985), 617—656.

[60] G. A. Edgar, "Kieswetter's fractal has Hausdorff dimension 3/2," *Real Analysis Exchange*, Vol. 14(1988-89), 215—223.

[61] G. A. Edgar, *Measure Theory, Topology, and Fractal Geometry*, Springer Verlag, New York, 1990.

[62] G. A. Edgar and R. D. Mauldin, "Multifractal decompositions of digraph recursive fractals," *Proc. Math. Soc.* **65**(3)(1992), 604—628.

[63] R. Engelking, *Dimension Theory*, North-Holland Publishing Company, Amsterdam, 1978.

[64] K. J. Falconer, *The Geometry of Fractal Sets*, Cambridge University Press, Cambridge, U. K., 1985.

[65] K. J. Falconer, *Fractal Geometry - Mathematical Foundations and Applications*, John Wiley & Sons, Michester, U. K., 1990.

[66] K. J. Falconer, "The Hausdorff dimension of self-affine fractals," *Math. Proc. Camb. Phil. Soc.* **103**(1988), 339—350.

[67] K. J. Falconer and D. T. Marsh, "The dimension of affine-invariant fractals," *J. Phys. A: Math. Gen.* **21**(1988), 121—125.

[68] K. J. Falconer, "Dimensions and measures of quasi self-similar sets," *Proc. Amer. Math. Soc.* **106**(2)(1989), 543—554.

[69] J. D. Farmer, E. Ott and J. A. Yorke, "The dimension of chaotic attractors," *Physica D* **7**(1983), 153—180.

[70] H. Federer, *Geometric Measure Theory*, Springer Verlag, New York, 1969.

[71] S. R. Foguel, *The Ergodic Theory of Markov Processes*, Van Nostrand Mathematical Studies, New York, 1969.

[72] M. Frazier, B. Jawerth and G. Weiss, *Littlewood-Paley Theory and the Study of Function Spaces*, CBMS **79**, American Mathematical Society, Providence, Rhode Island, 1990.

[73] P. Frederickson, J. L. Kaplan and J. A. Yorke, "The Lyapunov dimension of strange attractors," *J. Diff. Equations* **49**(1983), 185—207.

[74] O. Frostman, "Potential d'équilibre et capacité des ensembles avec quelques applications à la théorie des fonctions," *Meddel. Lunds Univ. Math. Sem.* **3**(1935), 1—118.

[75] J. S. Geronimo and D. P. Hardin, "An exact formula for the measure dimensions associated with a class of piece-wise linear maps," *Constr. Approx.* **5**(1)(1989), 89—98.

[76] J. S. Geronimo and D. P. Hardin, "Fractal interpolation surfaces and a related 2-D multiresolution analysis," *J. Math. Anal. and Appl.* **176**(2)(1993), 561—586.

[77] J. S. Geronimo, D. P. Hardin and P. R. Massopust, "Fractal functions and wavelet expansions based on several scaling functions," *to appear in J. Approx. Th.*

[78] J. S. Geronimo, D. P. Hardin and P. R. Massopust, "Fractal surfaces, multiresolution analyses, and wavelet transforms," *NATO ASI Series F*, Vol. 106 (1994), 275—290.

[79] J. S. Geronimo, D. P. Hardin and P. R. Massopust, "An application of Coxeter groups to the generation of wavelet bases in $\mathbb{R}^n$," *Fourier Analysis: Analytic and Geometric Aspects*, Lecture Notes in Pure and Applied Mathematics, Vol. 157 (1994), 187—196.

[80] S. Gibert and P. R. Massopust, "The exact Hausdorff dimension for a class of fractal functions," *J. Math. Anal. and Appl.* **168**(1) (1992), 171—183.

[81] R. Glowinski, W. Lawton, M. Rachavol, and E. Tenenbaum, "Wavelet solutions of linear and nonlinear elliptic, parabolic and hyperbolic problems in one space dimension," (1992), 55—120.

[82] T. N. T. Goodman, S. L. Lee, and W. S. Wang, "Wavelets in wandering subspaces," *Trans. Amer. Math. Soc.* **338**(2) (1993), 639—654.

[83] K. Gröchenig, "Analyse multiéchelles et bases d'ondelettes," *C. R. Acad. Sci. Paris* **305**, Série I(1987), 13—17.

[84] K. P. Grotemeyer, *Topologie*, B·I-Hochschultaschenbücher, Band 836, Bibliographisches Institut, Mannheim, Germany, 1969.

[85] D. P. Hardin and P. R. Massopust, "The capacity for a class of fractal functions," *Commun. Math. Phys.* **105**(1986), 455—460.

[86] D. P. Hardin and P. R. Massopust, "Fractal interpolation functions from $\mathbb{R}^n \to \mathbb{R}^m$ and their projections," *Zeitschrift für Analysis u. i. Anw.* **12**(1993), 535—548.

[87] D. P. Hardin, B. Kessler, and P. R. Massopust, "Multiresolution analyses based on fractal functions," *J. Approx. Th.* **71**(1)(1992), 104—120.

[88] J. Harrison, "Denjoy fractals," *Topology* **28**(1989), 59—80.

[89] J. Harrison and A. Norton, "Geometric integration on fractal curves in the plane," *Indiana Univ. Math. J.*, **40**(2)(1991), 567—594.

[90] H. Heijmans, "Discrete wavelets and multiresolution analysis," *CWI Quaterly* **5**(1)(1992), 5—32.

[91] D. Hilbert, "Über die stetige Abbildung einer Linie auf ein Flächenstück," *Math. Ann.* **38**(1891), 459—460.

[92] H. Hiller, *Geometry of Coxeter Groups*, Pitman, Boston, 1982.

[93] H. Holmann, *Lineare und Multilineare Algebra I*, B·I-Hochschultaschenbücher, Band 173/173a, Bibliographisches Institut, Mannheim, Germany, 1970.

[94] M. Holschneider, "On the wavelet transformation of fractal objects," *J. Stat. Phys.* **90**(5/6)(1988), 963—993.

[95] M. Hoffman and W. D. Withers, "Generalized Chebychev polynomials associated with affine Weyl groups," *Trans. Amer. Math. Soc.* **308**(1)(1988), 91—104.

[96] S. T. Hu, *Elements of Modern Algebra*, Holden-Day, San Francisco, 1965.

[97] W. Hurewicz and Henry Wallman, *Dimension Theory*, Princeton University Press, Princeton, 1941.

[98] J. E. Hutchinson, "Fractals and self similarity," *Indiana Univ. J. Math.* **30**(1981), 713—747.

[99] K. Jacobs, *Selecta Mathematica IV*, Springer Verlag, Berlin, 1972.

[100] R.-Q. Jia and Z. Shen, "Multiresolution and wavelets," *preprint*, 1993.

[101] G. A. Jones and D. Singerman, *Complex Functions*, Cambridge University Press, Cambridge, U. K., 1987.

[102] G. Julia, "Mémoirs sur l'itération des fonctions rationelles," *J. Math.* **1**(1918), 47—245.

[103] J. Kaplan and J. A. Yorke, "Functional differential equations and the approximation of fixed points," Procedings, Bonn, Springer Notes in Math. **730**(1978), 228—236.

[104] J. G. Kemeny, J. L. Snell, and A. W. Knapp, *Denumerable Markov Chains*, Springer Verlag, New York, 1976.

[105] K. Kiesswetter, "Ein einfaches Beispiel für eine Funktion welche überall stetig und nicht differenzierbar ist," *Math. Phys. Semesterber.* **13**(1966), 216—221.

[106] K. Knopp, "Einheitliche Erzeugung und Darstellung der Kurven von Peano, Osgood und Koch," *Arch. Math. Phys.* **26**(1917), 103—115.

[107] K. Kono, "On self-affine functions I. and II.," *Japan J. Appl. Math.* **3**(2)(1986), 252—269, and **5**(3)(1988), 441—454.

[108] R. Kultze, *Garbentheorie*, B. G. Teubner, Stuttgart, 1970.

[109] K. Kuratowski, *Introduction to Set Theory and Topology*, Addison-Wesley Publishing Company Inc., Reading, Massachusetts, 1962.

[110] S. Lalley, "The packing and covering functions of some self-similar fractals," *Indiana J. Math.* **37**(3)(1988), 699—709.

[111] H. Lebesgue, "Leçons sur l'intégration et la recherche des fonctions primitives," *Gauthiers-Villars*, Paris, (1904), 44—45.

[112] F. Ledrappier, "Some relations between dimension and Lyapunov exponents," *Commun. Math. Phys.* **81**(1981), 229—238.

[113] F. Ledrappier and L. S. Young, "The metric entropy of diffeomorphisms," *Bulletin Amer. Math. Soc.* **11**(2)(1984), 343–346.

[114] P. G. Lemarié, "Ondelettes á localisation exponentielle," *Journal de Math. Pure et Appl.* **67**(1988), 227—236.

[115] J. Liandrat, V. Perrier, and Ph. Tchamitchian, "Numerical resolution of nonlinear partial differential equations using the wavelet approach," *Wavelets and their Applications*, M. B. Ruskai, G. Beylkin, R. Coifman, I. Daubechies, S. Mallat, Y. Meyer, and L. Raphael, eds., Jones and Bartlett, Boston, 1992.

[116] T. Lindstrøm, "Nonstandard analysis, iterated function systems, and Brownian motion on fractals," in *Contemporary Stochastic Analysis*, World Scientific Publishing Co. Pte. Ltd, 1991, 71—108.

[117] T. Lindstrøm, "A nonstandard approach to iterated function systems," *preprint* (1992).

[118] T. Lindstrøm, "An invitation to nonstandard analysis," in N. J. Cutland, ed., *Nonstandard Analysis and its Applications*, Cambridge University Press, U. K., 1988, 1—105.

[119] H. Loïc, "Analyses multirésolutions de multiplicité *d*. Applications á l'interpolation dyadique," *preprint*.

[120] R. A. Lorentz and W. R. Madych, "Wavelets and generalized box splines," *preprint*.

[121] W. R. Madych, "Some elementary properties of multiresolution analyses in $L^2(\mathbb{R}^n)$," in *A Tutorial on Wavelets*, Vol. II, C. K. Chui, ed., Academic Press, Orlando, Florida, 1992.

[122] S. Mallat, "Multiresolution approximations and wavelet orthonormal bases of $L^2(\mathbb{R})$," *Trans. Amer. Math. Soc.* **315**, 69—87.

[123] B. Mandelbrot, *The Fractal Geometry of Nature*, W. H. Freeman and Company, New York, 1977.

[124] J. Marion, "Mesure de Hausdorff et théorie de Perron-Frobenius des matrices non-negatives," *Ann. Inst. Fourier*, Grenoble **35**(4)(1985), 99—125.

[125] E. Martensen, *Analysis V*, B·I-Hochschultaschenbücher, Band 768, Bibliographisches Institut, Mannheim, Germany, 1972.

[126] P. R. Massopust, "Dynamical systems, fractal functions, and dimension," *Topology Proc.* **12**(1987), 93—110.

[127] P. R. Massopust, "Fractal Peano curves," *J. of Geometry* **34** (1989), 127—138.

[128] P. R. Massopust, "Fractal surfaces," *J. Math. Anal. and Appl.* **151** (1)(1990), 275—290.

[129] P. R. Massopust, "Vector-valued fractal interpolation functions and their box dimension," *Aequationes Mathematicae* **42**(1991), 1—22.

[130] P. R. Massopust, *Space Curves Generated by Iterated Function Systems*, Ph.D. Thesis, Georgia Institute of Technology, 1986.

[131] P. R. Massopust, "Smooth interpolating curves and surfaces generated by iterated function systems," *Zeitschrift für Analysis u. i. Anwend.* **12** (1993), 201–210.

[132] P. R. Massopust, "A study of Wavelet-Galerkin methods for numerical solutions of differential equations using multigrid relaxation methods," *Final Report*, 1993 SFRP, AEDC, Arnold AFB, Tennessee.

[133] P. R. Massopust and P. van Fleet, "On the moments of fractal functions and Dirichlet spline functions," *preprint* (1993).

[134] R. D. Mauldin and S. C. Williams, "Random recursive constructions: Asymptotic geometric and topological properties," *Trans. Amer. Math. Soc.* **295**(1)(1986), 325—346.

[135] R. D. Mauldin and S. C. Williams, "On the Hausdorff dimension of some graphs," *Trans. Amer. Math. Soc.* **289**(1986), 793—803.

[136] R. D. Mauldin and S. C. Williams, "Hausdorff dimension in graph directed constructions," *Trans. Amer. Math. Soc.* **309**(2)(1988), 811—829.

[137] I. Meyer, *Ondelettes et Opérateurs*, Hermann, Paris, 1990.

[138] C. Micchelli, "Using the refinement equation for the construction of pre-wavelets VI: Shift invariant subspaces," in *Approximation Theory, Spline Functions and Applications*, S. P. Singh, ed., NATO ASI Series C, Vol. 356, 213—222.

[139] J.-P. Mongeau, G. Deslauriers and S. Dubuc, "Continuous and differentiable multidimensional iterative interpolation," *Linear Algebra and its Applications* **180**(1991), 95—120.

[140] P. A. P. Moran, "Additive functions of intervals and Hausdorff measure," *Proc. Camb. Phil. Soc.* **42**(1946), 15—23.

[141] K. Nagami, *Dimension Theory*, Academic Press, New York, 1970.

[142] J. Nagata, *Modern Dimension Theory*, John Wiley & Sons, New York, 1965.

[143] J. Naudts, "Dimension of discrete fractal spaces," *J. Phys. A: Math. Gen.* **21**(1988), 447—452.

[144] E. Neuman, "Dirichlet averages and their applications to special functions," *preprint* (1992).

[145] E. Neuman and P. Van Fleet, "Moments of Dirichlet splines and their applications to hypergeometric functions," *to appear in J. Comput. and Applied Math.*.

[146] Y. I. Oseledec, "A multiplicative ergodic theorem, Lyapunov characteristic numbers for dynamical systems," *Trudy Moskov. Mat. Obshch.* **19** (1968), 179—210 [English translation].

[147] J. Palis and W. de Melo, *Geometric Theory of Dynamical Systems. An introduction*, Springer Verlag, New York, 1982.

[148] G. Peano, "Sur une courbe, qui rempli toute une aire plane," *Math. Ann.* **36**(1890), 157—160.

[149] A. R. Pears, *Dimension Theory of General Spaces*, Cambridge University Press, Cambridge, U. K., 1975.

[150] J. Peetre, *New Thoughts on Besov Spaces*, Duke Univ. Math. Series, Durham, North Carolina, 1976.

[151] S. Pelikan, "Invariant densities for random maps on the interval," *Trans. Amer. Math. Soc.* **281**(2)(1984), 313—325.

[152] J. Pesin, "Characteristic Lyapunov exponents and smooth ergodic theory," *Russ. Math. Surveys* **32**(4)(1977), 55—114.

[153] F. Przytycki and M. Urbański, "On Hausdorff dimension of some fractal sets," *Studia Math.* **93**(1989), 155—186.

[154] A. H. Read, "The solution of a functional equation," *Proc. Roy. Soc. Edinburgh Sect. A*, **63**(1951-52), 336—345.

[155] A. Rényi, "Dimension, entropy and information," in *Transactions of the Second Prague Conference on Information Theory, Statistical Decision Functions, Random Processes*, Academic Press, New York, 1960, 545—556.

[156] C. A. Rogers, *Hausdorff measures*, Cambridge University Press, Cambridge, U. K., 1970.

[157] R. Ronan, "Buildings: Main Ideas and applications, I. Main ideas," *Bull. Lond. Math. Soc.* **24** (1992), 1—51.

[158] D. Ruelle, "Ergodic theory of differentiable dynamical systems," *Publ. Math. IHES* **50**(1979), 27—58.

[159] H. Sagan, "Approximating polygons for Lebesgue's and Schoenberg's space filling curves," *Amer. Math. Monthly* (1986), 361—368.

[160] I. J. Schoenberg, "The Peano curve of Lebesgue," *Elem. Math.* **28** (1973), 1—10.

[161] W. Sierpiński, "Sur une courbe dont tout point est un point de ramification," *C. R. Acad. Sci. Paris* **160**(1915), 302.

[162] J. Simon, "Sobolev, Besov and Nikolskiĭ Fractional Spaces: Imbeddings and comparisons for vector-valued spaces on an interval," *Annali di Matematica pura ed applicata* Vol. LCVII(IV) (1990), 117—148.

[163] P. Singer and P. Zajdler, "Wavelet series and self-affine fractal functions," *preprint* (1993).

[164] P. Singer and P. Zajdler, "Continuous nowhere differentiable functions constructed by wavelet series," *preprint* (1993).

[165] E. M. Stein, *Singular Integrals and Differentiablity Properties of Functions*, Princeton University Press, Princeton, 1970.

[166] G. Strang and G. Fix, "A Fourier analysis of the finite element variational method," in *Constructive Aspects of Functional Analysis*, C. I. M. E. II Ciclo 1971, G. Geymonat, ed., 793—840.

[167] R. Strichartz, "How to make wavelets," *The American Mathematical Monthly* **100** (6) (1993), 539—556.

[168] K. D. Stroyan and W. A. J. Luxemburg, *Introduction to the Theory of Infinitesimals*, Academic Press, New York, 1976.

[169] T. Takagi, "A simple example of the continuous function without derivative," *Proc. Phys. Math. Soc. Japan* **1**(1903), 176—177.

[170] S. J. Taylor, "On the connection between Hausdorff measures and generalized capacities," *Proc. Camb. Phil. Soc.* **57**(1961), 524—531.

[171] S. J. Taylor, "The measure theory of random fractals," *Math. Proc. Camb. Phil. Soc.* **100**(1986), 383—406.

[172] C. Tricot, "Douze définitions de la densité logarithmique," *C.R. Acad. Sci. Paris* **293**(1981), 549—552.

[173] C. Tricot, "Two definitions of fractal dimension," *Math. Proc. Camb. Phil. Soc.* **91**(1982), 57—74.

[174] M. Urbański, "Hausdorff dimension of the graphs of continuous self-affine functions," *Proc. Amer. Math. Soc.* **108**(1990), 921—930.

[175] H. von Koch, "Sur une courbe continue sans tangente obtenue par une construction géométrique élémentaire," *Arkiv für Matematik, Astronomie och Fysik* **1**(1904), 681—704.

[176] H. von Koch, "Une méthode géométrique élémentaire pour l'étude de certaines questions de la théorie des courbes planes," *Acta Math.* **30**(1906), 145—174.

[177] P. Walters, *An Introduction to Ergodic Theory*, Graduate Texts in Mathematics, Springer Verlag, New York, 1982.

[178] H. Wegmann, "Über den Dimensionsbegriff in Wahrscheinlichkeitsräumen," *Z. Wahrscheinlichkeitstheorie und Verw. Gebiete* **9**(1968), 216—221.

[179] H. Wegmann, "Über den Dimensionsbegriff in Wahrscheinlichkeitsräumen II.," *Z. Wahrscheinlichkeitstheorie und Verw. Gebiete* **9**(1968), 222—231.

[180] K. R. Wicks, *Fractals and Hyperspaces*, Lecture Notes in Mathematics, No. 1492, Springer Verlag, Heidelberg, 1991.

[181] P. Wojtaszczyk, *Banach Spaces for Analysts*, Cambridge University Press, Cambridge, U. K., 1991.

[182] W. Wunderlich, "Über Peano Kurven," *Elem. Math.* **28** (1973), 1—10.

[183] L.-S. Young, "Dimension, entropy and Lyapunov exponents," *Ergod. Th. and Dynam. Sys.* **2**(1982), 109—124.

[184] L.-S. Young, "Entropy, Lyapunov exponents, and Hausdorff dimension in differentiable dynamical systems," *IEEE Trans. on Circuits and Sys.*, **30** (8)(1983), 599—607.

[185] L.-S. Young, "Capacity of attractors," *Ergod. Th. and Dynam. Sys.* **1**(1981), 381—388.

Index